Introduction

When driving down a long hill on a crowded freeway, a situation can occur that demands good brakes. Many of the cars in this photo are going over 55 mph. If an emergency were to happen, the resulting chain-reaction braking would result in some drivers locking the wheels.

Most of us only think about brakes when a panic stop occurs ahead in traffic and all we see are brakelights and the undersides of cars. These near-emergencies illustrate how important brakes are to our safety. Brakes are also a vital part of high performance, as any racer can tell you.

Because everyone wants higher performance and safety, brakes deserve a great deal of attention. We not only want our car to go fast, but it should also stop quickly and safely. Any car with powerful and consistent brakes instills confidence in the driver. It also increases driving pleasure. Bad brakes are terrifying.

If racing is your game, you need to know more about brakes than the casual driver. No matter what type of racing you do, brake performance is vital. Road racing is most demanding on brakes, although drag racing and oval-track events have special problems, too.

I talk about brakes as a *system*. This includes fluid, lines, pedals, levers and linkages, as well as the brake units. Wheels, bodywork and even the frame structure become a part of the brake system when they affect brake performance. This book covers each part of the system and how it relates to overall brake performance. It will help in the selection of components if you prefer to design a brake system.

The first section of the book deals with particular parts of a brake system. The second section, starting with Chapter 9, describes how to design, install, test, maintain and modify a brake system for racing applications. Race-car concepts also apply to high-performance road vehicles. I also note where there are important differences between racing and road use.

Braking is essential in winning races. These stock cars are competing on a road course—the most severe duty for automotive brakes. Notice smoke from the right front tire on car that's braking hard in the corner.

Basics

1

Good brakes are essential to overall vehicle performance. Even though engine performance, suspension and body aerodynamics approach perfection, race cars, such as this GTP Corvette, will not be competitive without good brakes. Photo by Tom Monroe.

Most production sedans have drum brakes on the rear and disc brakes on the front. Drum brakes are preferred on rear wheels because a parking brake is easily adapted. This large drum is on the rear of a 2-ton sedan.

Brake systems are designed to do one thing—stop the vehicle. Sounds easy, but problems start when brakes must stop a vehicle from high speed in a short distance, and do it over and over again. We expect no failures or loss of control. All brake systems should stop a vehicle. The difference between a good system and a bad one is how well it will perform under the most adverse conditions.

All vehicles have brakes, and they always did. Ever since man discovered the wheel, stopping it was a problem. Carts, wagons and carriages had brakes, usually simple blocks rubbing on a wheel. This established a basic that has yet to change, even with the most sophisticated brake system: All brakes use *friction* to stop the vehicle.

BRAKE TYPES

When two parts rub together, the resulting friction generates heat. In brakes, the friction materials rub against metal surfaces. Different types of brakes are arranged differently, or use different methods of forcing rubbing surfaces together. There are also differences in dissipating heat once it is generated.

Either *drum brakes* or *disc brakes,* or a combination of the two, are used on most vehicles. These terms refer to how friction surfaces are designed and configured.

Drum Brakes—All early vehicles used drum brakes; many of today's vehicles still do. The rubbing surface is a metal cylinder called a *brake drum,* usually made of cast iron.

Early drum brakes were *external*—rubbing surface was *outside* of the drum. More modern *internal drum brakes* have the rubbing surface inside the drum. There are *shoes* inside the drum with friction material attached. This friction material is called *lining.* It

BRAKE HANDBOOK

Fred Puhn
Registered Professional Engineer

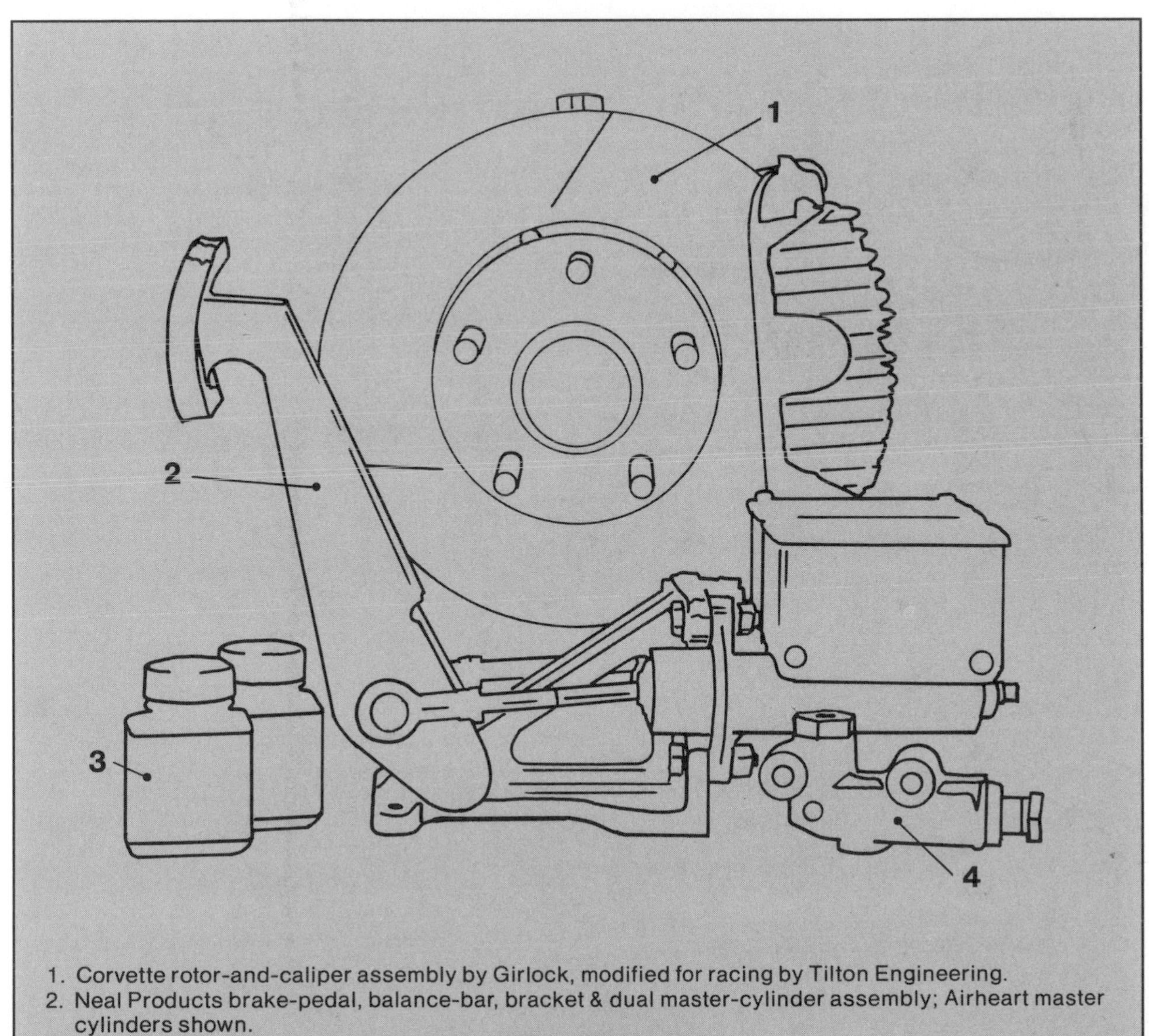

1. Corvette rotor-and-caliper assembly by Girlock, modified for racing by Tilton Engineering.
2. Neal Products brake-pedal, balance-bar, bracket & dual master-cylinder assembly; Airheart master cylinders shown.
3. Tempilaq temperature-sensing paint from Big-Three Industries.
4. Alston proportioning valve.

HPBooks
a division of
PRICE STERN SLOAN
Los Angeles

Contents

THANKS

Thanks to the many brake suppliers and experts who helped with technical information and photographs. Special thanks to Mac Tilton of Tilton Engineering, Bill Neal of Neal Products, and John Moore of AP Racing for their personal contributions. Thanks also to Carroll Smith of Carroll Smith Consulting for helpful suggestions given after reading the finished manuscript.

Special thanks to Garrett Van Camp, Van Camp Racing Enterprises, Inc., 25192 Maplebrooke, Southfield, MI 48034. His heroic efforts and great technical knowledge gained over the years as a Ford Motor brake-design engineer and race-car brake-design consultant made a significant contribution to the content and completeness of this book. It could not have been this quality without him. Thanks Garrett.

Photos: Fred Puhn, others noted; Cover Photo: Bill Keller

Published by HPBooks, a division of Price Stern Sloan, Inc.
360 North La Cienega Blvd., Los Angeles, California 90048
ISBN 0-89586-232-8 Library of Congress Catalog Number 84-62610

Printed in U.S.A.

7 6 5 4

Honda disc brake is typical of front brakes used on small sedans. Exposed rubbing surface of disc brake aids cooling.

Earliest automotive brakes were drum type with the rubbing surface on the outside of the drum. Because friction material surrounds the outside of the drum, little cooling air contacts the hot rubbing surface. External drum brakes are simple and easy to service, but have horrible cooling ability.

Prewar MG used mechanical brakes. Front brakes are operated by cables that flex as the wheels steer and move up and down. Finned aluminum drums give better cooling than plain cast iron.

is designed to rub against the drum without burning, melting or wearing rapidly. The shoes are forced against the inside surface of the drum when the driver pushes the brake pedal, creating friction between the lining and the drum surface. Drum brakes are covered in Chapter 2.

Disc Brakes—A modern brake design is the disc brake. The drum is replaced by a flat metal disc, or *rotor,* with a rubbing surface on each side. The rotor is usually made of cast iron. Friction materials are inside a *caliper,* which surrounds the rotor. Disc-brake friction material—one on each side of the rotor—is called a brake *pad, puck* or *lining.* This caliper is designed to clamp the pads against the sides of the rotor to create friction. Disc brakes are covered in Chapter 3.

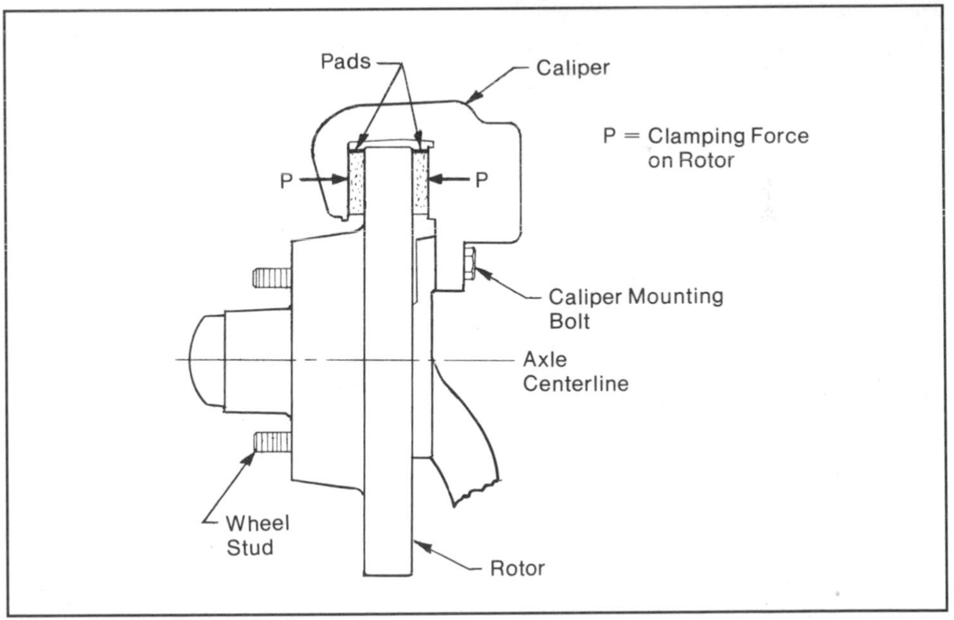

Disc brake operates by clamping rotor between two stationary pads. Rotor turns with the wheel; caliper is mounted to a fixed part of the suspension, usually the spindle or upright.

BRAKE-ACTUATING SYSTEM

Between the driver's foot and the wheel brakes are components that translate force from the driver into friction force at the brake-rubbing surfaces. I call this the *actuating system.* This system can be mechanical, hydraulic, pneumatic or a combination of these. Future vehicles could use electric systems. Whatever the type of actuating system, the result is the same: When the driver operates the system, brakes are applied.

Brake Pedal & Linkage—Brake pedals and linkages are integral parts of a brake system. The pedal is the familiar lever that the driver pushes with his foot to apply the brakes. Regardless of the type of brake-actuating system used, system application always begins with the driver operating a pedal—or lever in rare cases. Brake-pedal design determines the leg force required to stop the car. It is also a factor in determining how solid the brakes feel to the driver.

Trade-offs are made when designing brake pedals. Long pedals reduce the pedal force required to stop a vehicle. However, long pedals have long travel. They can also feel *spongy* to the driver. Brake-pedal design is detailed in Chapter 6.

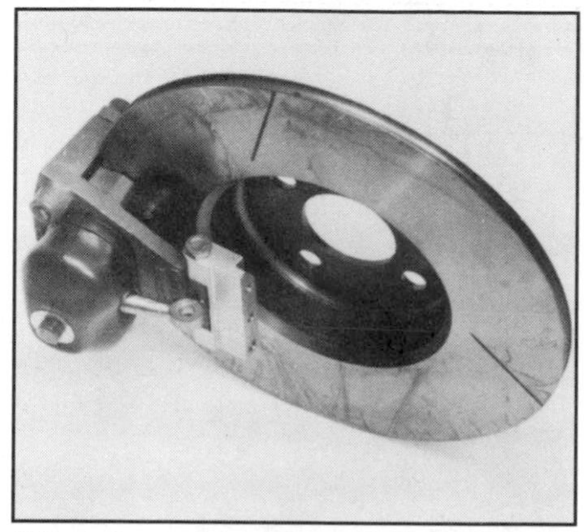

Although most mechanically actuated brakes are found on antique cars, some are still being used. Developed by AP Racing for use on competition rally cars, this modern caliper is mechanically actuated. Independent of the hydraulically actuated brakes, these calipers are used on the rear for high-speed control on slick surfaces. Photo courtesy AP Racing.

Early race cars had no front brakes. Although this Peugeot was a winner with its advanced high-speed dual-overhead-cam engine, it used cable-operated rear-wheel brakes.

In the '20s, Duesenberg introduced hydraulic brakes with large finned drums. This was probably the first road car with enough horsepower that required a great improvement in braking. It did not use flex hoses, but instead ran fluid through internal passages in the suspension.

The brake pedal is connected to a linkage that transfers force to the actuating system. This linkage can be as simple as a push/pull rod operating a single hydraulic master cylinder. Or, the linkage may be a complicated, adjustable *balance-bar* system for changing the balance between front and rear brakes. Early-design, mechanical-actuating linkage extends all the way to the brakes themselves. Brake-linkage design is discussed in more detail in Chapter 6.

Mechanical Brakes—The simplest brake-actuating system is a mechanical system. The brake pedal operates cables or rods that apply the brakes when the pedal is pushed. Early systems were mechanical and are still used for parking brakes on present-day vehicles. The mechanical linkage moves the shoes outward in a drum brake, or clamps the pads against a disc-brake rotor.

Hydraulic Brakes—Modern cars use hydraulic brakes. In a hydraulically actuated system, the cables or rods of the mechanical system are replaced by fluid-filled lines and hoses. The brake-pedal linkage operates a piston in a *master cylinder* to pressurize the fluid inside the lines and hoses. Fluid pressure in each wheel cylinder forces the friction material against the drum or rotor. See Chapter 5 for a detailed explanation of how a hydraulic system works.

Pneumatic Brakes—In a pneumatic, or air-brake, system the brakes are controlled by compressed air. Air brakes are generally used on large commercial vehicles and trucks. An advantage of the pneumatic brake system is safety. Small leaks cannot cause a total loss of braking because air is constantly supplied by a compressor and stored in large volume. Pneumatic brake-system operation is described briefly in Chapter 8.

BRAKE HISTORY

The earliest brakes were derived from those used on horse-drawn wagons. As cars became heavier and more powerful, these primitive brakes soon were improved to the early external-type drum brakes, all with mechanical-actuating systems.

Early brakes were on the rear wheels only. The major reason for this was the difficulty in designing an actuating system on wheels that are steered. Engineers avoided the problem by omitting front-wheel brakes. Another reason for not using front brakes was concern that the car might tip over on its nose if front brakes were applied hard!

Early external-type drum brakes used a band of friction material outside the drum. This type of brake was easy to design, but the friction material prevented the drum from cooling. Also, exposed friction materials were subject to dirt, oil and water contamination. When the brake shoes were relocated inside the drum, a modern drum brake was born. These were first used on the 1902 Renault.

Although four-wheel braking was tried early in the 20th century, most early cars had rear brakes only. Then, in the 1920s it was discovered that front brakes added greatly to a car's stopping ability, and they were judged safe. Four-wheel brake systems soon became universal.

Mechanical-actuating systems were still used on most cars until the late '20s when hydraulic systems came into use. Mechanical brakes were used for auto racing long after hydrau-

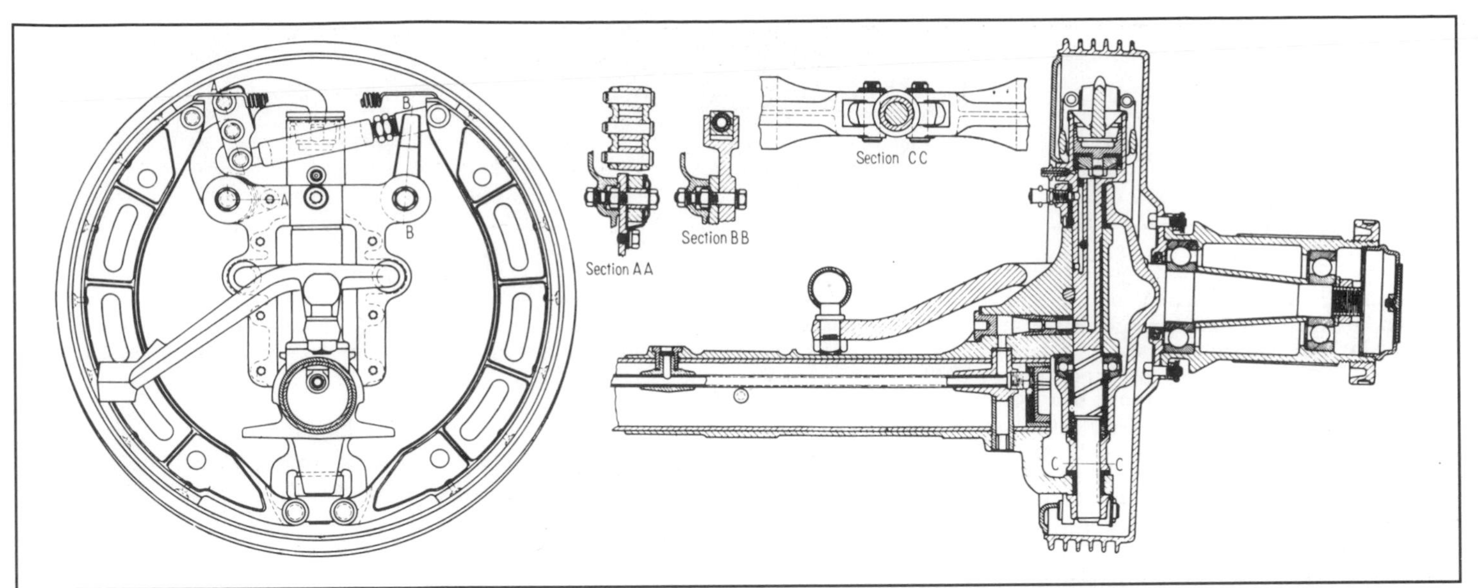

Internal details of early Duesenberg hydraulic drum brake: Notice fluid passages through axle, kingpin and spindle. Sealing was a problem eventually solved by use of flexible hydraulic lines.

This little disc brake started it all—first disc brake used on a mass-produced car. Goodyear-Hawley disc brakes were offered on Crosley Hotshot and Supersports roadsters in the early '50s. A Crosley Hotshot won the first Sebring 12-hour endurance race with these brakes. In later years, they were popular for small sports-racing cars.

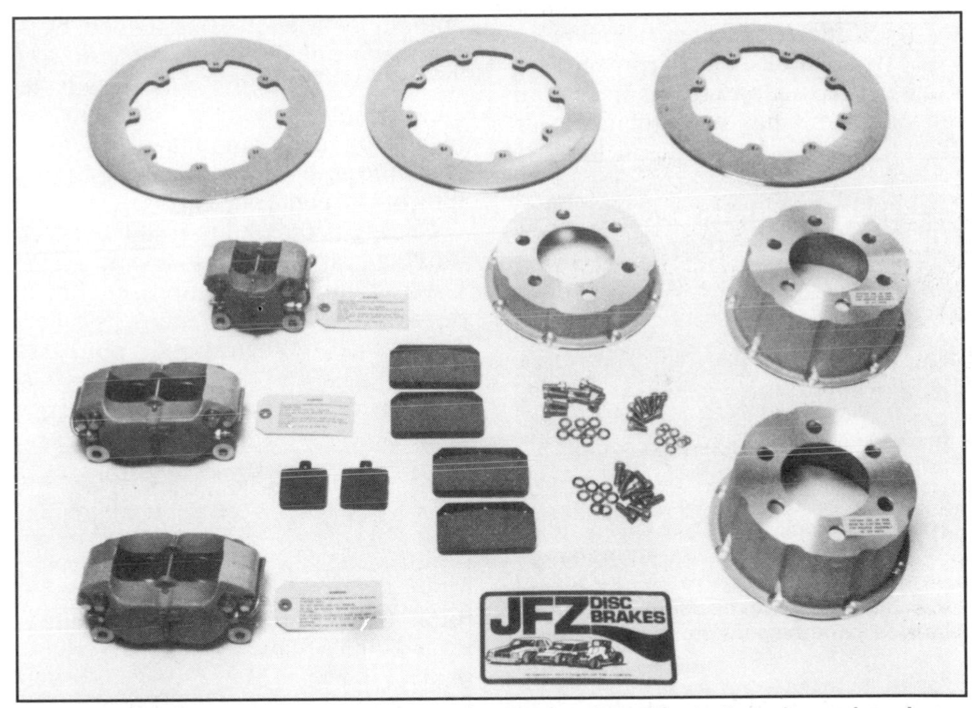

Since the mid-'50s, disc brakes have been highly developed for both racing and road use. Companies such as JFZ Engineered Products have taken brake development far beyond the early disc-brake concept. Rotors, pads and brake-mounting hardware have all benefited from rigors of faster race cars. Photo courtesy JFZ Engineered Products.

lic systems were developed for passenger cars. The simple mechanical brake system was reliable, easy to understand and maintain, and not subject to sudden loss of braking that could happen to a hydraulic system with a failed line or seal.

After World War II, disc brakes began to appear. The first production car with disc brakes was the 1949 Crosley Supersport. Disc brakes were used successfully on the 24-hours-of-LeMans-winning Jaguar in the '50s. Disc brakes soon became popular on many race cars. Indianapolis 500 cars used disc brakes early too, but they had little effect on the outcome of races on this fast track. Only at LeMans, where cars must decelerate from 180 to 30 mph every lap, and do it for 24 hours, were disc brakes tested to their limit. Even in racing, drum brakes are still used in certain classes, but most modern race cars use disc brakes.

Beyond the basic changes in brake design, there have been many important improvements. Brake systems today are very safe; and complete system failure occurs rarely. Modern brakes can go for years with little or no attention in highway use, but therein lies a problem: When really needed, performance may be marginal because of inattention. This book will help you keep that from happening.

FRICTION & ENERGY

Friction is *resistance* to sliding. Any

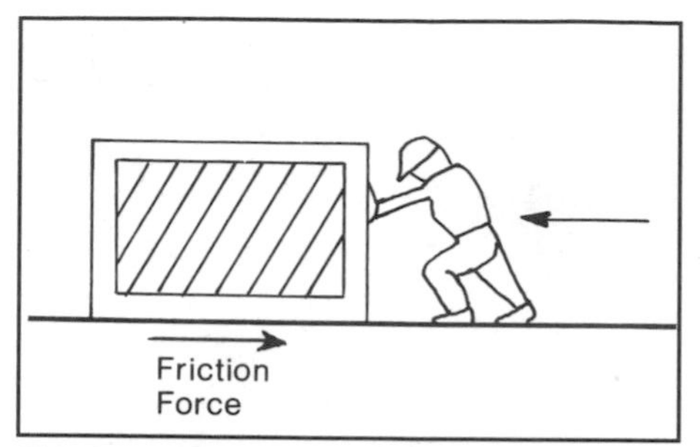

Friction between the box and floor is what makes the box difficult to slide. If box weight or friction between the floor and box increases, so must the force to slide it. Heat is developed on sliding surfaces as box is moved.

Type of Energy	Example
Heat	Energy stored in a hot brake rotor.
Sound	Noise from exhaust.
Light	Light from headlights.
Stored Mechanical	Energy stored in a compressed coil spring.
Chemical	Energy in a gallon of gasoline.
Electrical	Current from a battery turning a starter.
Radiation	Microwave energy in a microwave oven.
Kinetic	Energy stored in a speeding bullet.

Energy can be changed from one form to another, but it can't be created or destroyed. Here are some different forms of energy. Most forms are transformed into heat after energy does its useful work.

two objects in contact with and trying to move relative to each other have friction. It can be high or low depending on the types of surfaces in contact. Friction helps keep your feet from sliding out from under you. When you are standing on ice, friction is low and it is difficult to prevent slipping.

If two surfaces in contact are sliding, the friction creates heat. You can confirm this by rubbing your hands together rapidly back and forth. You can feel the warmth. This friction can help warm your hands on a cold day. In brakes, friction is used to create heat. The process of creating heat stops the car.

The amount of friction between two rubbing surfaces depends on the materials and their roughness. The amount of friction is described by a number called the *coefficient of friction*. A high number means a large amount of friction; a low number means a small amount of friction. Read more about friction at the beginning of Chapter 4.

Energy is the ability to do work. A moving car develops energy. The faster it moves, the more energy it develops. This type of energy is called *kinetic energy*. When speed is doubled, four times the kinetic energy is developed. That is, kinetic energy varies as the *square* of speed. To calculate kinetic energy of a car, use the following formula:

$$\text{Kinetic energy} = \frac{W_c S^2}{29.9}$$

in foot-pounds (ft-lb)
W_c = Weight of Car in pounds (lb)
S = Speed of car in miles per hour (mph)

Converting Energy—The first law of thermodynamics says: Energy can never be created nor destroyed. However, energy can be converted from one form to another.

Different forms of energy are heat, sound, light, stored mechanical, chemical, electrical and radiated. Stored electrical energy in a battery will convert energy into heat or light by connecting the battery to a light bulb. Stored mechanical energy in a spring can be converted to kinetic

KINETIC ENERGY OF ROTATION

Strictly speaking, an object with kinetic energy can be either moving in a straight line or rotating about its own center of gravity (CG). In a speeding car, kinetic energy is mostly in the moving car. Unless the car is spinning down the road, less than 10% of the total kinetic energy is stored in rotating parts of the car. Rotating parts include tires, wheels, brakes, engine and drive line. Additional kinetic energy stored in these rotating parts must be absorbed in the brakes. However, to make calculations simpler, I ignore the small amount of kinetic energy stored in rotating parts.

At high speed, kinetic energy stored in the rotating tire-and-wheel assemblies increases significantly. If you hit the brakes hard at high speed, the rotating parts must be stopped before the wheel can lock and slide. It takes time and pedal effort to stop this rotating weight, even if the car doesn't slow at all. Consequently, it is more difficult to lock the wheels when traveling at higher speeds. Ironically, this makes a car safer at high speed if the driver panics and hits the brakes too hard. However, in racing, it increases pedal effort as the driver tries to reach the traction limit of the tires.

TEMPERATURE & HEAT

The difference between temperature and heat may be confusing. We all are familiar with temperature, measured in either degrees Farenheit (F) or degrees centigrade (C).

Heat is a form of energy. When heat is added to a material, its temperature rises; when heat is removed, its temperature drops. Thus, *temperature is the effect of adding or subtracting heat energy.* When I say something *heats up,* I mean heat is added. When I say *cools off,* I mean heat is removed. In either case, the temperature of the object changes.

I measure kinetic energy in foot-pounds (ft-lb). However, it could be measured in *British Thermal Units* (BTU's), just as engineers do. One BTU is the amount of heat it takes to raise the temperature of one pound of water by one degree Farenheit. One BTU is equal to 778 ft-lb of energy, or one ft-lb equals 0.0013 BTU. Although one pound of water changes temperature one degree F when one BTU of heat is added, other materials do not react the same. Their temperature change is different when one BTU of heat is added.

The relationship between temperature change and heat-energy change is governed by a property called *specific heat.* Each material has its own specific heat as shown for typical materials in the accompanying table. Specific heat is the temperature rise for one pound of material when one BTU of heat is added. A material's specific heat is very important to a brake-design engineer for calculating brake-temperature change for each stop.

Ideally, brakes should be made from materials with a high specific heat. This would result in a small temperature rise for a given amount of kinetic energy put into the brakes. A small temperature rise means the brakes would have fewer problems.

Material	MELTING TEMPERATURE Degrees F	Degrees C	SPECIFIC HEAT BTU/1b/F
Water	32	0	1.00
Beryllium—pure	2340	1282	0.52
Beryllium—QMV	2340	[illegible]	[illegible]
Magnesium—AZ 31B-H24	1100	[illegible]	[illegible]
Aluminum—6061-T6	1080	[illegible]	[illegible]
Aluminum—2024-T3	940	[illegible]	[illegible]
Carbon—pure	6700	[illegible]	[illegible]
Titanium—pure	3070	[illegible]	[illegible]
Titanium—B 120VCA	3100	[illegible]	[illegible]
Magnesium—HK 31A-H24	1100	[illegible]	[illegible]
Stainless Steel—304	2600	[illegible]	[illegible]
Cast Iron	2750	[illegible]	[illegible]
Steel—C1020	2750	[illegible]	[illegible]
Copper—pure	1980	1082	0.09

Specific heats of various materials are listed from the highest to lowest. Specific heat is amount of heat energy required to raise one pound of material by one degree Fahrenheit. Material with highest specific heat is not necessarily the best for brakes. To be a good brake material, it must withstand high temperature, conduct heat rapidly and have a good rubbing surface.

Car parked at top of hill has zero kinetic energy. However, its position at top of the hill gives it *potential energy.* This potential energy is changed into kinetic energy as car coasts down hill.

energy when a wind-up toy car is released. Chemical energy stored in gunpowder is converted into sound, heat and kinetic energy when ignited.

A car moving down the road has kinetic energy. To stop the car, you must dispose of this kinetic energy. Because energy cannot be destroyed, it must be converted to another form. This kinetic energy could be converted into any of the forms listed on page 8, but conversion to heat is easiest. By forcing friction material against drums or rotors, heat is created and the car slows. If the brakes or tires squeal, some sound energy is also produced, but the amount of kinetic energy converted to sound is small compared to heat energy.

If you had an electric car using batteries, you could brake the car by converting the electric motor into a generator. This could be done by switching the connections. The motion of the car would turn the generator, putting electric energy into the batteries. The car would slow because power—kinetic energy in this case—turns the generator. Some electric cars maintain their battery charge by using this form of braking called *regenerative braking.*

It would be wonderful if fuel could be put in the gas tank by hitting the brakes. It can't, however, so all that kinetic energy is lost. On the other hand, if you drive more slowly or anticipate every stop, you could use the brakes less and, thus, conserve fuel. Try this while driving to work. The fuel energy used to move the car is lost in heat energy each time you hit the brakes. By minimizing braking, you increase mileage.

If you slow the car without using the brakes, the kinetic energy can be changed into two different forms. If you are driving on a flat road and take your foot off the throttle, the car will slow. Kinetic energy is lost in air drag, friction (heat) in the engine and drive line, and rolling resistance of the tires.

Because these items get hot, kinetic energy is converted into heat. However, because slowing takes longer and the items being heated are much larger, drive-line components and tires don't reach the high temperatures achieved by the brakes. But, the heat is there. Put your hand on your tires after a fast run on the highway and see how they feel. Touch the rear-axle housing and the transmission. A portion of the power of the engine was lost supplying the energy to heat those parts.

The other way you can slow a car without the brakes is by coasting up a hill. The car loses energy to drag the same as on a flat road, but it slows quicker. The kinetic energy is converted into *potential energy* as the car climbs the hill.

Potential energy is just another form of stored mechanical energy. It is increased when a weight is raised to a greater height. As the car climbs a hill, some of its kinetic energy is converted to potential energy. This potential energy can be changed back into kinetic energy by allowing the car to coast down the hill with the engine shut off. The speed at the bottom of the hill will be less than the original speed when you began to coast up the hill because some energy is lost in friction and drag going up the hill, plus the energy lost going down the hill.

After coasting half way down the hill, car has lost half of its potential energy, but has gained kinetic energy. Kinetic energy increases as car speed increases.

All potential energy of coasting car is converted into kinetic energy at bottom of the hill where car reaches maximum speed and kinetic energy.

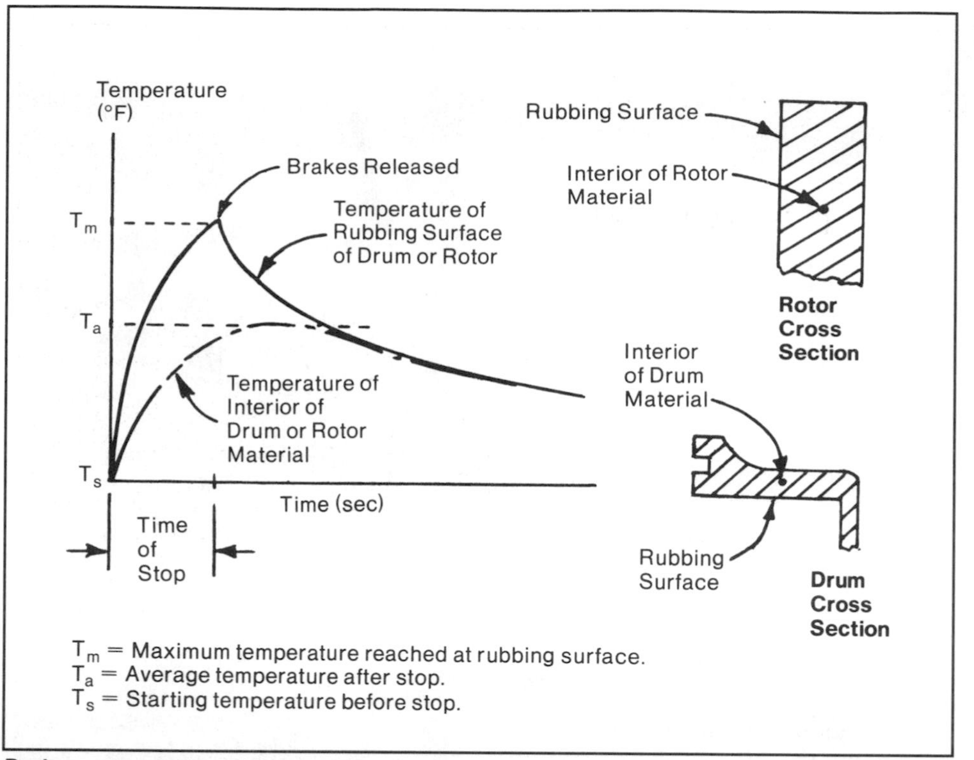

During a stop, brake rubbing-surface temperature increases more rapidly than interior temperature of a drum or rotor. Eventually, temperatures equalize *after* brake is released. Average brake temperature occurs at a point between rubbing-surface and interior temperatures before much cooling takes place.

The heavier a drum or rotor, the lower its temperature rise during a single stop. Designed for a 3000-lb sports car powered by a 300-HP engine, this large drum weighs 22 lb.

Potential energy is measured in units of foot-pounds (ft-lb). Potential-energy change equals the weight of the object multiplied by the change in height. If a 3000-1b car coasts up a 100-ft-high hill, it gains 300,000 ft-lb of potential energy. If all this energy is converted back into kinetic energy, the speed of the car can be calculated from the formula on page 8. By changing the formula using algebra, it comes out as follows:

$$\text{Speed} = \sqrt{\frac{29.9E_p}{W_c}} \text{ in miles per hour (mph)}$$

E_p = Potential energy of car in foot-pounds (ft-lb)
W_c = Weight of car in pounds (lb)

For our example,

$$\text{Speed} = \sqrt{\frac{(29.9)(300{,}000)}{3000}} = 55 \text{ mph.}$$

Obviously, this is much faster than would happen if you tried it with a real car. The difference between this sample problem and a real test is the kinetic energy lost in drag and friction.

How hot do brakes get during one stop?—Because the brake-rubbing surfaces are heated by friction, it is important to know what the temperature is after one stop. A high surface temperature can cause fade or damage to a brake. The problem is that heat is constantly being transferred from the friction surface of the drum or rotor to air and cooler interior metal. This makes the exact surface temperature difficult to calculate. However, it is easy to calculate the *average* temperature of a drum or rotor after one stop. The difference between surface temperature and average temperature of a drum or rotor is shown above.

In calculating the average temperature of a drum or rotor, you must make some assumptions. This makes the calculation easier and the answer more accurate. Tests have shown that the following assumptions are valid:

- Assume that all heat energy from the stop flows into the drum or rotor. In fact this is an accurate assumption because the friction material insulates the rest of the brake from heat and the metal drum or rotor is a very good heat conductor.
- Assume drag on the car from all sources is zero, including the effects of air drag, rolling resistance and engine braking. This is a good assumption for stops below 100 mph because air drag is small compared to braking forces.
- Ignore kinetic energy stored in rotating parts of the car. This assumption and the assumption of zero drag cause errors in opposite directions. Thus, the error in the brake-temperature calculation is small.
- Ignore cooling of brakes during stop. Heat flow into rotor or drum material is rapid compared to cooling time.

The first step is to calculate the temperature change of the drum or rotor. Any increase in temperature is known as *temperature rise*. Temperatures are always related to energy by change—not by the absolute values of energy or temperature. It is important to think of this as a *change* in energy causing a *change* in temperature. Temperature rise in the brake is caused by a kinetic-energy reduction in the moving car.

For a particular stop, figure the change in kinetic energy using the formula on page 8. This results in the following relationship:

K_c = Kinetic energy change in foot-pounds = K_B - K_A in foot-pounds
K_B = Kinetic energy before the stop in foot-pounds
K_A = Kinetic energy after the stop in foot-pounds

Obviously, if the car comes to a complete halt, K_A is zero. The *change* in kinetic energy is used to compute the temperature rise of the brake. For the weight of the brake, use only the weight of the drums and/or rotors. The temperature rise must be added to the temperature of the brake before the stop to obtain final brake temperature.

The temperature rise of the brakes is calculated as follows:

$$\text{Temperature rise} = \frac{K_c}{77.8\,W_B}$$

in degrees Fahrenheit (F)
W_B = Weight of all the rotors and drums in pounds

For example, let's compare the temperature rise in the brakes of a 3500-lb sedan stopping from 60 mph to the temperature rise of a 3500-lb stock-car's brakes slowing from 120 mph to 60 mph.

First let's figure the temperature rise for the sedan. Assume the brakes weigh 5 lb each, for a total brake weight of 20 lb. Calculate the change in kinetic energy slowing from 60 mph to stop. From page 8, the kinetic energy of a moving car is:

$$\text{Kinetic energy} = \frac{W_c S^2}{29.9}$$

For our sedan traveling at 60 mph:

K_B = (3500 lb)(60 mph)^2/29.9
= 421,000 ft-lb.
K_B = Kinetic energy before stop

Kinetic energy after the stop = 0, because car now has zero speed. Change in kinetic energy = 421,000 ft-lb. From the above formula temperature rise of the brakes is:

$$\text{Temperature rise} = \frac{(421{,}000\ \text{ft-lb})}{(77.8)(20\ \text{lb})}$$
$$= 270\text{F}\ (132\text{C})$$

For stock car slowing from 120 to 60 mph, change in kinetic energy is:

$$K_B = \frac{(3500\ \text{lb})(120\ \text{mph})^2}{29.9}$$
$$= 1{,}686{,}000\ \text{ft-lb}$$

$$K_A = \frac{(3500\ \text{lb})(60\ \text{mph})^2}{29.9}$$
$$= 421{,}000\ \text{ft-lb}$$

$K_C = K_B - K_A$
K_C = (1,686,000) − (421,000)
K_C = 1,265,000 ft-lb.

K_A = Kinetic energy after stop
K_B = Kinetic energy before stop
K_C = Kinetic-energy change

Notice how much greater this change in kinetic energy is compared to the sedan. Even though the speed reduction was the same 60 mph, the speed reduction occurring at a higher initial speed resulted in much greater energy put into the brakes. The temperature rise for the stock car in the race is:

Temperature rise = (1,265,000 ft-lb)/(77.8)(20 lb) = 813F (434C).

However, because of less cooling time, racing on a medium-speed track with short straights might result in higher brake temperatures than a high-speed track with long straights. Both temperature rise per stop and cooling time are critical to brake performance. Rotors weighing more than 5 lb would normally be used on a 3500-lb race car. Heavier rotors reduce the temperature rise per stop.

Tires on this car have reached their maximum coefficient of friction. Although a car stops faster if the wheels are not locked, most drivers hit the brakes too hard during a panic stop. This test, being conducted at Goodyear's San Angelo, Texas, test track, is to see how tires react during a panic stop. Photo courtesy Goodyear.

DECELERATION

Deceleration is a measure of how quickly a car slows. Deceleration means slowing the car—acceleration means speeding it up.

Both acceleration and deceleration are measured in units of *gravity—g's.* One g is the force exerted by an object due to gravity at the Earth's surface—how much an object weighs while at rest. One g is also a measure of acceleration or deceleration—22 mph per second. Zero g occurs in a weightless environment. Acceleration is positive; deceleration is negative.

Once the brakes are applied, a car is stopped by the friction force between the tires and the road. During braking, friction acts on the tires in a direction opposite to movement. The higher the deceleration, the greater this friction force becomes. Maximum possible deceleration occurs at the maximum *coefficient of friction* between the tires and the road. This happens just as the tires are about to skid. Once tires lose traction and skid, deceleration drops. Read on for an explanation of coefficient of friction, or simply *friction coefficient.*

INERTIA FORCES

More than 200 years ago, Sir Isaac Newton wrote the basic law relating a force on an object to its acceleration. Simply stated, Newton's Law is:

$$\text{Acceleration of an object} = \frac{F}{W}\ \text{in g's}$$

F = Unbalanced force on object in pounds
W = Weight of object in pounds

In this formula, the force F causes acceleration. If a car weighing 3000 lb has a braking force of −1500 lb, the stopping force is −1500 lb ÷ 3000 lb = −0.5 g.

If the force pushes in a direction to cause the object to speed up, the force is positive and acceleration is a positive number. If the force causes the object to slow, the force is negative and acceleration is negative. We call the negative acceleration *deceleration.*

The force referred to in Newton's Law is *unbalanced* force. That means if the force is not resisted by opposing force, the object is free to move, or accelerate. If I push on a tree with a force of 100 lb, the tree will not move. The 100-lb force is resisted by the tree roots with a 100-lb force, so there is no unbalanced force on the tree. To cause acceleration, either positive or negative, the force on the object must be unbalanced.

To understand how unbalanced forces work on a car, assume you are driving a car on a drag strip. At the start, you let out the clutch and push down on the accelerator, and the friction force between the tires and road pushes the car forward. At the start, air drag is zero, so the forward force is not resisted by anything but drive-line friction and tire drag, or rolling resistance. The forward force is almost unbalanced. Consequently, acceleration is maximum at the start.

As the car gains speed, air drag increases. It opposes the forward force on the tires that is trying to accelerate the car. The unbalanced force is the forward force minus the rearward force. As speed increases, the unbalanced force becomes smaller because of air drag. Consequently, acceleration also gets smaller.

Near the end of the drag strip, the car is moving so fast that the air-drag force approaches the forward force of the tires against the road. Assuming engine rpm and track length aren't limiting factors, speed will increase until no force unbalance exists and acceleration becomes zero. The car has reached its maximum speed. The only way to accelerate the car to a higher speed would be to increase the forward force (more engine power), or reduce the rearward force (less air drag). Every race-car designer knows that more power or less drag will increase car speed.

In addition to the basic relationship between force and acceleration, Newton came up with other important laws of nature. He discovered that every moving object has *inertia*. That is, an object always moves at the same speed and in the same direction until acted on by an unbalanced force. Inertia is what keeps the Earth moving around the Sun and keeps satellites in orbit. There is no air drag in outer space, so once an object is at speed, it keeps moving forever. Only a force can change its speed or direction.

A car also acts according to Newton's laws. Once it is moving, a car wants to continue in a straight line at the same speed. Every part of the car and its passengers also want to keep moving. When the brakes are applied, a force is applied to the car by the tires, causing deceleration.

Because passengers tend to continue at the same speed, they will move forward in the seat and strike the instrument panel unless held by restraint devices, legs or friction against the seat. This forward force that tends to "throw" a passenger forward during braking is called *inertia force*. Inertia force acts on everything in a car as its speed changes.

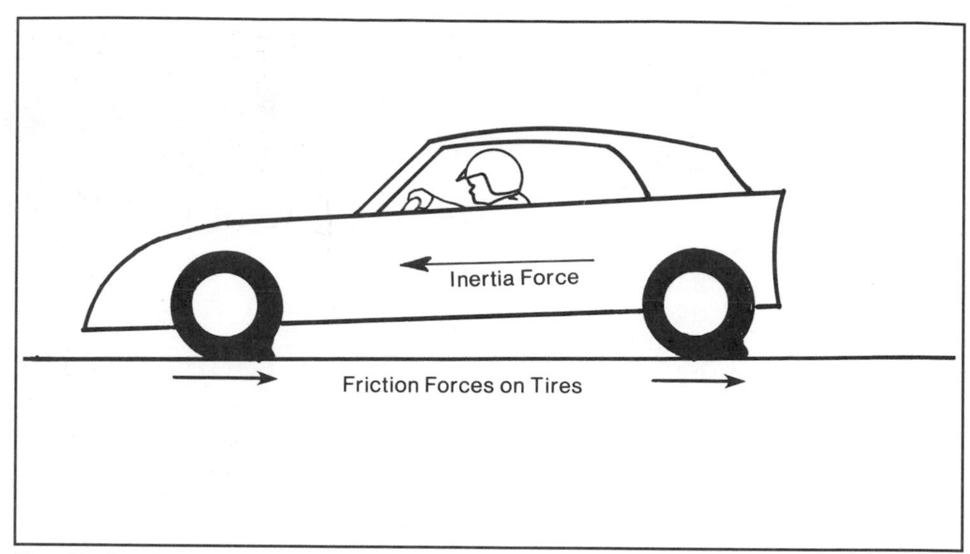

Tire friction forces are external forces that cause deceleration. The road pushes on the tires in the opposite direction of motion. Inertia force of the car acts in the same direction as motion and equals friction force.

Therefore, if a car accelerates, the inertia force acts toward the rear; if it decelerates, the inertia force acts forward.

Inertia force in g's is equal to car's acceleration or deceleration. Inertia forces are easy to calculate in pounds if you know the acceleration.

W = Weight of the object in pounds
a = Acceleration of the object in g's

Inertia force is measured in pounds. If the car is decelerating, the inertia force is *negative*—it acts in a forward direction.

Try visualizing what happens when a car decelerates. It is easier to visualize the forces if you imagine a driver trying to stop so quickly that he locks the wheels. Imagine a car skidding with all four wheels locked and smoke pouring off the tires. Assume the car weighs 3000 lb and the coefficient of friction between the tires and road is 0.7. The friction force on all four tires is:

Friction force = μF_N
μ = Coefficient of friction between two sliding surfaces
F_N = Force pushing the two surfaces together in pounds

In this example,

$\mu = 0.7$ and $F_N = 3000$ lb
Friction force = (0.7)(3000 lb) = 2100 lb.

This friction force on the tires tries to decelerate the car. The deceleration is easy to calculate from Newton's law, page 11:

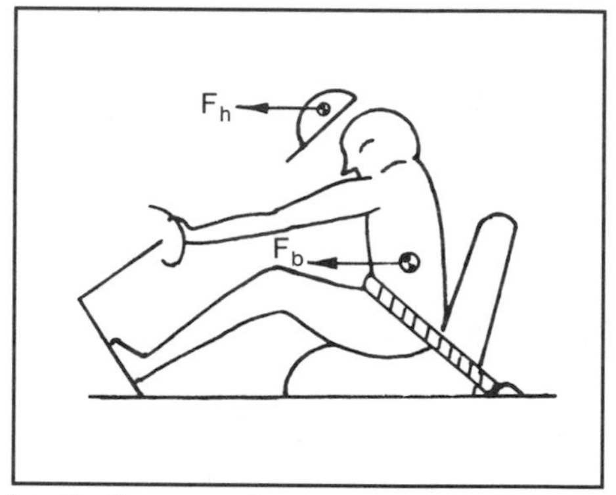

Inertia forces act on every part of a car. This passenger resists inertia force on his body with his arms against the dash, feet against the floor, and seat belt around his waist. Inertia force on his hat is not resisted by anything, so it flies forward into the windshield.

Acceleration of car = F/W_C
F = Unbalanced force on car in pounds
W_C = Weight of car in pounds

Acceleration of car = (-2100 lb)/(3000 lb) = -0.7 g.

The unbalanced force has a minus sign because it acts in a direction opposite to the car. Remember: Negative acceleration means deceleration; it causes the car to slow.

Because deceleration is 0.7 g, every part of the car has an inertia force on it of 0.7 g. The inertia force on a 200-lb passenger is:

Inertia force = Wa
W = Weight of the passenger in pounds
a = Acceleration of the passenger in g's

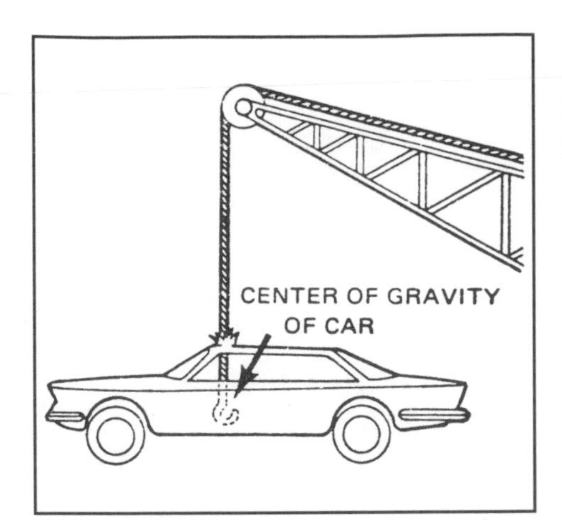

If there were some way to support a car at its center of gravity (CG), it would be balanced. Rotate the car to another position, such as on a side, and it would remain balanced. The CG is the *only* support point where balance exists regardless of position.

A front-wheel-drive car, such as this VW, has majority of weight at front. When brakes are applied hard, more weight is transferred to the front. The nose drops and tail rises during braking due to weight transfer.

Inertia force = (200 lb)(-0.7 g)
= -140 lb.

The minus sign means the force is forward. If there is little friction between the passenger and the seat, the 140-lb inertia force acts on his seat belt or legs.

In this example, the car decelerates at only 0.7 g. With the wheels locked and sliding, the coefficient of friction is lower than its maximum possible value. Maximum deceleration of a car is determined by many factors, including tires, aerodynamic forces, road condition and brake-system design. Most cars on street tires can reach about 0.8-g deceleration on dry pavement. Race cars can decelerate at well over 1 g.

WEIGHT TRANSFER

A car's inertia force acts at its *center of gravity,* or CG, of the *whole car.* The CG is the point about which the entire car is balanced. If you could hang the car by a cable attached at its CG, the car would balance in any position.

The CG is the center of all the weight. All the inertia forces on the individual parts of the car added together are the same as a single inertia force for the whole car acting at its CG. Because a car's CG is always above the road, inertia force from braking always tries to load the front tires and lift the rears. This effect is called *weight transfer.*

Weight transfer means that the front tires are loaded more during a stop, and the rear tires are unloaded. The weight of the whole car does not change. Weight added to the front tires is subtracted from the rear tires during weight transfer. The forces acting on a car during braking are shown in the accompanying drawing. In this simple illustration, aerodynamic forces are not shown. Aerodynamic forces change the amount of the forces, but not the basic principle of weight transfer.

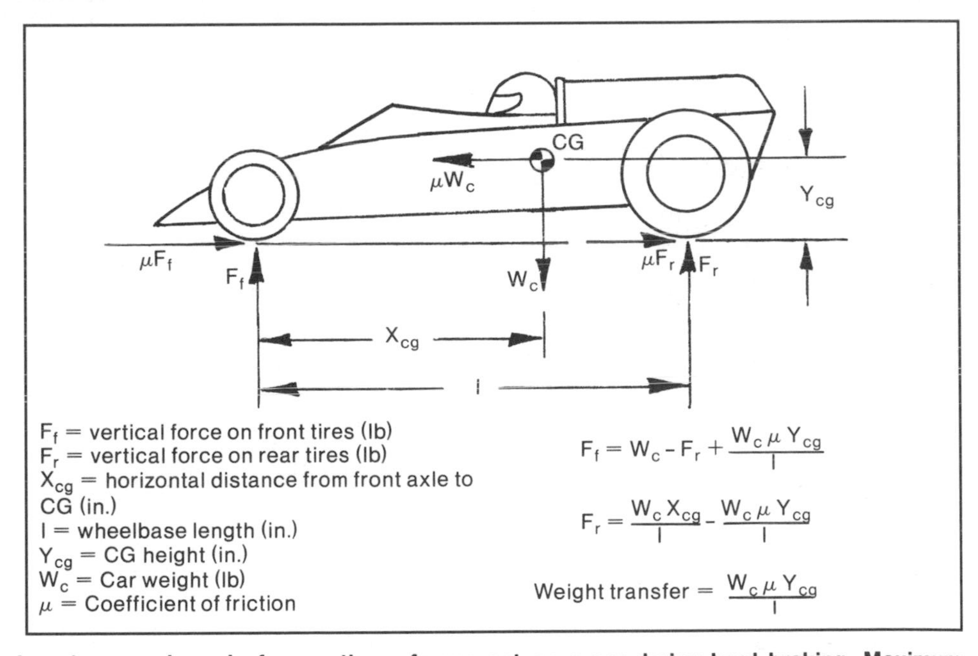

Ignoring aerodynamic forces, these forces act on a car during hard braking. Maximum weight transfer can be calculated if tire coefficient of friction is known.

Because weight transfer loads the front tires, additional friction force can be developed by the front tires before they skid. To produce this extra friction force, front brakes have to work harder than if there was no weight transfer. At the same time, the rear brakes can do less work. Typically, front brakes supply about two-thirds of the total braking force in a hard stop. The ratio is even higher on an extremely nose-heavy car. Because the front brakes do most of the work, they need to be larger than rear brakes. The forces are usually higher on front brakes, so they must absorb more heat energy.

BRAKE FADE

Brake fade is loss of braking due to overheating. It can cause longer pedal travel as the brakes get hotter—maybe to the point where the pedal goes to the floor. Pedal effort may also increase as heat builds up, even to the point where pushing with maximum force won't lock the wheels! Many times, fade causes a combination of both longer pedal travel and increased pedal effort.

Whatever fade is, the driver is faced with a panic situation. Brake fade typically occurs at the worst possible moment—going down a long hill pulling a trailer, at the end of a long race, or during a panic stop from freeway speed in heavy traffic. This is when you need your brakes the most. Consequently, brake fade is always scary, and often dangerous.

Geoff Brabham is at the traction limit of his Toyota Celica in this turn. Combination of braking and cornering forces causes tires to slip and point at extreme angles to the direction of travel. When car is at its traction limit, any attempt to corner, accelerate or brake harder will throw it into a skid.

BRAKING LIMITS

With a modern brake system, how good can brakes be? What determines the limits to brake performance? What makes your car stop quicker than the next car, or vice versa?

There are limits that determine how quick a car can stop. Some of these limits can be altered by design or maintenance, so only the basic laws of nature limit a car's stopping ability. Brake-performance limits are:

1. Force
2. Deflection
3. Wear
4. Temperature
5. Tire traction

A brake system should be designed and maintained so that tire traction determines how quickly your car can stop. If any of the other four limits keep you from stopping quicker, your brakes are not adequate.

Force limit means the driver pushes as hard as possible with his foot and the car can't stop any quicker. In other words, if the driver could push harder, the car would stop quicker. This limit can be altered by reducing master-cylinder size, putting on different lining, using power-assist brakes, or other methods. I discuss how to reduce the force a driver has to exert later in the book. In some cases, a force limit occurs when the brakes get hot. This is called *brake fade.* The answer here is dissipating heat. Perhaps the force limit you've encountered is really a temperature limit. How to handle excess heat is discussed in Chapters 10 and 12.

Deflection limit is reached as the brake pedal stops at the floor or stop. This means the pedal is moving too far to get maximum efficiency from the brakes. A deflection limit can be eliminated by design changes such as stiffening the pedal-support structure, increasing master-cylinder size, installing stiffer brake hoses, changing to stiffer calipers, or other modifications. Maintenance can eliminate a deflection limit if air is trapped in the brake lines.

Wear limit won't happen when brakes are new. However, if friction material is worn excessively, it may be worn out just when you need the brakes most—such as at the end of a long race. Wear limits can be eliminated or reduced by changing linings, using larger brakes, or by dissipating heat. Brake wear is discussed in Chapters 4 and 12.

Temperature limit: Brakes cannot absorb the full power of an engine continuously without some time to cool. When the temperature limit is reached, you can reach a force limit, deflection limit, or greatly increase the wear at the same time. Other things can happen, too, such as complete destruction of the brakes or total collapse of a structural part. Excessive temperature is a common cause of brake problems.

Traction limit: If brakes are properly designed and maintained, and don't get too hot, the only stopping limit is tire traction. If you try to stop quicker than the traction limit allows, the wheels lock up and the tires skid. The traction limit is *always* the limit with good brakes, but it can be increased through correct adjustment of *brake balance.* Adjusting brake balance is discussed in Chapter 10. Modifications to allow brake-balance adjustment are described in Chapter 12.

Drum Brakes

2

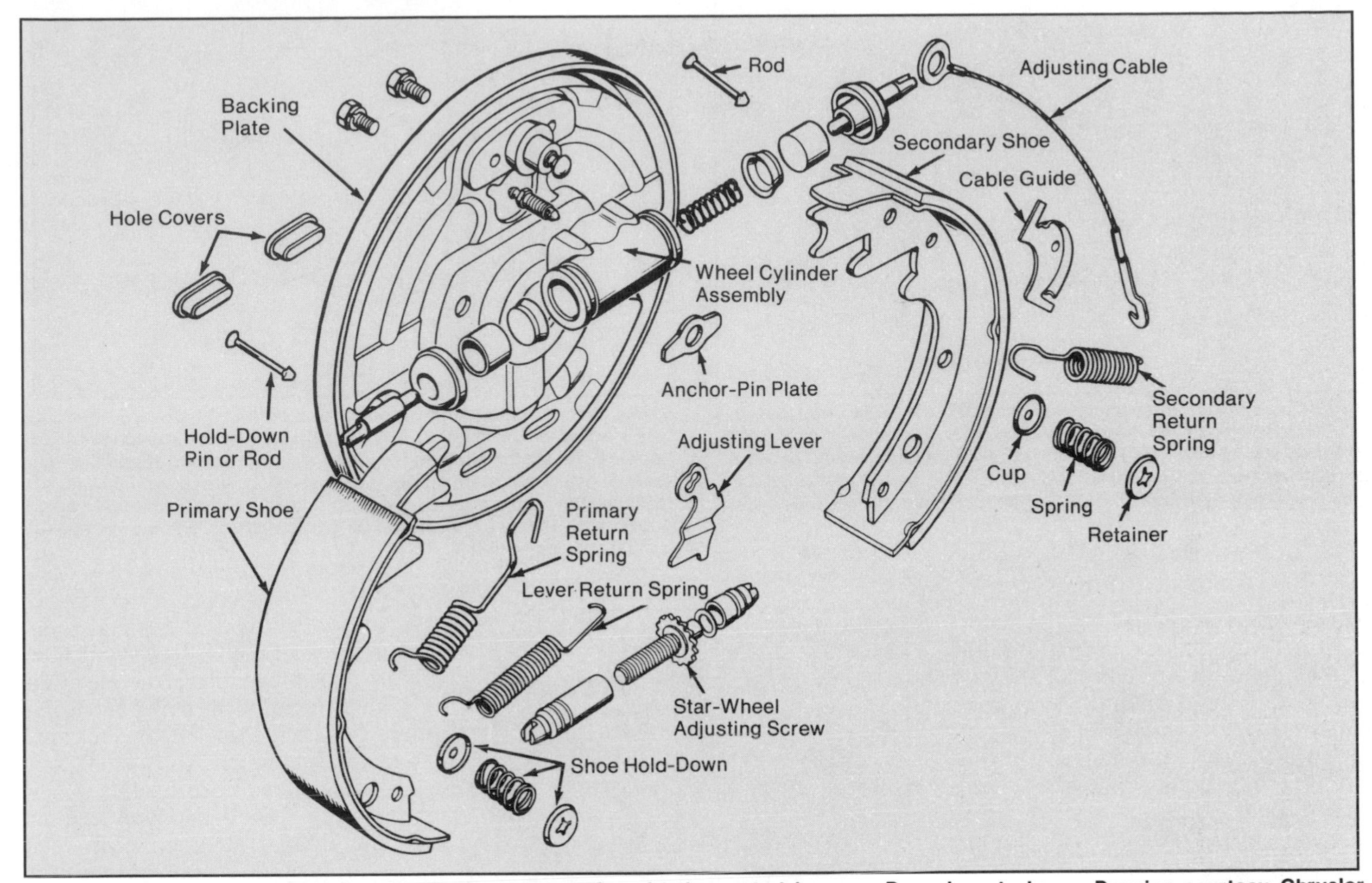

Modern drum brake automatically adjusts shoes outward as friction material wears. Drum is not shown. Drawing courtesy Chrysler Corporation.

Most cars have used internal drum brakes over the years. They continue to be used on the rear of most road cars.

Even though drum brakes share common features, details may differ. Each has a metal drum, usually cast iron. The drum rotates with the wheel. Within the drum are *brake shoes* lined with friction material. This material, consisting of various organic and metallic compounds, is the *brake lining.* The brake shoes are moved against the inside of the drum by pistons inside the *wheel cylinders.* Hydraulic fluid under pressure in the wheel cylinders moves the pistons. Wheel cylinders and brake shoes are mounted on a metal *backing plate.* This backing plate is bolted to the car's axle housing or suspension upright.

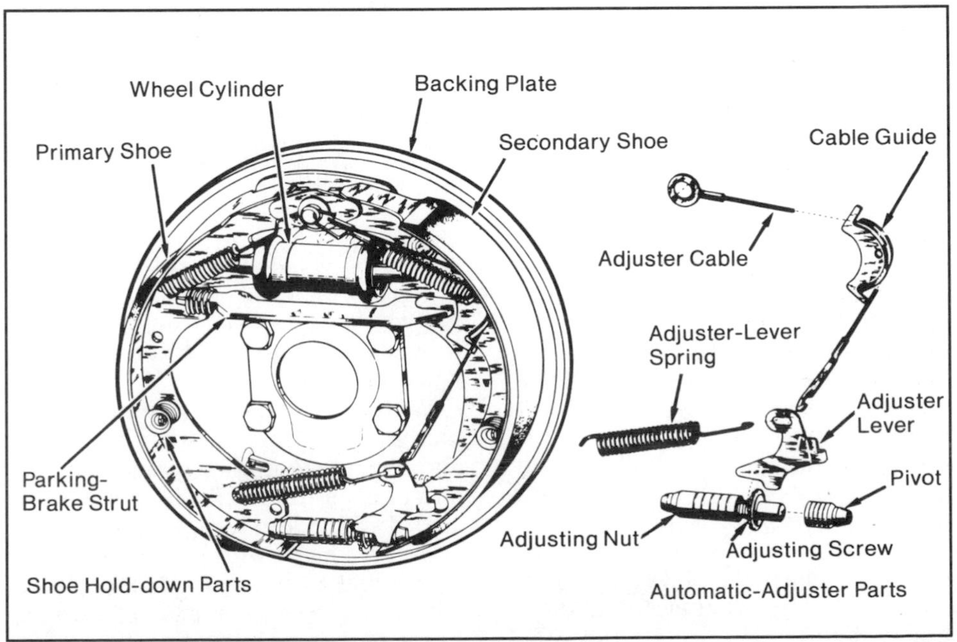

Bendix duo-servo rear brake is typical of drum brakes found on American cars. Brake features automatic adjuster and high servo action. Included is linkage to operate shoes from the parking-brake cable. Drawing courtesy Bendix Corp.

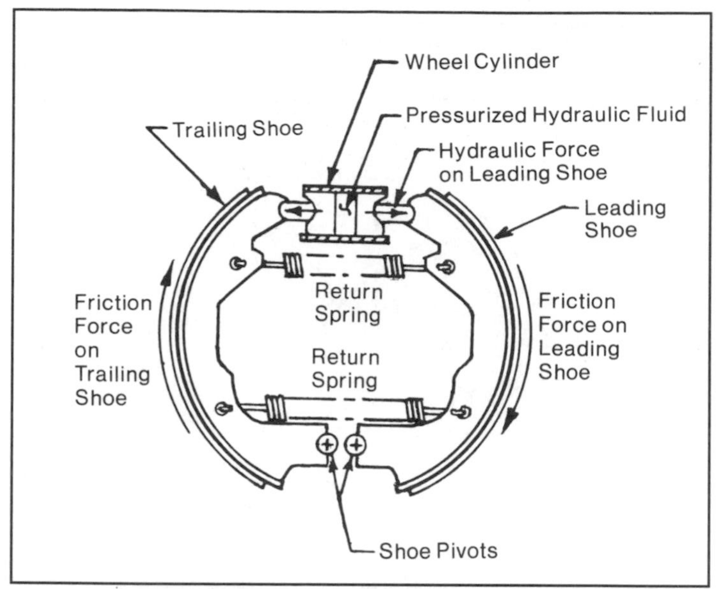

Single leading-shoe drum brake—sometimes called a *leading-and-trailing-shoe* brake—has one leading and one trailing shoe. As drum rotates clockwise, friction force on leading shoe forces it against drum, creating servo action, or force multiplication. Drum rotation tends to reduce shoe-to-drum force of trailing shoe.

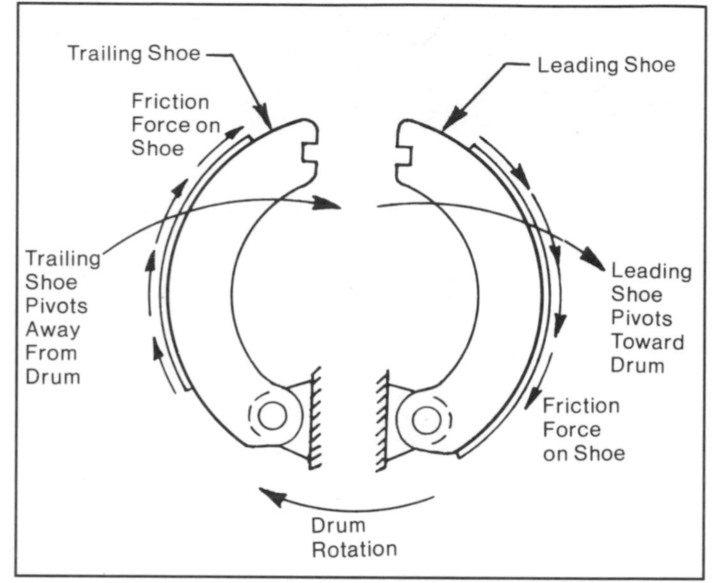

Shoes in this single leading-shoe brake both pivot toward the right when acted on by friction forces. Although wheel-cylinder force keeps both shoes against the drum, friction forces modify pressure exerted by each shoe. Leading-shoe pressure increases; trailing-shoe pressure decreases. There is little overall servo action with this type of brake, as friction-force effects on shoes cancel out each other.

SERVO ACTION

There are many variations of simple drum-brake design. Designs differ in the amount of *force-multiplication,* or *servo action.* Imagine driving one car with standard brakes and another equipped with power-assist brakes. The amount of pedal force is greatly reduced with power-assist brakes. Servo action acts much like power assist—it reduces the force required on the brake pedal for a given amount of braking. However, servo action occurs within the brake itself.

Leading or Trailing—To help understand servo action, let's look at how brake shoes are mounted. Imagine a brake shoe pivoting at one end and a wheel cylinder pushing on the other. There are two ways to mount a brake shoe, as *leading shoe* or as *trailing shoe.* This depends on which end of the shoe pivots in relation to drum rotation. If the brake drum rotates from the free (wheel-cylinder) end of the shoe toward the pivoted end, it is a leading shoe. If the brake drum motion is from the pivot end toward the free end, the shoe is trailing.

Now let's look at the forces applied to each type of brake-shoe arrangement. The friction force on a leading shoe tends to rotate the shoe around its pivot and against the drum. This assists the wheel cylinder in applying the brake shoe. A trailing shoe is just the opposite—the friction force moves the shoe away from the drum, thus counteracting the force of the wheel cylinder.

Single leading-shoe brake is generally found on rear wheels only. Although brake has low servo action, it works equally well in both directions.

With a leading shoe, drum rotation increases pressure between the shoe and drum, giving increased friction and braking force. This is *servo action.* The driver doesn't have to push the pedal as hard with a leading-shoe drum brake. It is just the opposite with a trailing-shoe brake. The driver has to push very hard.

When a car with leading-shoe brakes on all wheels is backed up, the reversed direction of the wheel rotation changes all the leading shoes into trailing shoes. The driver notices this as a huge increase in pedal force required to stop the car. For this reason, drum brakes have a mixture of leading and trailing shoes—two leading shoes on the front wheels and one leading and one trailing shoe on each rear wheel. This gives some amount of servo action and still allows one set of shoes to work as leading shoes when braking in reverse.

Duo-Servo—Another type of brake, which differs from leading- or trailing-shoe types, is the *duo-servo* drum brake. Its features are shown in the accompanying drawing.

The duo-servo brake does not have a simple pivot on its brake shoes. The shoes are connected to each other at the end opposite the wheel cylinder by a floating link. This link transmits the force and motion of one shoe to the other shoe. An anchor pin next to the wheel cylinder keeps the shoes from rotating with the brake drum. The two shoes are called the *primary shoe* and the *secondary shoe.* The primary shoe pushes on the secondary shoe through the link; the secondary shoe pushes on the anchor pin. When braking in reverse, the action reverses and the shoes change roles. The primary shoe then becomes a secondary shoe in the way it works.

Typical duo-servo front brake used on large front-engine, rear-drive American cars; anchor pin is above wheel cylinder. Adjuster joins lower ends of shoes. Only rigid connection between backing plate and shoes is at anchor pin.

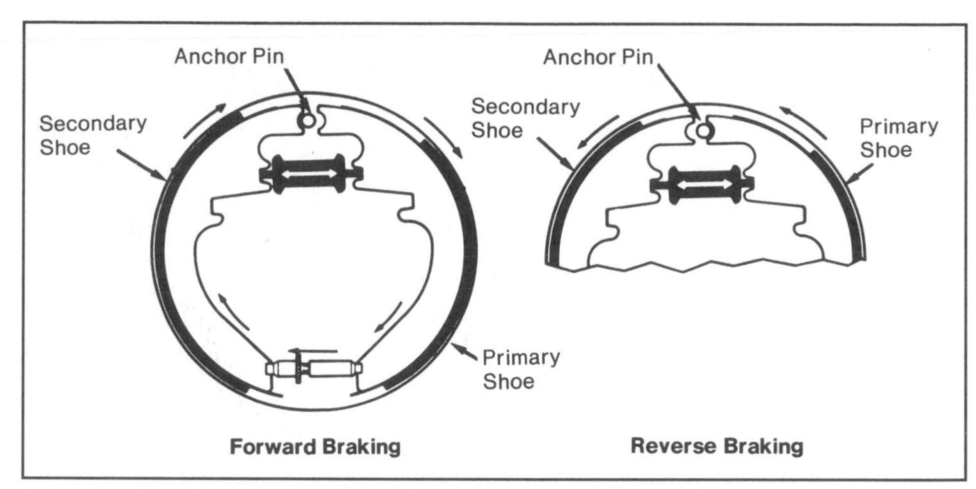

Duo-servo drum-brake operation: Notice that primary shoe is pushed away from anchor pin in forward braking. It moves until it is stopped against the drum. When backing up, rotation reverses and other shoe acts as primary shoe. Drawing courtesy Bendix Corp.

Duo-servo brakes have servo action regardless of rotation. The brake is designed for mostly forward motion or rotation. Lining wear is equalized between the primary and secondary shoes by putting more lining on the face of the secondary shoe. However, the driver can feel little difference in servo action when braking in reverse. Duo-servo brakes were used on most American cars with rear-wheel drive and drum brakes before the use of disc brakes. Because duo-servo brakes have the most servo action, they work well on heavy cars.

Pedal Effort—Let's compare the three types of brake-shoe arrangements and see how the servo action affects *pedal effort.* Pedal effort is the force the driver applies to the pedal. In the accompanying chart, it is obvious that the duo-servo brake has less pedal effort for a given rate of deceleration.

Now comes the bad part—we never get something for nothing. Let's see what happens when brakes get too hot. With most brake-lining material, friction decreases with increasing temperature. As an example, assume the friction coefficient between the brake drum and lining drops from 0.5 to 0.4 when it gets hot. The comparison is shown in the accompanying table for each type of drum brake. The duo-servo drum brake loses the most braking force. Thus, the brake that has the greatest servo action is also the one affected the most by a decrease in friction.

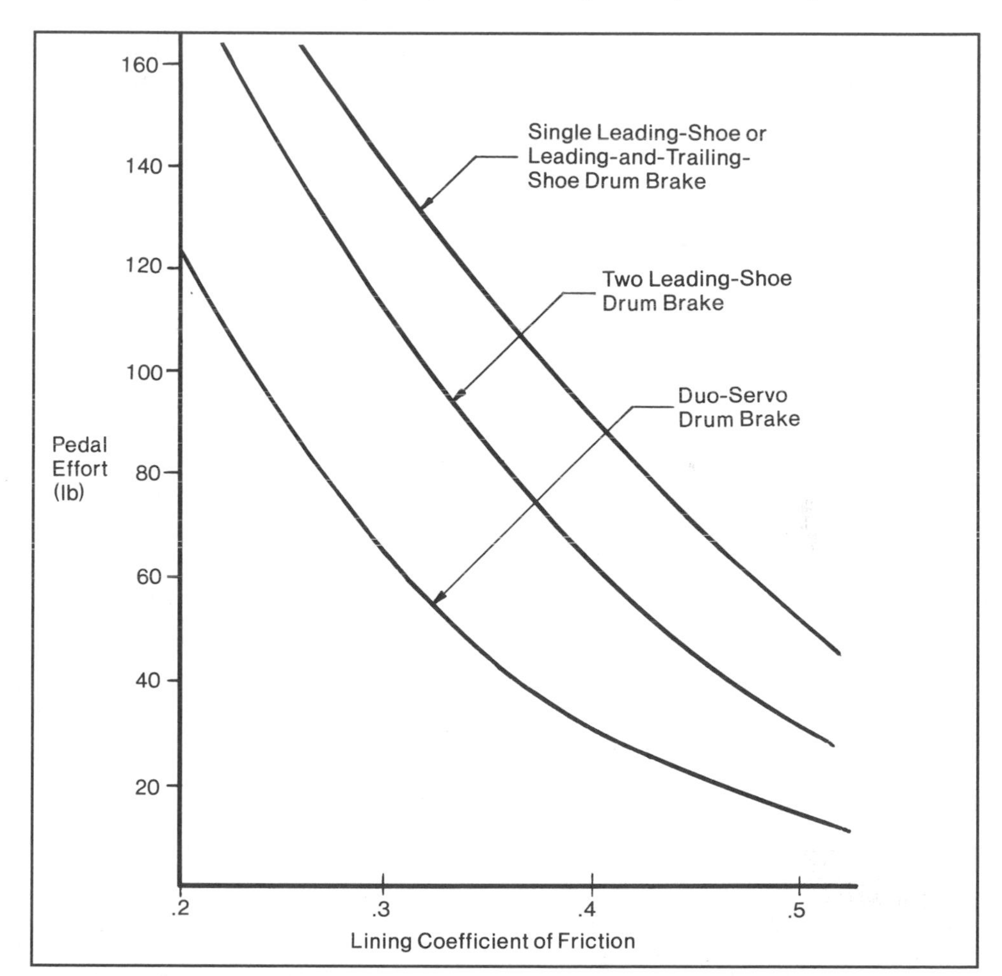

Duo-servo drum brake gives lowest pedal effort for all practical friction coefficients.

Brake Type	Pedal Effort at CF = 0.5	Pedal Effort at CF = 0.4	Percent Increase in Pedal Effort
Duo-servo	50 lb	107.6 lb	115.2%
Two-leading shoe	50 lb	99.8 lb	99.8%
Leading-trailing	50 lb	88.0 lb	76.0%
Disc brake	50 lb	62.5 lb	25.0%

A 0.1 friction-coefficient drop could occur from brake fade. Pedal effort more than doubles with duo-servo drum brakes. Disc brakes have no servo action, so effect of fade is considerably less.

Cast-iron brake drum is large and heavy compared to disc-brake rotor of equal effectiveness. Many '60s and '70s American cars have drums of this type. This front-brake drum has integral wheel hub and bearings.

Duo-servo brakes are most susceptible to the type of brake fade where a force limit is reached. Thus, if they are ever used in a racing or high-performance application, care must be taken to use linings that do not have a drastic drop in friction with increasing temperature. Cars fitted with duo-servo front brakes have a tendency to pull to the right or left if the friction in one brake is slightly higher than the other.

More is said about linings in Chapter 4.

BRAKE DRUMS

The brake drum is a large, critical part of a brake system. If the drum is too small or flexible, the brake will perform poorly under severe use, no matter how good the system may be. Let's see what constitutes a good brake drum. The important properties are:

- Must have a hard wear-resistant rubbing surface and the surface finish must not damage the lining.
- Must be strong enough to withstand the hardest braking, while at high temperatures.
- Must be stiff and resistant to distortion and warping.
- Must dissipate heat rapidly and withstand excessive temperatures.

Most brake drums are made of *grey cast iron,* because it is hard and wear-resistant. Cast iron contains carbon, which prevents galling and seizing when hot. It's also a good *dry* rubbing surface without lubrication.

Cast iron is very rigid, compared to most metals. Therefore, the drum resists distortion under load. Because the drum is cast at high temperature, it resists warpage when repeatedly heated and cooled.

Racers soon discovered that bigger brakes are better. This old Talbot Grand Prix car used 18-in. wheels—brake drums are bigger! Fitting drum into wheel limits its size.

However, cast iron is not the strongest metal in the world; it tends to crack if overstressed. To avoid this, thick sections are used in a drum. This extra metal not only strengthens the drum, but it increases the stiffness of the drum. It also reduces temperature buildup during hard use. When fins are used, they help reduce temperatures quickly by exposing more metal surface to the cooling air. When all design tricks are used, a drum will have adequate strength. In addition, heat dissipation is improved with the fins and extra metal.

Drum Cooling—High-performance brake drums must be *BIG; bigger the better* is the key for reducing temperatures and eliminating fade. A large drum diameter has the added advantage of reducing pedal effort because of the increased "leverage" of the friction material—similar to a longer pry bar. Racing drum brakes are usually the largest possible diameter that will fit inside the wheel.

The size and weight of a brake drum are important in determining how much heat energy it can absorb. Brake drums are measured by the inside diameter of the drum and width of the linings. The drum rubbing surface is slightly wider than the lining to allow clearance. Don't measure a drum to determine lining width.

During later days of race-car drum-brake development, a great deal of work was spent on cooling-fin design. This type of radial-finned drum pumps cooling air between the wheel and drum.

Swept Area—The brake-drum inside circumference multiplied by lining width is the brake *swept area,* an important measure of how effective a brake is. Usually a car's brake specification includes a ratio of brake swept area to the car weight. A figure of square inches (sq in.) per ton is typical. A car with a high swept area per ton will have long-wearing, fade-resistant brakes, with all other factors being equal. A typical road car has about 200 sq in. per ton of swept area; a race car may have twice this amount.

Calculate drum-brake swept area by this formula:

Swept area = 3.14 DL in square inches
D = Brake-drum inside diameter in inches
L = Lining width in inches

To calculate swept area per ton, divide swept area of all four brakes by the car's weight in tons.

To help remove heat and cool the drum, fins are necessary. If the fins and wheel are designed as a unit to promote airflow, they can act as an air pump. Some older race-car drum brakes were highly sophisticated in forcing air around the hot parts, as the photos show.

Bimetallic Drums—To further improve cooling, a high heat-conducting material is used outside the cast-iron rubbing surface. This is usually aluminum, although copper has been tried. Such drums are called *bimetallic.* Aluminum carries heat from the rubbing surface to the cooling fins more rapidly than iron. An additional bene-

Vintage sports car is powered by fuel-injected Chrysler hemi. Because of its high weight and a top speed of over 160 mph, it needed all the brakes it could get. Buick aluminum/cast-iron bimetallic drums were some of the best ever produced for passenger-car use.

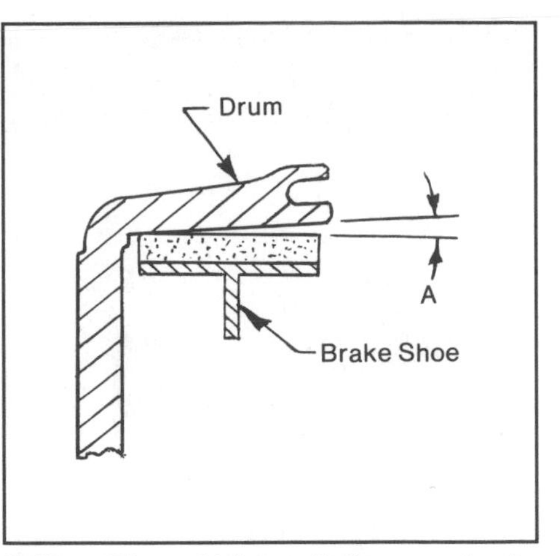

Bellmouthing, which results from overheating and hard use, creates angle A between drum and shoe. This angle causes shoe distortion, extra pedal movement, and faster lining wear on inside edges.

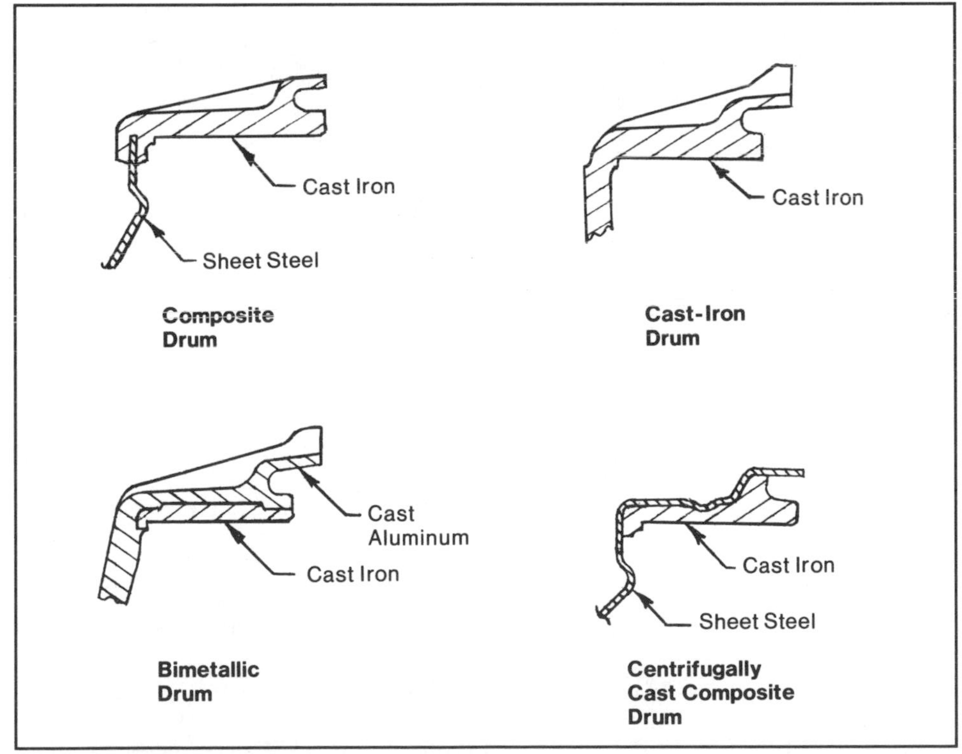

Of commonly used designs, bimetallic drums cool best due to high heat transfer of aluminum. Bimetallic drums are also the lightest.

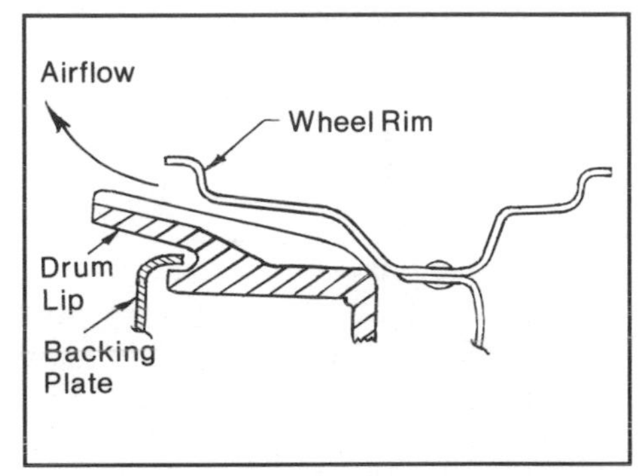

Angled lip on drum can help cooling. If lip has fins, drum rotation can pump air between the wheel and drum. If lip extends inboard past wheel and backing plate, additional cooling is realized.

fit is that aluminum is lighter.

For an aluminum/iron drum to work properly, the aluminum must be attached tightly to the iron. The drum will not cool sufficiently if there are air gaps between the two metals. A once-popular type of bimetallic drum used a special process for casting aluminum around the iron, known as the *Al-fin process.*

There are four basic types of brake-drum construction. The simplest is cast in one piece from iron. Another, lighter drum style, is the *composite* drum. This type uses a sheet-steel hub with an iron rim cast to it. A variation of the composite drum uses a stamped sheet-steel drum with a cast-iron rubbing surface inside it. Last is a bimetallic drum, made of aluminum and cast iron.

Drum Design—The design of the open edge of the brake drum is critical to a drum's performance. The stiffness of the lip at the edge will keep the drum from going *out-of-round* (egg shaped) or *bellmouthing* (diameter increasing at open end) under severe loads and high temperatures. Also, the shape of the lip can determine how well a drum cools. An angled lip can help promote airflow around the drum. Because airflow around a drum is complicated and affected by other parts on the car, drum cooling can only be determined by tests.

Besides stiffening and cooling the drum, the lip also mates with the backing plate. This helps keep dirt and water out of the drum. The usual design has a groove at the edge of the brake drum. The outer edge of the

Fins on Alfa-Romeo aluminum/cast-iron drum are designed to pump air between drum and tight-fitting wheel. Drum goes with brake assembly pictured on page 22. Photo by Ron Sessions.

Early race-car drum brake has finned backing plate to help cooling. Unfortunately, the designer neglected to use vents. And, because the backing plate doesn't get very hot, these cooling fins don't help much.

Large forward-facing scoop was used on 520-HP Auto Union Grand Prix cars from the mid-'30s. These cars used huge drums and hydraulic actuation, but brake cooling with that much power was a real problem.

backing plate is flanged. This flange fits into the groove, but does not touch the drum. This is a seal for dirt and water, making it difficult for foreign matter to blow or splash into the drum. The brake-lining surfaces are thus protected from contamination. A problem with this seal is it also prevents cooling air from entering the interior of the brake.

BACKING PLATE

The backing plate is a bracket on which the brake shoes and wheel cylinders mount. It also serves to protect the interior of the drum from contamination, as just discussed. Braking torque is transmitted from the shoes to the suspension of the car through the backing plate. A good backing plate must be stiff and strong so the shoes stay in alignment with the drum. When a backing plate deflects excessively under load, the driver feels this as a spongy pedal or as excessive pedal travel.

Most backing plates are stamped from heavy sheet steel. Ridges, bumps and edge lip stamped into the backing plate increase its strength and stiffness. Critical areas for strength are the brake-shoe-pivot and wheel-cylinder mounting points. Any cracks or other weakness in those areas can cause a dangerous situation.

Some backing plates on older race cars are cast aluminum or magnesium for lightness. These are not good materials for this application because they lack sufficient stiffness and strength. By the time an aluminum or magnesium casting is sufficiently beefed up to equal the strength of a steel backing plate, it will be nearly as heavy.

The backing plate is a precision part. Brake-shoe mounting points must be aligned with the brake drum for proper operation. A bent or twisted backing plate is useless, so inspect each one carefully for damage when working on your car. You should never pry on a backing plate to remove a brake drum. This will bend the backing plate and misalign the brake shoes.

A backing plate is usually bolted to a suspension upright or axle flange with tight-fitting, high-strength bolts. It is important that this joint is tight and cannot shift under load. This brings up an important point. When working on brakes, make sure the bolts are the correct length and strength for the application.

Drum-Brake Cooling—The backing plate has little brake-cooling effect. It never gets as hot as the other brake parts. Therefore, fins or air blowing on the backing plate is ineffective. Efforts to cool a drum brake should always be directed to the drum.

Some backing plates are vented or have scoops mounted on them. The purpose is to direct cooling air into the interior of the brake drum. This setup is found on most drum brakes used for racing, if the rules allow it. The problem with *ventilated* backing plates is that the venting may allow entry of water, dirt or other contamination. This creates braking problems, particularly if one side of the car is affected more than the other. Grabbing or pulling to the side while braking can result.

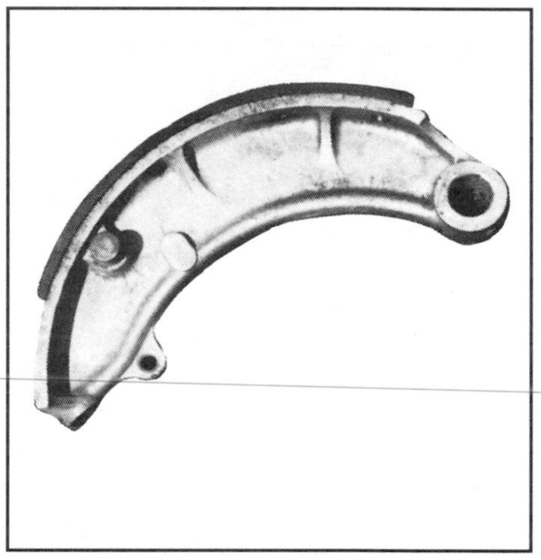

Although most brake shoes are made of steel plate, cast aluminum has been used. Aluminum is lighter, but weakens at extreme temperatures. If you are building a car with aluminum shoes, make sure they won't be subjected to extremely high temperatures. Never use metallic racing-brake linings on aluminum shoes.

A scoop should be designed to keep out contamination. To do this, some people place a screen flush with the scoop intake. This restricts airflow to the brake, thus defeating the purpose of the scoop. A good solution is to use a long air duct with a coarse screen re-

Return springs and hold-downs have been removed from these brake shoes. Spring-loaded hold-downs keep the shoes against flats on the backing plate. The shoe edges rub against these flats as they move in and out.

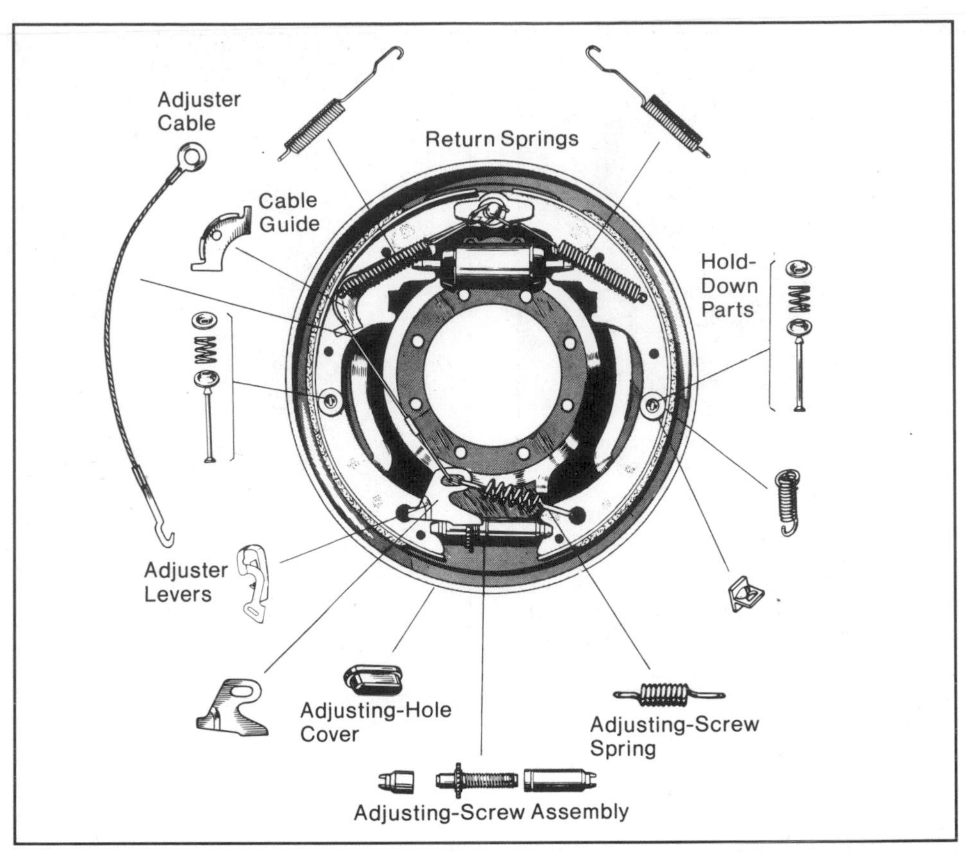

Various types of springs are used in the typical drum brake. This is a duo-servo rear brake with an automatic adjuster. Drawing courtesy Bendix Corp.

cessed into the duct inlet. Thus, only fine particles are likely to reach the brake. The screen will tend to repel stones and other large objects entering the duct, but won't block airflow as much. However, be aware that dust will enter the brake and may increase lining wear and drum-surface scoring.

BRAKE SHOES

Brake shoes are rigid metal assemblies to which friction material is attached. Friction material, or *brake lining,* is riveted or bonded to the brake shoe. Brake shoes are usually sold on an exchange basis so they can be rebuilt by installing new lining on the old shoes. Brake shoes can be relined and used repeatedly unless they have been damaged.

Most brake shoes are fabricated of sheet steel with a tee-shaped cross section. Some are made of cast aluminum. The shoe is shaped to match the inside diameter of the drum, with new full-thickness lining installed. Accurate fit between the lining and drum is assured by machining the lining after it is attached to the shoe. This machining process is called *arcing* the lining. The exact radius of the arc is determined by the inside radius of the drum. If the drum has been turned to an oversize diameter, the lining-arc radius is increased to match.

Brake shoes have rubbing points that are contacted by wheel-cylinder mechanisms and brake adjusters. Also, there's a mechanical linkage connected to the parking brake, which operates the rear brake shoes. These rubbing points are subject to wear and should be lubricated with special high-temperature grease for maximum life.

Brake shoes have holes in them for locating pins and return springs. When buying new shoes, make sure the holes in the new ones match those in the old shoes. Shoes often get mixed up when relining is performed, so carefully comparing old to new is worthwhile. When installing the new shoes, pay attention to where the return springs connect. Some shoes have extra holes for various applications. If a spring is placed in a wrong hole, poor brake performance can result, in the form of grabbing or dragging brakes.

Brake shoes get very hot in racing use, but not nearly as hot as the drum. If the shoe gets too hot and the lining is still working, failure of the lining-to-shoe bonding could occur. Fortunately, most heat generated by a drum brake goes to the drum and not the shoes, so overheated shoes are rare. Brakes using bonded metallic linings are the type for which bonding strength is extremely important.

Lining-material design is more important than brake-shoe design. See Chapter 4 for an in-depth discussion of brake linings.

Usually, brake shoes are held to the backing plate with a spring clip or spring-loaded pin. This pin lightly presses the brake shoe against the backing plate. Flats are provided on the backing plate for the edge of the shoe to rest against. Pressure of the shoe hold-down springs is low, so friction is minimal.

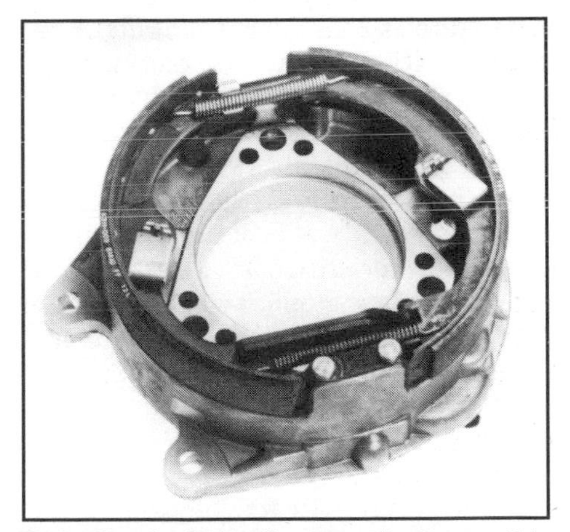

Girlock of Australia manufactures this lightweight mechanically-operated brake for use as a parking brake on '84-and-later Chevrolet Corvettes. It has all the features of a standard drum brake, including an adjuster and return springs. Drum installs in the disc-brake-rotor hat. Photo courtesy Girlock Ltd.

RETURN SPRINGS

Drum-brake shoes are pulled away

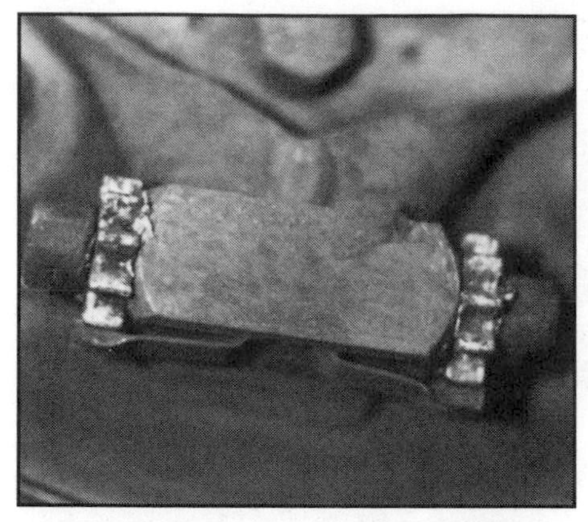

Brake adjuster is operated by turning *star wheels.* This type adjuster is used between the brake shoes on a leading-and-trailing-shoe drum brake. Adjuster housing is mounted to backing plate.

Prior to disc brakes, some drum-brake designs got pretty exotic. This three-leading-shoe drum brake was used on some Alfa Romeo sports cars. Although they did a good job of stopping the little sports cars, they were heavy and complex. Notice that each shoe has its own wheel cylinder.

from the drum—retracted—after application by a set of return springs. Because the force exerted by these springs is high, a special tool is required to install them. There are various arrangements and mounts for the return springs, depending on the specific design.

If the brake has any servo action, a loss of return-spring force can cause the brake to lock or grab. This may be as dangerous as a loss of braking. For instance, if only one wheel loses its return-spring force, that side can lock up and cause a sudden swerve or skid.

Some manufacturers provide different return springs that give a different force when installed. Springs are usually identified by color. Check with your parts supplier to be sure you have the correct ones on your vehicle. This is particularly important if the car is fitted with heavy-duty or racing brakes.

BRAKE ADJUSTERS

The return springs pull the brake shoes away from the drum a *minimum* specified distance. If the distance the shoes must travel before contacting the drum is too great, excess brake-pedal travel results. Thus, some adjustment must be provided to minimize shoe clearance and to compensate for lining wear.

There are two types of brake adjusters, *manual* and *automatic.*

Both manual- and automatic-adjuster designs use a cam or screw-thread mechanism to move the brake shoes toward or away from the drum. The duo-servo brake uses a turnbuckle-like threaded link between the primary and secondary shoes. There are right-hand threads on one end and left-hand threads on the other. A star-shaped wheel, which is rotated manually with a tool through a hole in the brake drum or backing plate, turns the threaded link. This rotation makes the link shorter or longer, increasing or decreasing shoe-to-drum clearance.

Automatic adjusters have a linkage that rotates the star wheel as the vehicle is backed up and brakes are applied. This is a ratchet-type mechanism—it only works in one direction. Automatic adjusters are not actuated in forward driving. There is a limit to the stroke of the actuating linkage. When clearance between the shoe and the drum reaches a certain value, the actuating linkage jumps over the next tooth on the star wheel. That feature prevents the automatic adjusters from overtightening the brakes.

Other types of brake adjusters use a cam or an eccentric bolt for adjusting the shoes. These are actuated manually by turning a bolt that protrudes through the backing plate or by using a screwdriver through a hole in the brake drum.

Adjusting brakes is covered in Chapter 11.

WHEEL CYLINDERS

Brakes are applied by hydraulic fluid inside *wheel cylinders.* One or two pistons in each cylinder are moved outward by brake-fluid pressure when the driver pushes on the brake pedal. Piston movement is transmitted to the movable end of the brake shoe by a pushrod or other linkage.

There are many types of wheel cylinders. Cylinders are made from aluminum or iron. There are single- or dual-piston cylinders, depending on brake design. Duo-servo brakes have one cylinder operating two shoes—two pistons are used with this design. Others have one cylinder on each shoe—one-piston cylinders are used.

Wheel cylinders and other hydraulic components are discussed in more detail in Chapter 5.

Disc Brakes 3

Racing-brake setup on this March Indy Car uses four-wheel disc brakes with dual master cylinders pressurized by remotely adjustable balance bar. Brake cooling is critical at Phoenix International's one-mile oval. Photo by Tom Monroe.

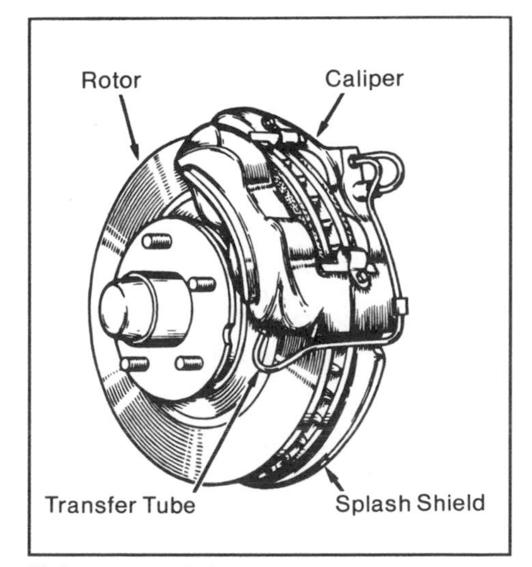

Major parts of disc brake are caliper, rotor and splash shield. A transfer tube connects hydraulic cylinders on opposite sides of this fixed caliper. Many other fixed calipers have internal fluid passages. Photo courtesy Bendix Corp.

Disc brakes are used on the front of most modern road cars, and on all four wheels of most race cars. Disc brakes were first introduced on cars in the late '40s. By the early '70s, disc brakes had replaced drum brakes on the front.

Two common types of disc brakes are single and multiple disc, both with *rotating discs* or *rotors.* Single-disc types have a rotor, which is clamped on by friction material called *brake pads.* Multiple-disc types, commonly used on aircraft, have a number of rotating discs separated by *stators,* or stationary discs. Operation is by a large-diameter hydraulic piston in the backing plate moving outward, clamping the rotors and stators together. Multiple-disc brakes are 100% metallic, while single-disc types use organic/metallic friction material.

ADVANTAGES OF DISC BRAKES

The auto industry changed from drum brakes to disc brakes for a number of reasons:

- More resistant to brake fade.
- Better cooling.
- Water and dirt resistant.
- Less maintenance.
- Greater surface area for a given weight of brake.

The main advantage disc brakes have over drum brakes is their increased resistance to fade. The reasons for this are:

- Friction surfaces directly exposed to cooling air.
- Drum deflection is eliminated.
- Disc brakes have no servo action.

The lack of servo action is a disadvantage with a heavy car. Resulting pedal effort is too high. As a result, disc brakes almost always require

Adapted to a hot rod, early Chrysler Imperial disc brake looks like a drum brake and works like a clutch. Although neat looking, it doesn't work as well as a modern disc brake. Rubbing surfaces on brake are on inside of finned housing.

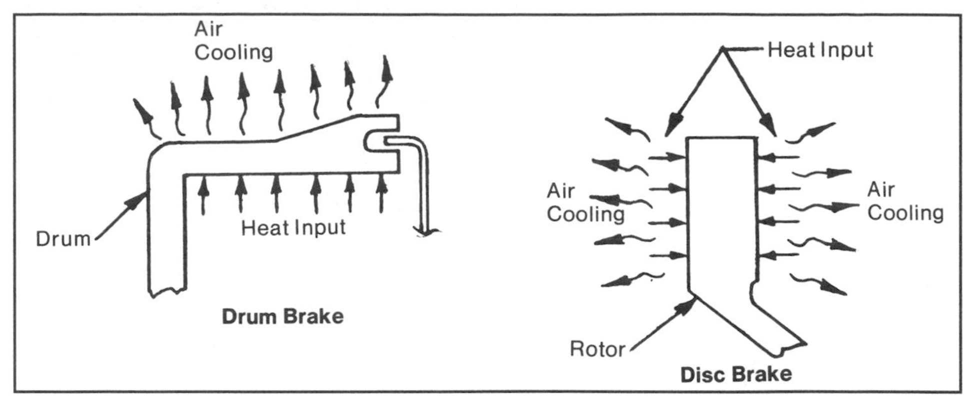

Why a disc brake has superior cooling is illustrated here. Heat generated by brake drum must flow through drum before air can cool the brake. On a disc brake, hot rubbing surfaces are directly exposed to cooling air; heat-to-air transfer begins immediately upon brake application.

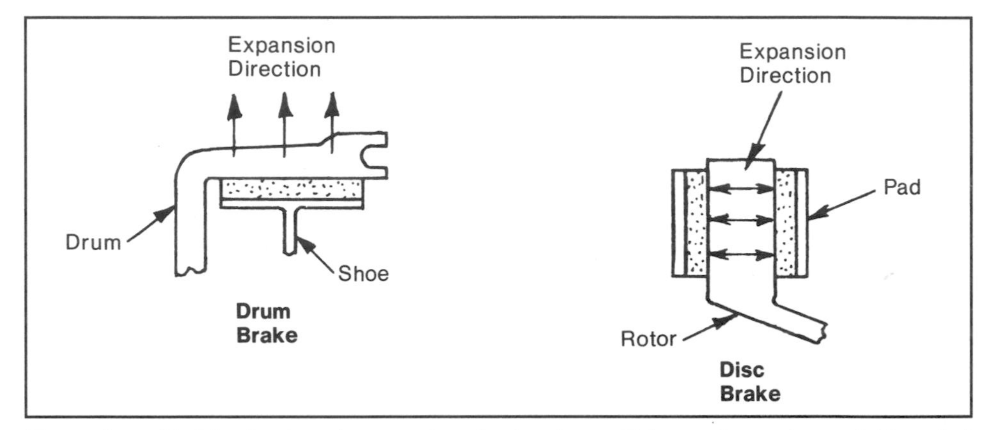

Expansion of hot brake drum is away from brake shoes. This requires extra pedal travel. A disc-brake rotor expands slightly toward the pads. Therefore, extra pedal travel due to heat expansion is not a problem with a hot disc brake.

Disc brakes were used on many race cars in the mid-'50s. This car, built in 1955, is powered by a 300-HP Chevy. Tiny solid rotors would never be used today on such a powerful car.

SPOT BRAKES & DISC BRAKES

The disc brake was first called a *spot brake* or *spot disc brake*. This distinguished it from a type of disc brake that has a clutch-like full circle of friction material. This unusual type of disc brake was used on the Chrysler Imperial in the early '50s, but was discontinued in favor of the traditional drum brake.

Full-circle multidisc brakes are used on large aircraft. Light aircraft typically use more conventional spot-type disc brakes. There's a brief discussion of aircraft brakes in Chapter 9.

To keep things simple, I call the standard automotive disc brake simply a *disc brake* because the term *spot brake* is no longer in common use. And, I call aircraft multidisc brakes simply *multidisc brakes*.

power assist, while drum brakes with high servo action may not. On race cars, weight is much less, so disc brakes usually do not require power assist.

The hottest part of a brake is the metal surface contacted by the friction material. On a drum brake, this surface is inside the drum; on a disc brake, it is the exterior of the rotor. For a drum brake to cool, the temperature of the entire drum must first rise. Then the drum is cooled at the exterior by surrounding air. A disc brake is cooled immediately by air blowing on the disc's rubbing surfaces.

A brake drum expands—ID increases—when it gets hot, increasing pedal travel. The drum can also warp from temperature or braking force. Brake-drum deflection decreases performance and causes fade. However, a disc-brake rotor is essentially a flat plate. Temperature expansion of the rotor is toward the friction material rather than away from it. Squeezing a disc cannot cause sufficient deflection to affect performance.

The lack of servo action means that a disc brake is affected little by changes in friction. As I mentioned earlier, drum brakes with maximum servo action are also affected the most by friction changes. A slight drop in friction, such as that caused by heat, is magnified by the servo action. The driver senses this as fade. Disc brakes are simply not as sensitive as drum

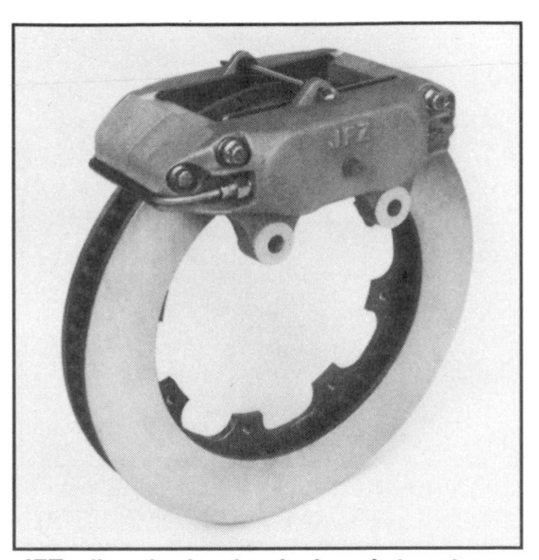

JFZ disc brake is designed to stop a powerful oval-track race car. Four-piston, lightweight caliper is massive to reduce deflections and minimize temperature. Photo courtesy JFZ Engineered Products.

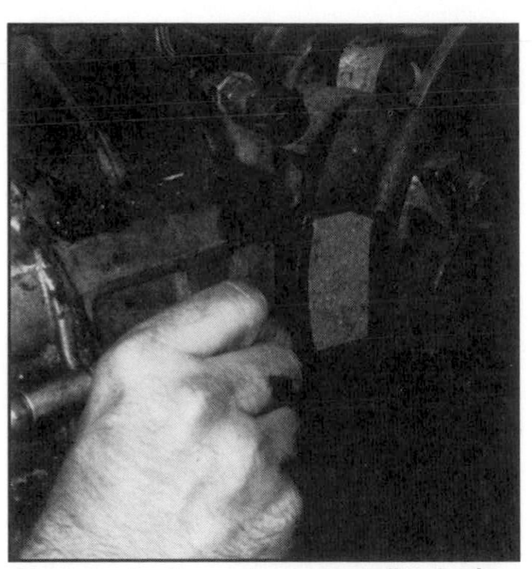

Changing brake pads on most disc brakes is easy. Pads are sometimes retained with quick-release pins. New pads are being installed in this transaxle-mounted caliper for bedding in.

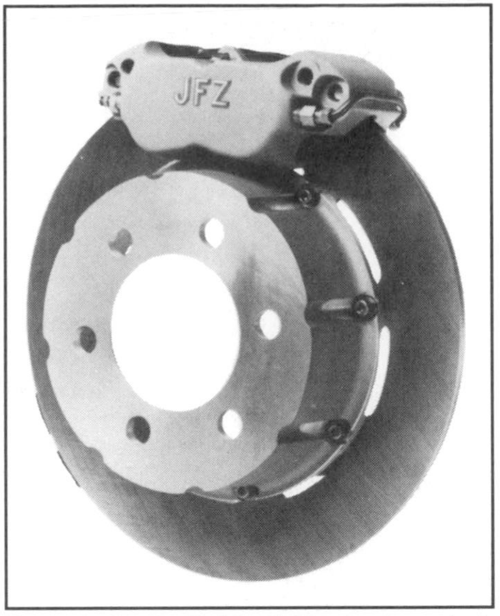

Rotor bolts to lightweight aluminum hat. Assembly then bolts to wheel hub or axle. Deep hat flexes when rotor expands from high temperature and minimizes brake heat transferred to the wheel hub and bearings. Photo courtesy JFZ Engineered Products.

brakes to changes in friction.

The shape of a drum brake also contributes to fade. Many organic brake linings expel gases when heated. These gases can act as a lubricant between the lining and drum, causing friction loss and severe brake fade. The cylindrical surface of a brake drum is similar to bearings in an engine. For example, high-pressure oil between a main bearing and crankshaft journal keeps the surfaces apart and friction to a minimum. A brake drum and shoes act much like an oiled bearing and journal when hot gas is expelled between them. They are pushed apart by the high-pressure gas, reducing friction and causing brake fade.

The small surface area of a disc-brake pad and the flat face of the rotor do not simulate a bearing container. There may be some gas coming from the lining, but because it is not contained well, there is little effect. To repeat, any drop in friction on a disc brake affects it less than on a drum brake because the disc brake has no servo action.

Disc-brake cooling is better than drum-brake cooling because the rotor-contact surfaces are directly exposed to cooling air. However, this makes the contact surfaces "potentially" more vulnerable to damage from corrosive dirt or water contamination. Fortunately, the constant wiping action of the pads against the rotor keeps the surfaces clean. Also, centrifugal force tends to throw material off the rotor. Because water is wiped from the rotor, disc brakes are less sensitive to this contamination than drum brakes. On the other hand, water lubricates a drum brake well. And, more servo action means less braking.

Disc-brake pads are easy to change on most cars. Usually, disc pads can be changed after removing the wheel and a simple locking device. Some installations requires removing the caliper.

Another advantage of disc brakes over drums is adjustment. Drum brakes must be manually adjusted or must have the added complexity of an automatic-adjusting system. Disc brakes are designed to run with little clearance, and are self-adjusting each time they are applied.

The disc-brake rotor has a contact surface on each side; the brake drum has a contact surface only on the inside. Disc-brake swept area is larger when compared to a drum brake of the same diameter and weight. More swept area means better cooling.

ROTORS & HATS

The rotor, usually made of cast iron, is the largest and heaviest part of a disc brake. It is a flat circular disc with a contact surface on each side. The rotor may either be solid or it can be vented with cooling passages through it.

Hats—The rotor is usually attached to a *hat,* which in turn is attached to a wheel hub or axle flange. The hat gives a long path for heat to travel from the brake rubbing surface to the wheel bearings. This keeps wheel-bearing temperature down. The hat sometimes is cast integral with the rotor, and sometimes it is a separate part.

Hats are made of iron or lighter material, such as aluminum. Production-car hats are usually iron and integral with the rotor. Most race-car hats are separate and made of aluminum alloy.

With non-integral rotors, the hat-to-rotor attachment can be either of two types, bolted or *dog-drive.* A dog-drive is a series of radial slots in the rotor with metal driving dogs attached to the hat. They act as a spline to transmit torque between brake and hat. These dogs allow the rotor to expand freely while keeping the rotor centered on the hat. This freedom to expand eliminates loads the hat would otherwise get when the rotor gets hot.

If the hat is bolted to the rotor, it must expand with the rotor as it gets hot. If the hat is relatively thin and flexible, it works without breaking. The bolted connection between the

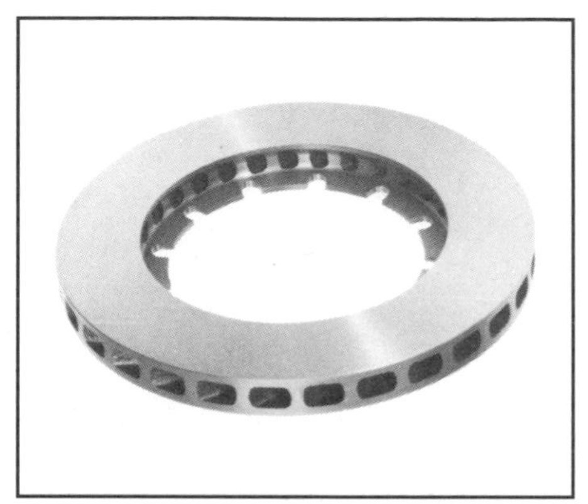

Racing rotor from AP Racing uses a *dog-drive* rather than bolts to transfer braking action to the hat. Dogs on hat fit in slots machined in edge of rotor ID. Slight clearance between hat dogs and rotor slots allow rotor to expand uniformly, but remain centered. Clever mounting arrangement is called *castellated* drive by AP Racing. Photo courtesy AP Racing.

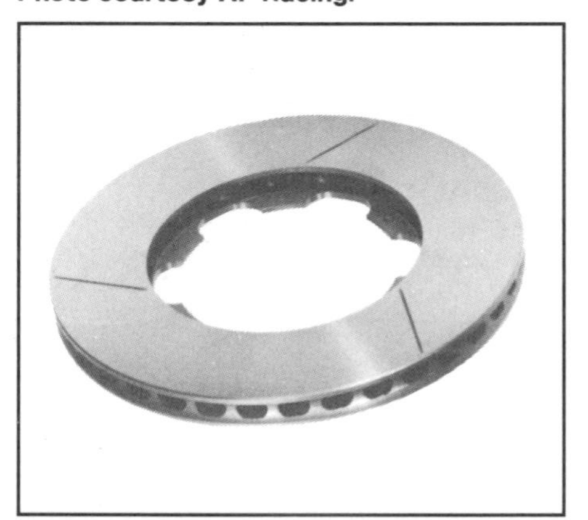

On this racing rotor, metal is removed at outer edge to reduce weight and slots are cut in rotor faces to reduce possibility of fade. Photo courtesy AP Racing.

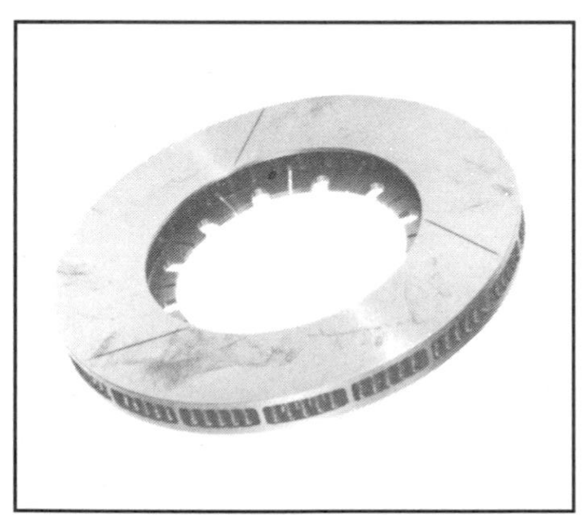

Highly developed AP Racing *Sphericone* rotor—notice projections cast *inside* vent holes—transmits heat to cooling air more efficiently than standard vented rotor. Design is used on Grand Prix cars, where every ounce of rotor weight counts. Photo courtesy AP Racing.

two items must use high-strength bolts with close-fitting, unthreaded shanks. Aircraft bolts are often used for this purpose, because they come

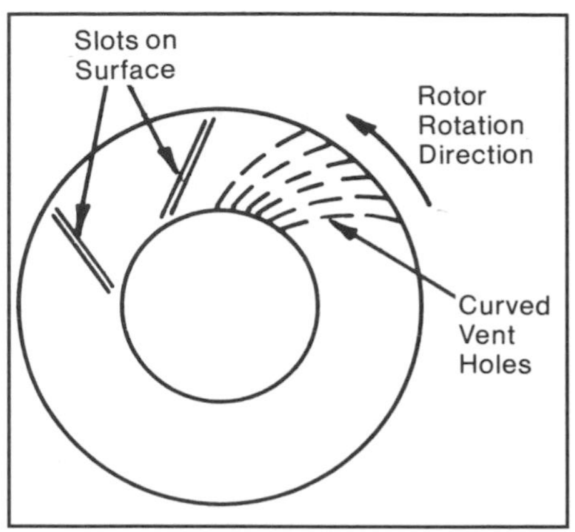

A rotor that has curved vent holes or angled slots must rotate in a particular direction to be effective. Correct rotation relative to vents and slots is shown.

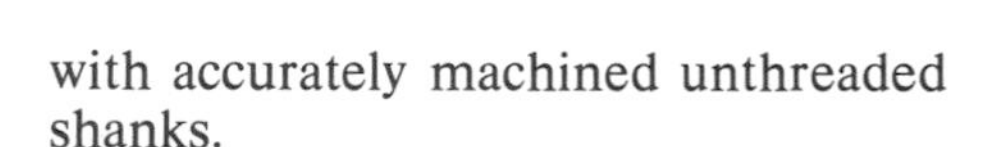

with accurately machined unthreaded shanks.

Aluminum hats reduce weight and thermal stress on the rotor. Because aluminum alloy expands at a greater rate than iron, it tends to equalize the expansion difference caused by extreme temperature of the rotor. Additionally, aluminum alloy is about one-third as rigid as iron, so there is much less force put on the rotor for a given amount of expansion. The lighter aluminum hat aids the car's performance and handling. One disadvantage of the aluminum-alloy hat is that it conducts heat better than iron. This causes the wheel bearings to run hotter.

Rotor Design—A disc-brake rotor has many features common with that of a brake drum. The material is usually grey cast iron for the same reasons it is popular for brake drums. Cast iron has good wear and friction properties, is rigid and strong at high temperatures, and is inexpensive and easy to machine.

A rotor is measured by its outside diameter and its total thickness across the two contact surfaces. A vented rotor is always thicker than a solid rotor. The diameter of the rotor is usually limited by the wheel size.

Swept area of a brake is an important measure of its effectiveness. The swept area of a disc brake is the total area contacted by both the brake pads in one revolution.

The combined swept area of all brakes divided by the weight of the car is one way of indicating how effective its brakes are likely to be. With

Here's what eventually happens to a rotor when thermal stresses are high. Because race-car rotors undergo such a beating, they should be inspected frequently and replaced if cracks are detected.

Tiny cracks on race-car-rotor surface are caused by thermal stresses. Rotor should be replaced before larger cracks form. Small surface cracks can often be removed by grinding.

good brake design, high swept area per ton indicates a high-performance brake system. Brake swept area for various cars is given on page 28.

Some rotors have slots or holes machined in their contact surfaces. These reduce hot-gas and dust-particle buildup between pad and rotor. Although fade caused by gas buildup is less for a disc brake than for a drum brake, some fade can still occur. This is more prevalant with large brake pads, because the hot gas has a harder time escaping than with small pads. Therefore, slots or holes have greater effect in racing, where pads are large and temperatures are very high. Holes increase the tendency for cast-iron rotors to crack, so they should be used only when required.

Vented Rotors—Many rotors are cast with radial cooling passages in them. These act as an air pump to circulate air from the rotor center through the

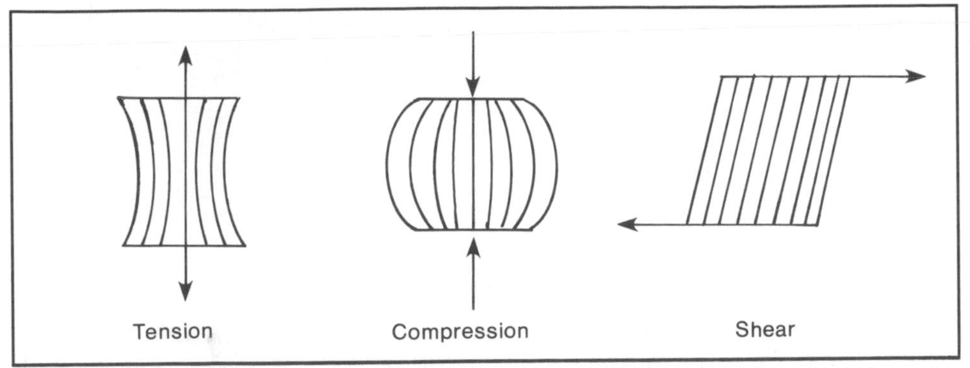

Three ways stress act on a material: Tension pulls; compression crushes; and shear rips sideways. Each stress occurs simultaneously in a hot brake component, particularly if one area of the part is hot and another is cool.

Lightweight race-car rotor has no hat. Rotor heat flows directly from rotor to aluminum hub. Such a design should only be used on light cars with small engines, where little heat is generated. Wheel bearings in this hub have to be replaced several times a season.

THERMAL STRESS

Stress is a measure of the internal force in a solid material, usually given in pounds per square inch (psi)—the same units for measuring hydraulic pressure. Stress acts in compression, tension or in a sliding direction called *shear.* Typically, only tension stress—pulling—causes cast iron to crack. The stress at which a material breaks in tension is called its *tensile strength.*

The tensile strength of grey cast iron is about 25,000 psi.

When a metal is heated, it expands. If the metal is restrained so it can't expand freely, it presses against its restraint with a force. The restraining force causes an additional stress in the metal. This stress caused by a change in temperature is called a *thermal stress.*

Thermal stress occurs when metal is hotter in one area. If a cold brake rotor is suddenly used for a hard stop, the rotor surfaces suddenly get hotter. The rotor interior remains cold until the surface heat has time to *soak* in. The hot surface tries to expand, but the cold interior keeps it from doing so. Therefore, the surface is in compression and the interior in tension. Thermal stresses gradually reduce as the surfaces cool and the interior warms. When the exterior and interior temperatures equalize, the surfaces go into tension because of being permanently compressed when heated. If high thermal stress is repeated over and over, eventually cracks form on the rotor surface.

Brake drums and rotors are designed to withstand the worst possible thermal stress for each specific application, but thousands of stress applications can cause *fatigue* cracks. Inspect used parts for cracks, particularly if they have been used hard.

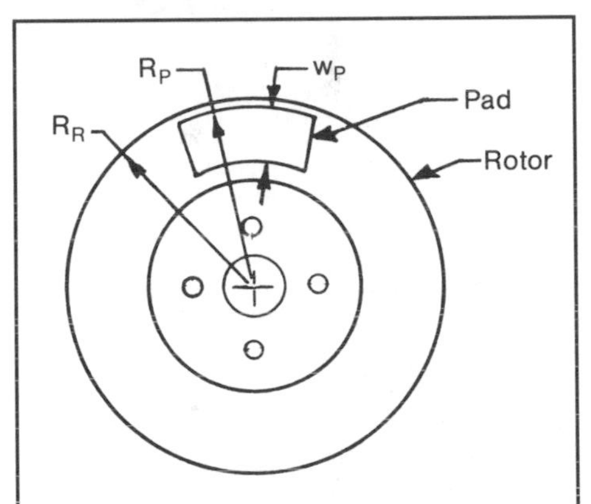

R_P = radius to outside of pad

R_R = radius to the outside of the rotor

w_P = pad width

If R_P nearly equals R_R, use the following formula to find swept area of brake:

Disc-brake swept area = $6.28\, w_P\, (2R_R - w_P)$

Disc-brake swept area is area on both sides of rotor rubbed by pads. It is more accurate to use R_p instead of R_R. However, because the two radii are nearly equal in most brakes, R_R is used for convenience. R_R is easier to measure than R_P.

passages to the outside of the rotor. This type of rotor is referred to as a *vented rotor.* Vented rotors are used on most heavy cars requiring the largest possible rotors. Lighter cars usually use solid rotors.

Powerful race cars use vented rotors. Sometimes there is a difference in the thickness of the rotor sidewalls. On many race-car brakes, the rotor side closest to the wheel is thicker than the side away from the wheel. This is done to try to equalize the temperature on each side of the rotor. The wheel tends to prevent cooling air from reaching the outboard rubbing surface of the rotor, making it run hotter than the inboard surface. The extra thickness next to the poorly cooled outboard surface tends to equalize temperatures.

Racing rotors often have curved cooling passages, which increase air-pumping efficiency. Because these vents are curved, there are different rotors for the right and left sides of the car. If your brakes use curved-vent rotors, don't get them mixed up!

A problem with vented rotors used in severe service is cracking. This is caused by thermal stresses and by brake-pad pressure against the unsupported metal at each cooling passage. Thermal stresses on a rotor with a cast- or bolted-on hat are caused by the contact portion of the rotor being hotter than the hat portion. Because heat causes expansion, the outer portion of the rotor expands more than the cooler hat. This distorts and bends the rotor into a conical shape. Repeated expansion and contraction cause cracks to appear. Even though cracking may not occur, coning causes uneven lining wear.

Rotor cracking is greatly reduced for two reasons with curved vents: First, curved vents provide better support for the sides. And, because curved-vent rotors cool better, thermal stresses are less.

Composite Rotors—The latest rotor material was developed for heavy aircraft brakes and is being

TYPICAL SWEPT AREAS PER TON OF VEHICLE WEIGHT

Car Make and Model	Swept Area Sq In./Ton	Car Make and Model	Swept Area Sq In./Ton
Alfa Romeo Spyder Veloce	259	Mercury Lynx RS	188
Audi 5000 Turbo	245	Mitsubishi Cordia LS	196
Audi Quatro	254	Mitsubishi Starion Turbo	236
BMW 528e	259	Nissan Sentra	272
BMW 745i	302	Nissan Stanza	174
Chevrolet Camaro Z28	176	Peugeot 505 STI	269
Chevrolet Citation X11 HO	180	Pontiac Firebird Trans-Am	277
Chevrolet Corvette	284	Pontiac Grand National Racer	236*
Datsun 200SX	266	Pontiac J2000	173
Delorean	346	Porsche 944	303
Dodge Aries Wagon	145	Renault Alliance	190
Dodge Challenger	230	Renault Fuego Turbo	185
Dodge Charger 2.2	161	Renault LeCar ISMA GTU Turbo	348*
Ferrari 308GTSi	263	Renault 5 Turbo	175
Fiat Brava	154	Renault 18i	189
Fiat Turbo Spyder	221	Saab 900	275
Ford Mustang GT 5.0	162	Subaru GL	169
Honda Accord	177	Toyota Celica	208
Honda Civic	171	Toyota Celica Supra	224
Isuzu I-mark LS	209	Toyota Starlet	196
Jaguar XJ-S	219	Volkswagen Quantum	183
Lamborghini Jalpa	227	Volkswagen Rabbit GTI	216
Lancia Monte Carlo GP.5 Turbo	537*	Volkswagen Scirocco	198
Mazda GLC	174	Volkswagen Scirocco SCCA GT3	304*
Mercedes-Benz 380SL	237	Volvo GLT Turbo	242
Mercury LN7	173		

*Indicates race car.

Listed are brake swept areas per ton for typical 1981 and '82 cars as published in *Road & Track* magazine. Notice that high-performance cars tend to have higher swept areas per ton than economy sedans. Those for race cars are indicated by asterisks. Race-car brakes are larger and race cars are generally lighter. The 537 sq-in./per ton area for the Lancia Monte Carlo GP 5 Turbo long-distance racer indicates long-wearing powerful brakes.

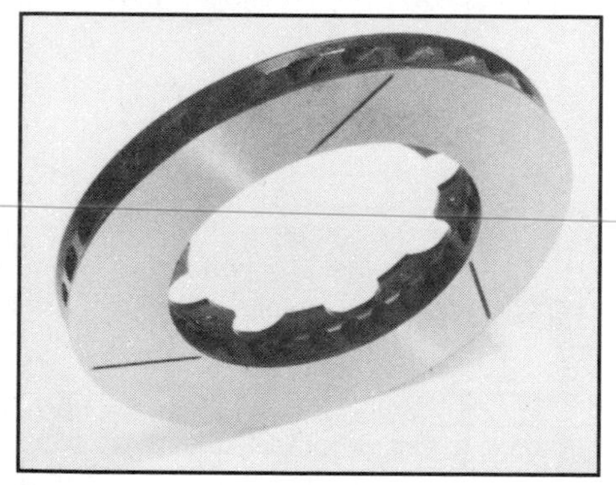

Curved-vent AP Racing rotor has slots angled so rotor rotation will tend to throw off dust. As viewed, rotor should be installed so it rotates counterclockwise. Photo courtesy AP Racing.

used on Grand Prix cars. These new rotors are made of carbon-graphite, boron composite. A composite rotor is lighter than a cast-iron rotor—about 30-lb less total weight on the Brabham F-1 car—and it can operate at much higher temperatures. Temperatures up to 1700F (927C) have been observed without damage. Graphite-composite pads are used with composite rotors due to high operating temperatures.

Because composite rotors are experimental for racing use, there is no certainty they will ever find their way into production cars. However, if the car manufacturers continue to work on weight reduction and the cost of composites become reasonable, there may be composite brakes on your car some day. The problem is that composite brakes experience extreme wear due to road-dirt contamination.

Brabham Grand Prix car is fitted with graphite-composite brake rotors. Exotic material from the aerospace industry can operate at extreme temperature. Brakes work while rotors glow red!

CALIPERS

Disc-brake calipers contain brake pads and hydraulic pistons that move the pads against the rotor surfaces. There are many types of calipers. They differ in material, structural design and piston arrangement. Regardless, all disc-brake calipers operate on the same principle—when the driver pushes the brake pedal, brake fluid forces the pistons against the brake pads, causing them to clamp the rotor.

Calipers used on most production cars are made of high-strength nodular cast iron. This low-cost material is suited to mass production, and makes a rigid caliper. However, cast-iron calipers tend to be heavy. Race cars or high-performance cars usually use aluminum-alloy brake calipers. A light aluminum-alloy caliper weighs about half as much as cast-iron caliper of the same size.

Fixed & Floating—There are two

Rotor on dragster rear axle has been given the swiss-cheese look by cross-drilling. Care must be exercised when drilling rotors to avoid weakening them or ruining their balance. Drilling reduces rotor weight and, therefore, the heat it can absorb. This is no problem for a dragster because the brakes are cool for each stop. Such drilling is more critical on oval- or road-racing cars. Be careful!

JFZ racing caliper is typical fixed-caliper design; caliper is fixed and only pistons move. Two pistons on each side are housed in cylinders bored in caliper body. Caliper assembly is bolted to axle housing or suspension upright through ears at bottom. Fluid-transfer tube is sometimes insulated to prevent fluid boil. Photo courtesy JFZ Engineered Products.

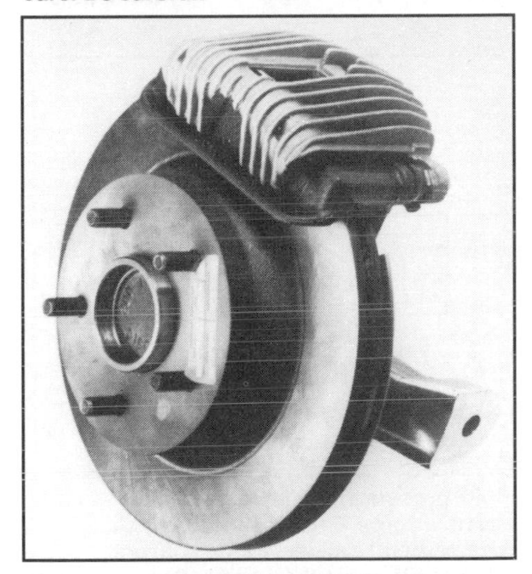

1984 Corvette floating caliper uses finned-aluminum caliper housing with one piston on inboard side of rotor. Mounting bracket is darker portioı to right of finned housing. Housing slides on two guide pins when brakes are applied to clamp rotor between pads. See separate parts in following drawing. Photo courtesy Girlock Ltd.

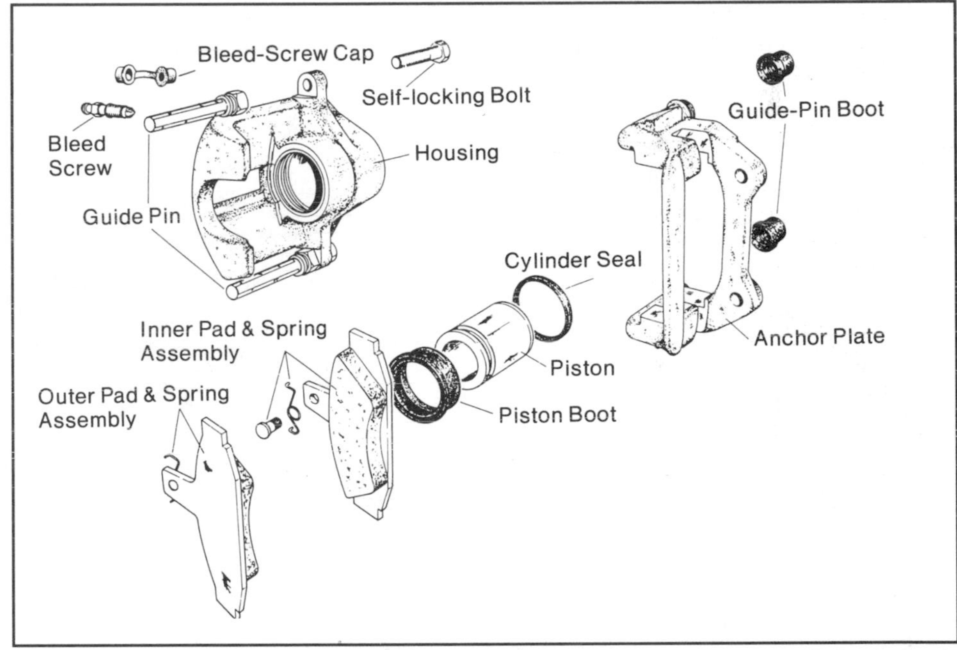

Blowup of '84 Corvette floating caliper. Guide pins are retained in close-fitting holes in anchor plate. Pins are in line with piston to minimize twisting of housing as it slides. Drawing courtesy Girlock Ltd.

basic types of calipers: *fixed* and *floating*. The difference is obvious when you look at them. A fixed caliper has one or two pistons on both sides of the rotor; a floating caliper usually has one piston on only one side of the rotor. The housing of a fixed caliper is bolted rigidly to the spindle or axle housing. A floating caliper is mounted so it moves in the direction opposite the piston(s). Because a floating caliper only has the piston(s) on the inboard side of the rotor, the entire caliper must shift inward for the outboard pad to contact the rotor.

Floating-caliper mountings vary. Some are mounted on guide pins and are held between machined surfaces to take braking torque. Others have a flexible mount or are mounted on a linkage. Depending on their mounting, they are also called *sliding* or *hinged calipers*.

Floating calipers are used on most production cars; fixed calipers are preferred for racing applications. This does not mean one type is good and the other is bad—each has advantages.

The fixed caliper has more pistons (two or four) and is bigger and heavier than a floating caliper. In severe use, it will take more hard stops to make the fixed caliper overheat. And, a fixed caliper generally will flex less than a floating caliper.

However, a floating caliper is more compact, thus packages better in a wheel. With the piston and fluid on the inboard side of the rotor, a floating caliper cools better. The floating caliper has fewer moving parts and seals, so is less likely to leak or wear out. On the negative side, the floating feature may cause the pads to wear at an angle due to caliper motion. One advantage

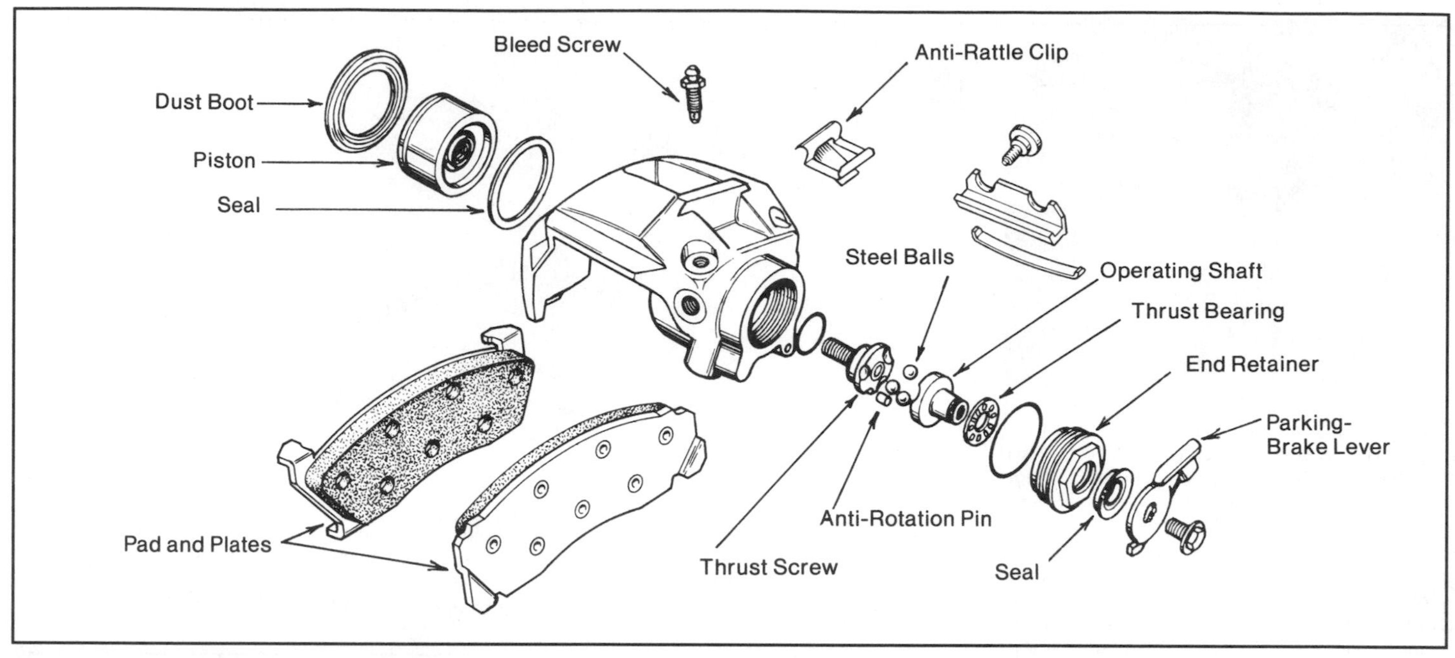

Floating caliper used on rear of some larger Ford cars features an integral parking-brake mechanism. As parking-brake shaft rotates, steel balls roll up ramps and push on piston to apply brake. Drawing courtesy Bendix Corp.

One-piece AP Racing caliper is used on small race cars using 10-in.-diameter, 0.375-in.-thick solid rotors. Caliper is available with 1.625 or 1.750-in. pistons. Photo courtesy AP Racing.

Two-piece caliper is for small race cars using 10.5-in.-diameter, 0.44-in.-thick solid rotors. Caliper uses 2.00-in. pistons. Photo courtesy AP Racing.

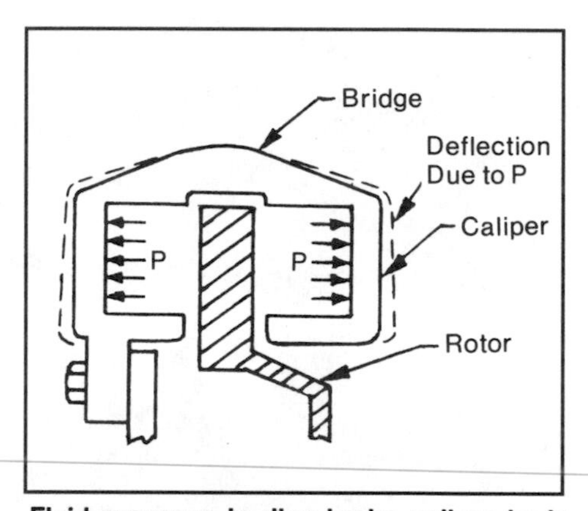

Fluid pressure in disc-brake-caliper body exerts a force P on each side of caliper and tries to bend caliper bridge. Caliper-bridge stiffness is critical part of caliper design. If caliper bridge is too thick, wheel may not fit over it. On the inside, most car designers want largest possible rotor diameter, so bridge thickness is limited.

of the floating caliper on road cars is the ease of using a mechanical parking brake. The single-piston design is easily operated by a parking-brake cable, while a fixed caliper with pistons on both sides of the rotor is more difficult.

A caliper body can be one piece of metal or it can be several pieces bolted together. Bolted designs use only high-strength, high torque-value bolts because a caliper must not flex during heavy brake application. Because it spans the rotor, the caliper body acts as a curved beam loaded by forces against each pad. If the caliper flexes, the driver senses this as a soft brake pedal. It will also tend to cock the pistons. This reduces braking consistency and makes the pads wear in a taper.

You can visualize caliper flex using a C-clamp. The frame of the C-clamp is similar to a disc-brake-caliper body. As the C-clamp is tightened, it simulates the clamping action of a disc-brake caliper. After the C-clamp is tight, continued turning of the handle will distort the C-frame. This is what happens to a flimsy caliper when the brakes are applied hard.

The portion of the caliper body that spans the outside of the rotor is called the *bridge*. Bridge stiffness determines overall caliper stiffness. Bridge-design requirements call for stiffness, which requires a thick cross section and weight. Because the caliper must fit between the outboard side of the rotor and inside the wheel rim, space requirements dictate a thinner cross section. Unfortunately, this may allow caliper flexing.

Because bridge design is a compromise, most calipers have about the same bridge thickness. But racing calipers are designed with a wider bridge to gain additional stiffness.

A disc-brake piston is installed in a machined bore in the caliper body. Usually, floating calipers have a single piston on one side of the rotor. Fixed calipers have one or two pistons on each side of the rotor.

When big calipers are required on heavy, powerful cars, a single piston will not do the job. It is difficult to

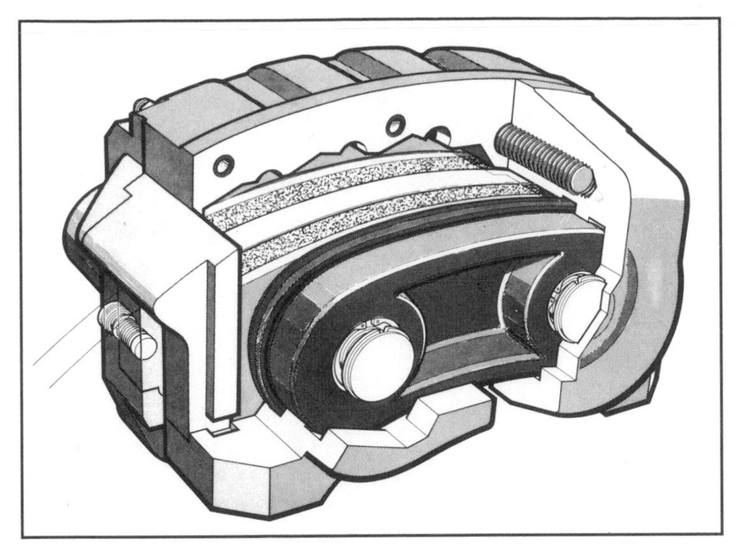

Frank Airheart, working for Alston Industries, designed this racing caliper with noncircular pistons. Kidney-shaped pistons closely match the pads, resulting in more even pad-to-rotor clamping pressure. Drawing courtesy Alston Industries.

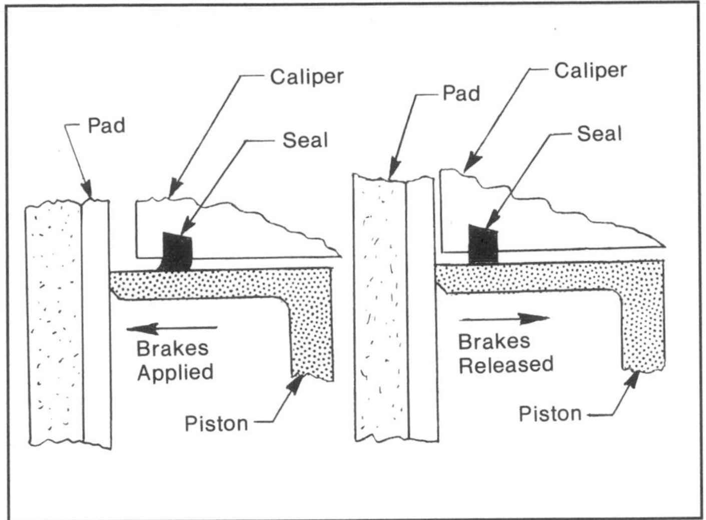

Cross section through caliper-piston seal shows position of seal with brakes applied and released. When brakes are released, seal retracts piston from rotor. As pad wears, seal slides on piston to compensate.

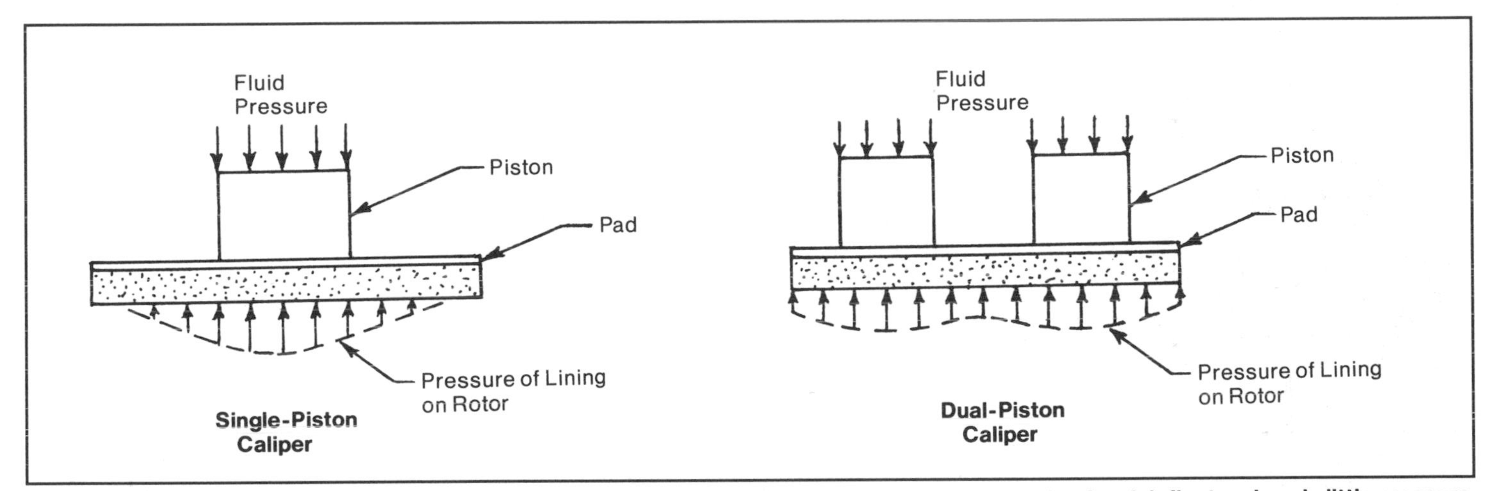

Wide brake pad used with large single piston results in uneven pressure on rotor. Unsupported ends of pad deflect and apply little pressure. Wide pads should be used with multiple-piston calipers. More-even pad-to-rotor pressure resulting from two smaller pistons gives improved braking.

make a large single-piston caliper that fits inside a wheel. As the caliper gets larger, it gets wider. If a single piston were used with a wide rectangular pad, the pad surface would not contact the rotor evenly. If the pad overhangs the piston, it will bend when the brakes are applied. Only that pad portion directly behind the piston is loaded against the disc. Multiple pistons minimize this.

In an effort to spread caliper-piston load evenly against the pad and achieve maximum piston area for a given-size caliper, Frank Airheart developed a racing caliper with a noncircular piston for Alston Industries. This unusual kidney-shaped piston roughly matches the shape of the brake lining. Consequently, there is little piston overhang and lining distortion with this design.

Multiple-piston calipers are used on most large and powerful cars. Most race cars with over 150-HP engines use multiple-piston calipers. These usually have four or more pistons. The largest calipers are found on racing stock and GT cars. These calipers cover nearly half the rotor. Usually larger, multiple-piston calipers use a fixed mount. Floating calipers using multiple pistons are rare in racing, although BMW uses a big ATE two-piston floating aluminum caliper on the M1 coupe.

Most caliper pistons are made of steel, aluminum or cast iron. Molded-phenolic plastic is becoming popular on road cars because of its low cost and low thermal conductivity. However, phenolic hasn't found its way into many racing applications yet, even though it should work quite well where fluid boil is a problem.

A disc-brake piston is usually sealed with a square cross-section O-ring. This seal stretches as the piston moves toward the rotor. When the brakes are released, the stretched rubber seal retracts the piston, eliminating the need for return springs. Because the pad and rotor surfaces are flat, only a slight movement is needed to obtain pad-to-rotor clearance. Consequently, disc brakes never need adjusting. The piston moves to take up wear.

Some calipers are designed with return springs on the pistons. Airheart racing calipers have this mechanical retraction feature to make sure the pad never drags on the rotor. With

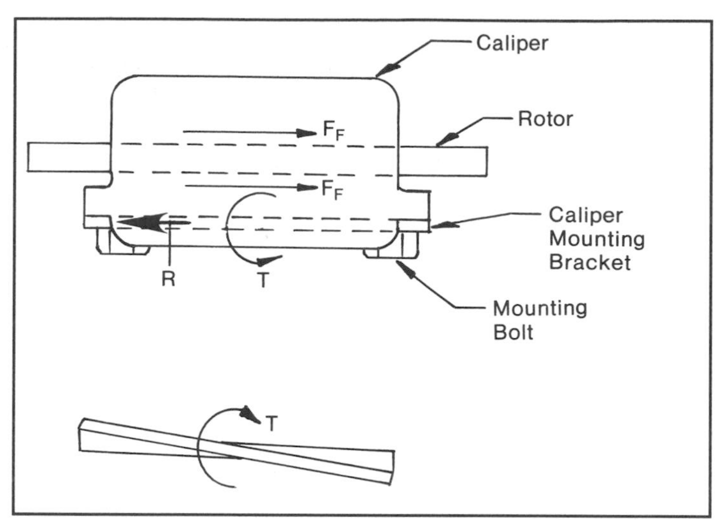

Forces on brake pads, indicated by F_F, are resisted by a force R at caliper bracket. Because force R is not in line with the rotor, twisting torque T is also applied to the bracket. Effect of twisting torque on the bracket is shown in lower drawing. If mounting bracket is not sufficiently rigid, caliper cocks against rotor, causing uneven pad wear, spongy pedal, and excessive pedal movement. Use a stiff caliper bracket and this will not happen.

Husky bracket attaches caliper to spindle on this sprint car. Because spindle was not designed to mount a disc-brake caliper, mounting bolts are too far from caliper. Risk of bracket twisting excessively is reduced by heavy plate.

Racing stock cars use the largest calipers. This is the popular Hurst/Airheart caliper with two pistons on each side of rotor.

Tilton Engineering sells sturdy caliper-mounting brackets for popular race-car applications. Bracket adapts caliper to four-bolt pattern normally used to mount drum-brake backing plate. Disc brakes can be fitted in place of drum brakes with a bracket such as this.

mechanical-retraction calipers such as this, a slightly different master-cylinder design is used. This is discussed in Chapter 5.

Caliper Mounting—It is essential that a caliper not move or flex as the brakes are applied—other than the lateral movement of a floating caliper. If the caliper-mounting structure is flexible, the caliper can twist on its mount. This causes uneven pad wear, a spongy pedal and excessive pedal movement.

A common race-car design error is using thin, flexible brackets to mount the caliper to the spindle. Because the rotor and caliper-mounting bracket lie in two different planes, the caliper bracket has a twisting force on it when the brakes are applied. If the bracket is too thin, it will twist, cocking the caliper against the rotor. Generally, a mounting bracket at least 1/2-in. thick should be used, particularly on large race cars such as stock cars.

SPLASH SHIELDS

The inboard side of the rotor on most production cars is shielded by a steel plate that looks similar to the backing plate on a drum brake. This is the *splash shield.* This shield only prevents mud, dirt and water from being thrown up against the inboard surface of the rotor. It does not have a structural function. If splash shields are not used, road debris may cause excess scoring and wear on the inboard side of the rotor.

Unfortunately, splash shields tend to block cooling air from reaching the rotor. This is why splash shields normally are not used on race cars. Race-car brakes usually need all the cooling they can get. Road-debris contamination is a secondary problem. Therefore, deleting the splash shield is a compromise between brake cooling and rotor protection.

BRAKE PADS

Disc-brake friction material is mounted on a *brake-pad backing plate,* which is usually a steel plate. This assembly is called a *brake pad.* Friction material is usually bonded to the backing plate during the lining-molding process, but sometimes it is joined in a separate operation with rivets and conventional bonding adhesives.

Brake pads are sold with friction material attached as are brake shoes. Unlike drum-brake shoes, pad backing plates usually are not reused. To make sure, check with your parts-supply house before discarding used disc-brake pads.

The piston in the brake caliper often does not bear directly against the brake pad. On many cars, there is an anti-squeal shim between the piston and brake pad. This sheet-metal shim is supposed to reduce the noise caused when the pad chatters against the rotor. Follow manufacturer's instructions when servicing the brake pads. These shims may require replacement along with the pads.

On some race cars, insulators have been used between the brake pads and caliper pistons. This is done to reduce brake-fluid and piston-seal temperature. Excessive temperature can cause seal failure or fluid boiling. If an insulator is used, it must be stiff, strong, and not too large. These conflicting requirements make the design of a good insulator very difficult.

Friction materials are covered in detail in the following chapter.

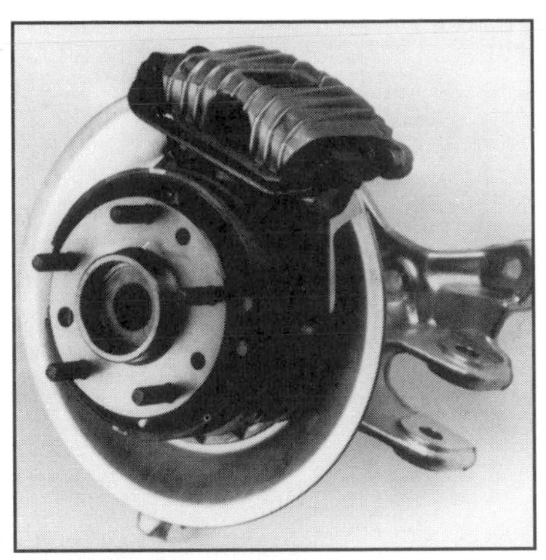

1984 Corvette rear brake less rotor shows splash shield. Thin circular plate mounted behind rotor protects it from road dirt and water. Road cars use splash shields to provide good braking in a dirty environment. Most race cars are better off without splash shields because they block cooling air. Photo courtesy Girlock Ltd.

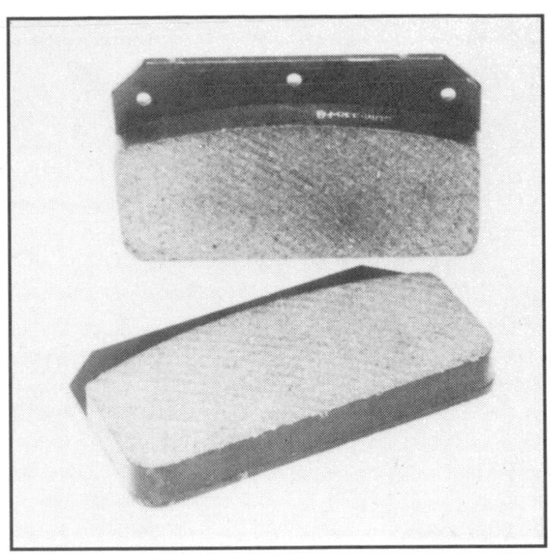

Racing brake pads look similar to those for road use. Pads come with lining bonded to a steel backing plate. These pads are designed for JFZ racing calipers for use on dragsters. Photo courtesy JFZ Engineered Products.

Friction Material 4

Stock car running on road course is hard on brakes. Stock cars are the heaviest road-racing vehicles, and are powerful. Brake ducts and heavy-duty linings are a must on hot day at Riverside Raceway.

The material that rubs on a brake drum or rotor is called *friction material* or *brake lining*. Drum-brake friction material is attached to the brake shoes. In a disc brake, friction material is attached to steel backing plates. Friction material composition is similar in both types of brakes, but it is designed for higher temperatures in disc brakes.

WHAT IS FRICTION?

Friction exists when two contacting surfaces either try to or do slide against each other. There is always resistance to sliding. This resistance is called *friction force*. The friction force acts on the sliding surfaces at their point of contact in a direction opposite to movement.

In a brake, there is friction between the friction material and the drum or rotor rubbing surface. The problem is controlling the amount of friction force and using it to stop the car.

The friction force depends on two things: types of surfaces in contact and amount of force pressing the surfaces together.

Each of these has its own name. The force pressing the two surfaces together is called *normal force*, or *perpendicular force*. Because only engineers use the word *normal* for *perpendicular*, I'll use perpendicular to avoid confusion. It should be obvious that a higher perpendicular force causes a higher friction force. If sliding surfaces are slippery, friction force is low. If the surfaces resist sliding, friction force is high. The "slipperiness" of a surface is described by a number called its *coefficient of friction*. The value of this friction coefficient varies, usually between 0 and 1. The friction coefficient for surfaces trying to slide is given by the following formula:

$$\mu = \frac{\text{Friction Force}}{\text{Perpendicular Force}}$$

If the coefficient of friction is low, the surfaces are slippery—the friction force is low. If the friction coefficient is high, the surfaces grip each other better—the friction force is high. Tires on dry pavement have a high friction coefficient, but on ice they have a low friction coefficient.

To make life complicated, the friction coefficient has two different values. It is somewhat higher if there is no sliding. As soon as the surfaces start to move relative to each other, the friction coefficient drops to a lower value. This is why it is harder to

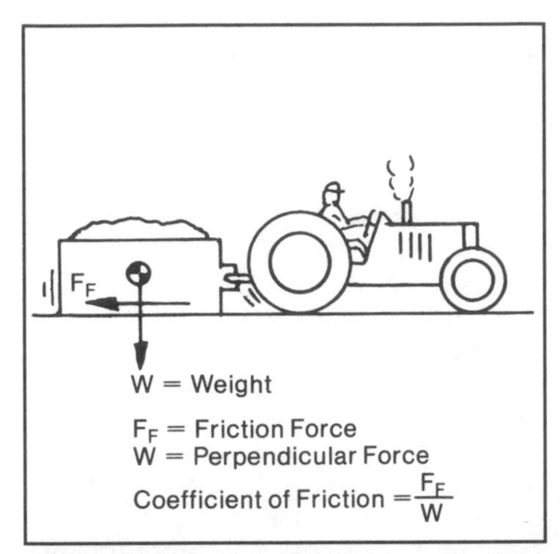

To slide box, tractor has to pull with a force equal to friction force F_F. Friction force is related to perpendicular force pushing two sliding surfaces together. In this example, the perpendicular force is box weight W. Friction coefficient is friction force divided by perpendicular force, or F_F/W.

As first car brakes in turn, inside front wheel locks. Tire smokes as it slides on pavement due to heat from friction. Following car can slow quicker because its wheels aren't locked. *Static*—nonsliding—coefficient of friction is higher than *dynamic*—sliding—coefficient of friction.

start something sliding than it is to keep it sliding. The friction coefficient before sliding occurs is called *static* coefficient of friction. After sliding starts, it's called *dynamic* coefficient of friction.

Either type of friction coefficient is calculated by the previously given formula. Just remember that the value drops once sliding starts.

In a brake, we are concerned about the coefficient of friction between the friction material and the drum or rotor surface. For most brake materials, the friction coefficient is about 0.3. I'll discuss later how the coefficient of friction affects brake-pedal effort.

Tire-to-Road Friction—The friction coefficient between the tires and road determines how fast a car can stop. This value varies from less than 0.1 on wet ice and 0.5 on rain-slick roads to more than 1.0 for racing tires on dry pavement. The friction coefficient for tires on pavement is critical for performance under all conditions. It determines how fast a car can accelerate and corner as well as how fast it can stop.

FRICTION-MATERIAL CHARACTERISTICS

A good friction material must have these characteristics:

- Friction coefficient must be high.
- Friction coefficient must not change with increasing temperature.
- Should not wear rapidly.
- Should not damage surface it rubs against.
- Must withstand high temperature without failing.
- Should not be noisy when brakes are applied.

Unfortunately, there is no perfect friction material. Instead, there are many different types, each one made to emphasize a particular characteristic or several of them. The ideal friction material for severe or racing use is not the same as that for lowest-cost passenger-car use.

Popular friction materials fall into one of these basic categories:

- *Organic material* held together by resin binders. Contains asbestos, glass or synthetic fibers, and may contain metallic particles.
- *Metallic material* fused or *sintered* together. Almost entirely made of

FRICTION & SURFACE AREA

For sliding surfaces, the friction force depends only on the coefficient of friction and perpendicular force. The amount of sliding or rubbing surface area does not affect friction. This holds true for all conditions of sliding where material *isn't removed from the sliding surface.*

Skid marks result from locking up the brakes on dry pavement. Each skid mark is rubber that was torn from the tires and deposited on the pavement. Because material is removed from the tires during sliding, surface area *does* affect the friction force. Consequently, the greater the area in contact, the greater the friction force. To make the friction formula work for tires on pavement, engineers choose to change the coefficient of friction rather than the friction formula.

For tires with all but width being equal, the coefficient of friction increases as width increases. This is a major factor that makes the friction coefficient greater than 1.0 for wide racing tires. So, the friction coefficient cannot be greater than 1.0 unless material is removed from the sliding surface.

When discussing handling, I call the coefficient of friction of a tire its *grip.* This differentiates grip from ordinary coefficient of friction, because tire area affects "coefficient of friction." If the subject of tires and grip interests you, read my HPBook "How to Make Your Car Handle."

Metallic linings for 1962 Corvette drum brakes were highly successful for racing. Prior to the use of disc brakes, drum brakes with these linings were relatively fade-free, even on a 3000-lb, 300-HP sports car.

metal particles.
- *Semi-metallic material* held together by resin binders. Contains steel fibers and metallic particles.

ORGANIC MATERIALS

Most brake friction material on the market is *organic*. This means that the friction material, such as petroleum-based products and plastics, was once a living plant or animal. Early friction materials were wood, leather or cotton impregnated with asphalt or rubber. These old-fashioned linings did not have many desirable characteristics. If used severely, they tended to catch on fire!

Superior friction material was obtained by weaving asbestos-impregnated fibers together with organic material to glue them together. Today, asbestos is still used in most brake linings, along with various *binders* and *friction modifiers*. Binder, usually a plastic resin, holds the friction material together. The fibers provide strength and resist wear. Glass and synthetic fibers are beginning to replace asbestos in some organic-lining material. Friction modifiers tune the friction of the material to the desired level.

Molded brake linings were first produced in the '30s. Woven lining material with small fibers was manufactured using a molding technique. In this type of friction material, the binder, modifiers and fibers are pressed together in a mold under heat and pressure to produce a curved brake-shoe lining.

In a molded lining, other materials such as mineral or metallic particles were added to change the characteristics of the friction material. Later, pads for disc brakes were developed from molded material, as well as semi-metallic high-temperature friction materials. Woven linings gradually became less popular because the range of materials was limited. Today, woven material is found in clutch discs, but not in brakes.

Organic friction material has a variety of ingredients. Manufacturers keep the exact specifications a secret. Generally, a typical organic brake lining contains these components:
- Asbestos fibers for friction and heat resistance. Friction material is usually more than 50% asbestos.
- Friction modifiers, such as oil of cashew-nut shells, give the desired friction coefficient.
- Fillers, such as rubber chips, reduce noise.
- Powdered lead, brass or aluminum improves braking performance.
- Binders, such as phenolic resin, hold the material together.
- Curing agents provide proper chemical reactions in the mixture during manufacturing.

The friction coefficient and wear properties of friction material can be modified by changing the components in this magic brew. When you change lining material, you are probably getting a different mixture. What is actually in it is a secret. The actual performance of the material is also difficult to determine. You must resort to a trial-and-error method *based on experience* to find the best friction material for a particular car.

Conventional organic materials work well at low temperatures, but their friction coefficient tends to drop when hot. They also usually wear rapidly at high temperature. Above 400F (204C), most organic materials have fade and wear problems. However, because of their good wear properties when cool, low noise and low cost, organic linings are used on most road cars.

High-Temperature Organic Friction Material—A special organic lining, called *high-temperature organic material,* has improved wear properties at high temperatures. However, it has poorer wear and low friction at low temperatures. It is also harder on the drums and rotors than *low-temperature* organic materials. High-temperature organic materials usually have a high concentration of metai particles in them to increase the friction coefficient and resist heat. Most popular racing linings are high-temperature organic materials. The best ones work well up to 1250F (677C).

High-temperature organic material is stiff and brittle compared to low-temperature organic friction material. As a result, it is harder to attach it to a shoe or pad. The old-fashioned riveting method won't work well because the material tends to crack at the rivet holes. And ordinary bonding techniques have too low a temperature limit. Therefore, great care must be used in attaching high-temperature organic friction material. Trust only an expert. The best attachment is one in which the friction material is *mold bonded* to the metal backing during manufacture. A combination of bonding and riveting also works well.

High-temperature organic lining may require a special finishing operation on the drums or rotors. The manufacturer of the lining material will specify what should be done. Always follow his instructions, or you won't achieve the superior performance of this material.

METALLIC MATERIALS

In the mid-'50s, Chevrolet was competing in road races with its new V8 Corvette. Using drum brakes to stop a 3000-lb car with its 300-HP engine was a big technical problem, particularly in the 12-hour endurance race at Sebring. To make the Corvettes competitive with the lighter foreign cars, the company developed a *ceramic-metallic* brake-lining material that could operate at extremely high temperatures.

These *cerametallic* linings worked well, but at the brutal Sebring 12-hour race they tended to wear through the brake drums! After these early experiments, less-harsh materials were developed, using an iron base rather than abrasive ceramic. Iron-base metallic linings were offered by Chevrolet for street use prior to the introduction of disc brakes on the Corvette. The Chevrolet metallic linings were excellent for the time. Only the superior characteristics of the disc brake has made them obsolete.

Metallic linings are designed to operate at higher temperatures than

Grand Prix cars frequently run on courses with long straights and tight turns, which is extremely hard on brakes. Friction-material hardness is selected to match the course. "Hard" linings are required when running a race like this one on the streets of Long Beach.

organic friction materials. All organic materials, such as plastic resin, disintegrate at high temperatures reached in severe braking conditions. The asbestos fiber in organic linings can withstand high temperature, but with no binder to hold it together it has no strength. The metallic materials solve this problem.

Sintering—Metallic linings are manufactured using a *sintering* process. A metal-powder mixture is compressed in a mold at high temperature and pressure. The metal partially fuses together into a solid material.

Sintered friction material can be attached to the shoe or pad by brazing, or it can be fused on during the sintering process. Sometimes, the material is sintered to a thin plate and the plate is then riveted to the shoe or pad. This processes eliminates the need for organic adhesive.

Metallic friction material that is glued on loses some of its high-temperature capability. This is because the organic adhesive will fail at temperatures that the friction material can withstand. When such failure occurs, the friction material comes off its backing, and braking is severely affected.

Metallic friction material does not fade at high temperature, but it does have other bad properties. It costs more than organic material, works worse when cold, wears the drum or rotor more, and is harder to attach to the shoe or pad. Many metallic racing linings are dangerous if not warmed up. When cold, some metallic materials either grab viciously or have very low friction. In racing, warming up the brakes prior to using them may be acceptable, but not for street usage.

Sintered-metallic friction material is used extensively in heavy aircraft brakes, and some high-performance racing clutches and brakes. Some motorcycle disc brakes use stainless-steel rotors and sintered linings.

SEMIMETALLIC MATERIALS

To solve some of the problems with organic and sintered-metallic material, *semimetallic* friction material was developed. Semimetallic linings contain no asbestos. This material is made of steel fibers bonded together with organic resins to give characteristics of both organic and sintered-metallic materials. Semimetallic linings operate well at temperatures of up to 1000F (538C). The steel fibers in the material tend to melt and weld to the rotor when operated at higher temperatures.

FRICTION-MATERIAL HARDNESS

Racers often talk about *hard* or *soft* brake pads. This is slang for how high the coefficient of friction is. "Soft" friction material has a high friction coefficient; "hard" material has a low friction coefficient.

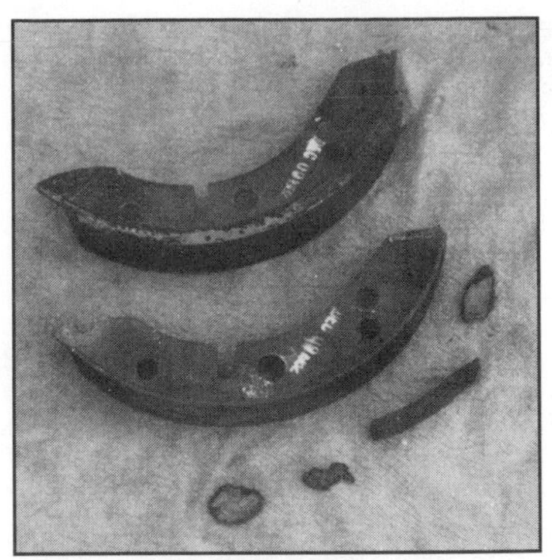
Organic friction material attached by bonding only didn't stay attached long under racing conditions.

Why does a racer want to use a "harder" friction material? The reason is wear and fade characteristics of the material. Unfortunately, "soft" friction material tends to wear rapidly; it also fades easier. "Soft" material is usually organic. "Hard" material may have a higher percentage of metallics. A racer usually chooses soft material for short races or on tracks that are easy on brakes. Hard materials are used where wear is high, or where very hard braking and little brake cooling occurs.

FRICTION-MATERIAL ATTACHMENT

Friction material is attached to a metal backing. On a drum brake, the backing is the brake shoe. On a disc brake, a metal backing plate is used. The friction material is either riveted or bonded to the shoe or backing plate. This attachment is very important. If it comes apart, the brakes can fail suddenly.

Riveted—Although riveting is an old way of attaching friction material, and the method is very reliable, rivet holes can cause cracks in the friction material when it gets hot. The rivets seldom fail—usually it is the friction material itself. Also, if the friction material wears out, the rivets can contact the drum or rotor. This will score the surface and usually ruin it.

Bonded—A more modern attachment is bonding with high-temperature adhesive. There is less tendency for the friction material to crack because it has no holes. And, if the friction material wears out, a bonded attachment does less damage to the rotor or drum surface. In addition, bonded-

Brake-specialty shop is best place to get facts for comparing lining materials. Because they often work with repeat customers, shop personnel see long-term results of using different materials. Stores selling brake linings typically don't see results of actual use.

Passenger car used for towing should have heavy-duty brake linings. Cheap linings may work in light-duty use, but will have high wear and may fade when towing heavy loads. Consider vehicle use when selecting linings.

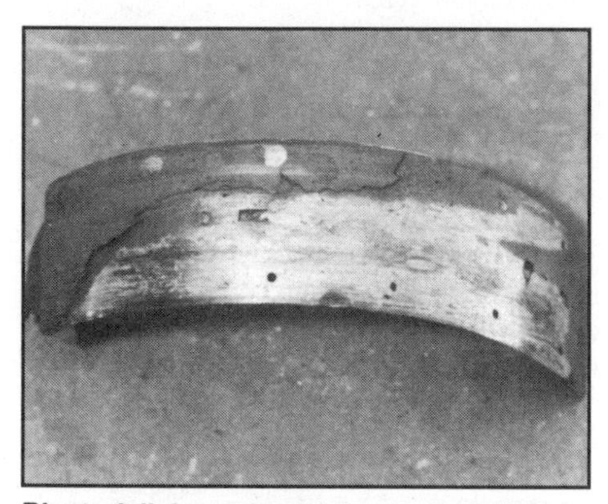
Riveted linings should be replaced when rivets get close to touching the drums. Here's what happens when warning signs are ignored. Most of lining and rivets are gone and shoe is worn part way through! Drum was also ruined.

only linings last longer because they have more *usable* thickness; there are no rivet heads to contend with.

The disadvantage of bonding is a lower resistance to high temperature. Sometimes, riveting and bonding are combined to secure brittle material for high-temperature operation. Rivets are much stronger at extreme temperatures.

The best attachment is achieved by bonding during the lining-molding process. This is called *mold bonding* or *integral molding*. During this process, holes are punched in the brake-pad backing plate. A high-temperature adhesive is then applied to the plate and the plate is put into the lining mold. During the molding process, pressure forces the lining material into these holes and high temperature activates the adhesive. This process combines the advantages of bonding with the positive mechanical-attachment feature of riveting. Most racing disc-brake pads are made this way. Unfortunately, due to their curved shape, it is difficult to produce drum-brake shoe-and-lining assemblies with this process.

SELECTING FRICTION MATERIAL

Selection of the proper friction material depends on the following:

- Vehicle use—easy street driving, hard street driving, or racing.
- Cost versus budget.
- Maintenance you are willing to put up with.

Easy Street Driving—For street use, most people choose the lowest-cost friction material. However, you get what you pay for. Cheap linings or pads may squeal and wear out faster than the more expensive ones. Consequently, they may end up costing more. Worse yet, they may fade in some unexpected hard usage. So, even though most street driving is easy, it is best to get high-quality friction material.

An auto-parts store usually can't judge friction-material wear or fade resistance. Therefore, consult a specialty shop that deals in brakes. There are usually a number of brake-repair shops in every city or locale. They deal with all sorts of brake failures. And their repeat customers allow them to judge the long-term performance of various materials. The shop foreman or service manager is usually the person to ask. Take his advice rather than that of an over-the-counter general parts salesman.

Shoes are fitted with different types of lining. Upper shoe is darker due to extra metallics in lining. This "harder" lining is used for severe use.

The problem with easy street driving is getting both good low-temperature brake performance and fade resistance. Even the most leisurely street driver will occasionally have to make a hard stop from high speeds. The brakes must be capable of a hard stop. Organic linings are the best for this type of service, as they work best at low temperature. The one hard stop should be within their capability, but not necessarily repeated stops from high speed.

If you've selected a lining correctly matched to easy street driving, *don't* use it for racing or hard use. If you have to pull a trailer in the mountains, either change friction material or go very slowly down hills in low gear without riding the brakes.

Hard Street Driving—Use heavy-duty friction material for hard street driving or hauling heavy loads. This is usually a semimetallic lining. Heavy-duty material has greater wear resistance and better braking performance at high temperature. It will usually cost more initially, but will be more economical overall.

Heavy-duty linings should be purchased only after consulting an experienced brake specialist. Select a brand that he is familiar with and has confidence in. Ask him what he would use on his own vehicle for similar driving conditions. Go to several shops, if possible, to see if the advice you are getting is consistent. Be aware that many American cars produced after 1979 are fitted with semimetallic disc-brake linings. Most European and Japanese cars use high-temperature organic linings.

For hard street driving, you may be tempted to use a high-temperature, organic racing friction material. This may work well under severe conditions, but be prepared for trouble during normal driving. They actually have some of the same characteristics when cold as organic linings when hot.

Racing friction material cannot be kept hot enough to work on long trips on the highway. This would require repeated brake applications to keep them warm. A panic stop caused by an unexpected emergency could have disastrous results with cold brakes.

Racing—Selecting the correct friction material for racing is easier than for the street because driving conditions are more predictable—always severe. Exact brake use depends on the particular track, driver's technique and race length. Friction material must meet these requirements:

- No fade during race.
- Will not wear out during race.
- Correct composition to withstand temperatures encountered.

Because road racing is usually the hardest on brakes, let's look at how to select friction material for this application.

It turns out that most road-racing cars use the same material, Ferodo DS11. And, nearly all road-racing cars use disc brakes. DS11 is so common because road-racing brakes are designed to be as light as possible, relative to the power and weight of the car. Big, fast cars have bigger brakes than light, slow cars. This means that most road-racing brakes operate at about the same temperature.

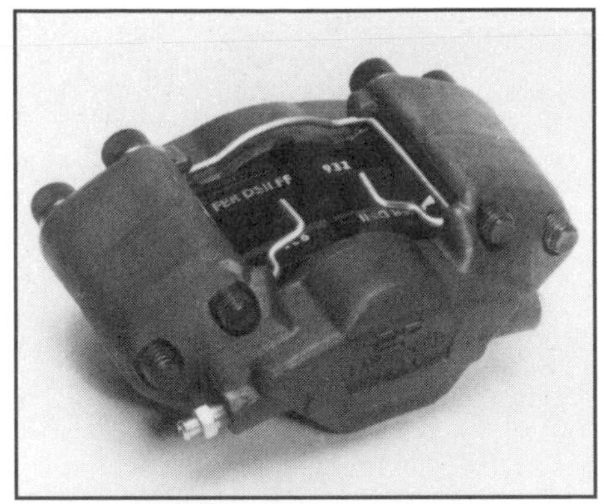

Racing caliper is fitted with Ferodo DS11 lining material. This lining is popular for all types of racing. Photo courtesy AP Racing.

Compare Indy Car brake with Super Vee brake in photo, above right. Indy car uses dual-piston aluminum caliper and vented rotor. Rotor diameter is slightly larger because 15-in. wheels are used. Indy car has 210-mph capability, and is hundreds of pounds heavier than Super Vee. Car speed and weight show in the brakes. Photo by Tom Monroe.

Super Vee brake is typical of those used on small race cars. Rules require using VW-brake components. VW cast-iron single-piston—one piston per side—caliper clamps on solid rotor. VW rotor bolts to special aluminum hat that allows quick-change wheel mounting. Design is adequate for light cars with 160-mph capability.

Scoop for Indy Car rear brake projects above body and into airstream. Because drag is created, the smallest-possible scoop should be used. Smaller scoops mean a faster car, but hotter brakes. Maximum ducting is necessary on short ovals such as Phoenix International Raceway. Photo by Tom Monroe.

Ferodo DS11 operates properly at a disc temperature of 780—1100F (416—593C). The friction coefficient rises to its operating value of about 0.3 and is reasonably constant in this temperature range. Outside this temperature range, the friction coefficient is lower.

Most race cars can use the same friction material because car designers size the brakes in proportion to the speed and weight of the car. Try the DS11 material first and see how it works. Check rotor temperature to be sure it is within the correct range, using a *temperature-indicating paint*. Read Chapter 10 for more information on brake testing.

If the linings wear too fast or the brakes fade, and cooling cannot be improved, a harder friction material is required. There are materials that can withstand temperatures over 1200F (649C), but these have disadvantages.

Raybestos M19 and Hardie-Ferodo 1103 are ultra-high-temperature materials. These friction materials tend to wear out rotors and also take a little time to work after they cool on a long straight. Consequently, Ferodo DS11 is more popular than higher-temperature material. It is better to use bigger brakes or improve brake cooling than to live with extremely high brake temperatures.

If brakes run cool, that's good. You have two options: Use softer linings to give lower pedal effort or reduce airflow to the brakes. However, because changes occur during a race that can overheat the brakes, reducing cooling air is a better choice. Cooling airflow increases aerodynamic drag, so slightly smaller ducts or none at all will reduce drag and make the car go faster. This will also increase brake temperature. Usually, the only valid excuse for running soft linings is a very high pedal effort. If you can run soft linings, Hardie-Ferodo Premium works well. It is a *low-copper* organic material, with a high friction coefficient.

Always run the same friction material on the front and rear. If you use DS11 friction material at one end, use it at the other end, too. Brakes warm up during a race and reach steady-state temperature after a few laps. If the brakes change friction *differently* from front to rear during the warmup period, front-to-rear brake balance will also change. This can cause loss of control. If you find a great difference in temperature between the front and rear brakes, modify the cooling airflow or the brake size, not brake friction material.

STOCK-CAR BRAKE LININGS

Because of their high weight, speed and horsepower, a racing stock car is the hardest on brakes. The most difficult track for brakes is one consisting of medium-length straightaways and slow corners. And, racing stock-car brakes require more attention on a road course than on a superspeedway. Certain medium-speed ovals are also hard on brakes.

Disc Brakes—Before stock cars were used on road courses, most used drum brakes. Their huge drums with metallic linings were adequate for superspeedways and most smaller oval tracks. Things have changed in stock-car racing. They now use disc brakes. The calipers are huge, even when compared to the high-horsepower Indy cars. The 200-mph speed potential of a stock car combined with a weight over 3000 lb, makes large brakes necessary.

As with road-racing cars, brake lining must be chosen according to the type of track and length of the race. Tracks can vary from short dirt ovals, to road courses, to 200-mph superspeedways. The type of lining that works best for each type of racing has been developed over years of experience. If you are new to the sport, consult your brake-lining supplier or other racers for advice.

Size of caliper used on a stock car indicates how much energy brakes must absorb. Usually, the biggest-possible brakes are used. Linings are varied to suit track and length of race.

Many grades of high-quality racing linings are available for stock cars. Many racing-brake suppliers offer Hardie-Ferodo linings. Again, Premium is the softest lining. Hardie-Ferodo Premium is recommended for high-banked ovals and tracks, such as superspeedways, where little braking is required. These soft pads are easiest on rotors and require no warm-up time for full effectiveness. If brake temperatures are low enough to avoid fade or excess pad wear, Premium is the best choice.

For running at higher temperatures, try the Hardie-Ferodo DP11 pads. These are the medium-hardness racing material. The DP11 material has more copper content than Premium. Consequently, it has increased temperature resistance and less wear.

For road racing, a hard lining material will be required on most stock cars. Hard materials will withstand high temperatures best of all. They will operate above 1200F (649C) while the popular Ferodo DS11 compound loses effectiveness at about 1100F (593C).

Two hard materials are the Hardie-Ferodo 1103 and Raybestos M19. Both have high copper content, but are still considered organic linings. Not only do they withstand high temperatures. they wear long enough to last in a long hard race. The drawback is that 1103 or M19 material tends to wear rotor surfaces more than other compounds. Also, pedal effort will be higher and they take longer to bed in. Therefore, use the hard compound if you have to, but better overall results will be obtained if you can use the softer DP11 material.

Other heavy-duty racing pads are available from various manufacturers. They also tend to be hard on rotors.

If you race on superspeedways, don't use a hard material. Such material cools off too much and the brakes don't work well when an emergency occurs.

Drum Brakes—Although racing stock cars now use disc brakes, there are still some restricted-class machines using drums. If you race in a class that is so limited, you might be required to use original-equipment drum brakes at the rear.

There are several metallic or semi-metallic materials for use in stock-car racing with drum brakes. One popular brand is made by the Grey Rock division of Raybestos Manhattan Corp. They make two metallic linings with high-temperature capability: 5262 and 5191. The 5262 compound was first developed for use on fast ovals and superspeedways with the characteristic of low change in friction under extreme-temperature conditions. The 5191 material is softer, but is said to wear faster.

For some types of oval-track racing, a semimetallic or heavy-duty street lining may be suitable. This assumes that speeds are low, corners are banked, and races are short. Here, brakes are used mostly for setting the car up for cornering and for unexpected emergencies. The Chevrolet semi-metallic linings or Bendix EDF heavy-duty service linings might be suitable. Careful testing is required to see if a medium-temperature material will work properly on your car.

BEDDING IN LININGS

New linings must be *bedded in* before they are raced to ensure proper brake operation. Bedding in brakes is similar to breaking in a new engine. If an attempt is made to race on new linings, *green fade* can occur. Green fade is severe brake fade caused by gas or liquid coming out of the friction material and lubricating the rubbing surfaces. The driver feels green fade as a sudden brake loss.

Oval-track cars running on dirt are not extremely hard on brakes. For light cars such as this late-model stock car, softer racing linings or heavy-duty street linings may work. Careful testing during a long practice session is desirable before trying new linings in a race.

Organic material in the lining tends to boil off rapidly when first heated. This material not only acts as a lubricant between the lining and rotor, but it can also *glaze* the lining surface. A glazed surface is caused by the organic material cooling and resolidifying on the pad surface. A glazed surface is hard and slick, and never has the right friction coefficient. Once glazed, the friction material must be refaced or replaced.

Bedding in brakes should be done during a practice session, not during a race. If the practice is long enough, you can bed in more than one set of pads and have them race-ready for later. It is wise to have a spare set of unworn bedded-in pads as spares.

The bedding-in procedure starts by taking the car out and gradually warming up the brakes. This is followed by using the brakes very hard several times. The car is then brought into the pits for a cooling-off period.

When Ferodo DS11 is properly bedded in, it changes from a black or dark grey to a lighter grey or brownish tone near the surface. Various brands of lining material have similar color-change characteristics, which you learn by experience.

Metallic particles are visible on a worn racing-lining surface. Check your brake linings frequently. Get familiar with the appearance of the lining surface when the brakes are working correctly. This will help you visually detect when a problem occurs. Make sure the surface is not polished, crumbling or cracking. Also

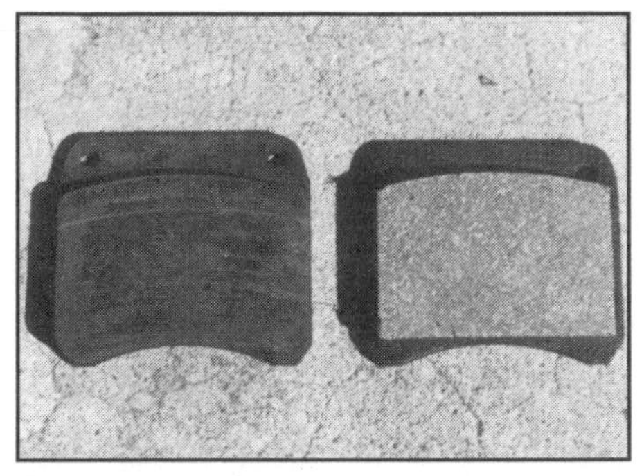

Brake-pad appearance changes greatly after bedding in. Pad on left is bedded in, but not raced. Pad at right is new. If new pad were used hard, *green fade* could result. Pads should always be bedded in first.

inspect the rotor surface. Friction material should not transfer from the lining to the rotor.

MACHINING FRICTION MATERIAL

When installing new drum-brake linings, you may need to turn or grind the drums. However, do not turn the drums unless they are out-of-round or deeply grooved. Regardless, brake linings should be machined to fit the drums for best braking performance.

Because brake-lining material usually contains asbestos fiber, machining results in dangerous asbestos dust. Due to recent discoveries of the hazards of asbestos dust to human health, many new safety precautions are required in brake shops. Consequently, many of them have stopped machining brake linings. You may have to search for a shop to do this, but it is worth it in terms of optimum brake performance.

The drum-brake-shoe machining process is called *arcing the shoes.* The drums must be measured, so bring them to the brake shop. The machine cuts the correct radius on the shoe so the shoes contact the drum over a large area. If a mismatch exists between the drum and shoes, brake friction may be too high or too low, depending on where the shoe contacts the drum. The worst possibility is when the right and left wheels have a different fit between the shoes and drums. This will cause dangerous pulling to the side.

Disc-brake pads for racing use are often machined to make them flat. This can be done on a milling machine or surface grinder. Be careful. Some types of brake lining dull milling cutters in a hurry.

As brake pads wear, they tend to

This machine was used to arc, or grind correct radius, on brake linings before asbestos was discovered to be a health problem. Consequently, many shops have stopped arcing brake shoes. Better braking is achieved with arced shoes if drums are turned.

get tapered. In that condition, the braking gets worse, with higher pedal effort and longer pedal travel required. To restore full braking effectiveness, machine the pads so the rubbing surface is parallel to the back of the pad. Deglazing and minor flattening may be accomplished by rubbing the lining on medium-grit sandpaper laid on a flat surface, such as a mill table or a piece of glass. Rebedding is necessary after machining or sanding.

Also, it sometimes helps to machine a groove in the surface of the brake pad to get rid of the lining dust. Most pads have a groove in them for this purpose. If it always seems to be filled with dust, a second groove may help. A groove machined 90° to the stock one is a reasonable plan. See Chapter 12 for details on modifying brake pads.

If you machine the linings, be careful not to breathe the dust and chips. Wear an approved *respirator* and do not blow dust with an air hose. Pick up asbestos dust with a vacuum cleaner. The dust in the shop must be cleaned up after the operation, so other people won't breathe the potentially deadly stuff.

See your local health authorities for safety regulations before you take a chance. Remember, many organic linings contain more than 50% asbestos. Most, if not all, new road cars will use asbestos-free semimetallic or non-asbestos organic linings by 1990.

Hydraulic Systems 5

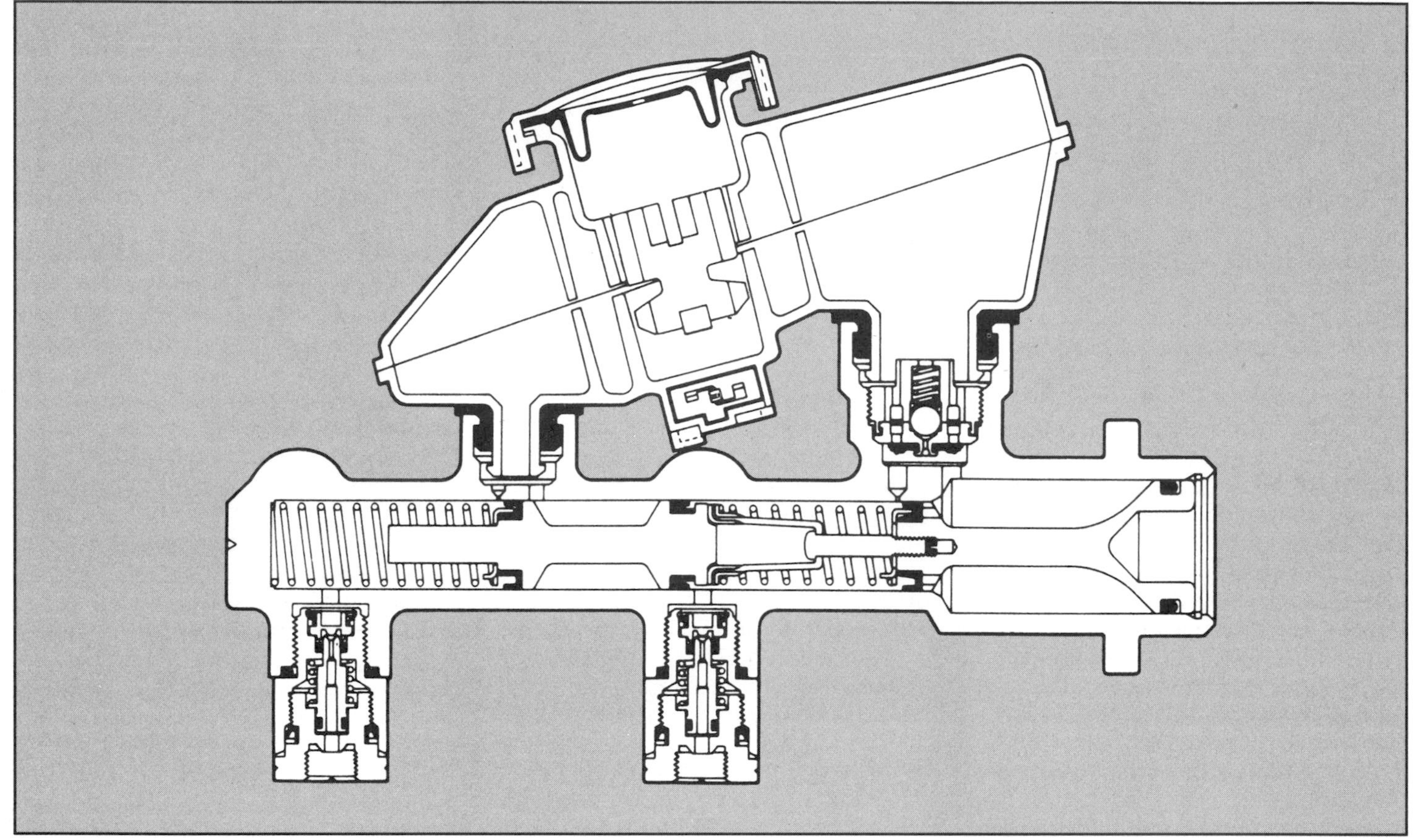

Modern fast-fill tandem master cylinder, page 50, has plastic reservoir and integral proportioning valves. Cylinder must be mounted so fluid is level in reservoir.

All modern cars use a hydraulic system to operate the brakes. This hydraulic system consists of cylinders, valves, hoses and tubing, all filled with *hydraulic fluid.* The type of hydraulic fluid used in brake systems is called *brake fluid.* The special properties of brake fluid are discussed later in this chapter.

A hydraulic system has two functions: move brake linings into position against drums or rotors; and apply a force to brake linings, creating friction force.

When you think about a hydraulic system, remember these important functions: *movement* and *force.* A hydraulic system has to do both jobs to operate the brakes. The movement must be enough to take up all clearances and deflections in a brake system. The force must be great enough to stop the car.

BASIC HYDRAULICS

To understand how a hydraulic system operates, we must first consider the basics. For a hydraulic system to function it must be closed and completely filled with fluid, and leak-free—no fluid out or air in.

In a closed, properly sealed hydraulic system, the following *laws* are true: Fluid cannot be compressed to a lesser volume, no matter how high the pressure. Pressure is equal over all surfaces of the containing system.

Turn to page 63 for more on hydraulic fluid for automotive use.

Pressure—Fluid is pressurized by the force of a piston. The piston moves in a cylinder and is sealed to prevent fluid from leaking out or air leaking in. The piston must move with little friction in the cylinder.

There is air on one side of the piston and hydraulic fluid on the other. The area of each piston is determined by its diameter.

Piston area $= 0.785\,D_p^{\,2}$
A_p = Area of piston in square inches
D_p = Piston diameter in inches

Force applied to the piston creates pressure in the hydraulic fluid. The pressure is the force divided by the area of the piston:

Hydraulic pressure $= \dfrac{F_p}{A_p}$
in pounds per square inch
F_p = Force on piston in pounds

Don't refer to pressure in *pounds*—it is pounds *per square inch (psi).*

Let's illustrate pressure with an example. See accompanying drawing of a simple hydraulic system with one piston. Fluid pressure is measured with a gage or calculated by a formula. The amount of pressure depends on

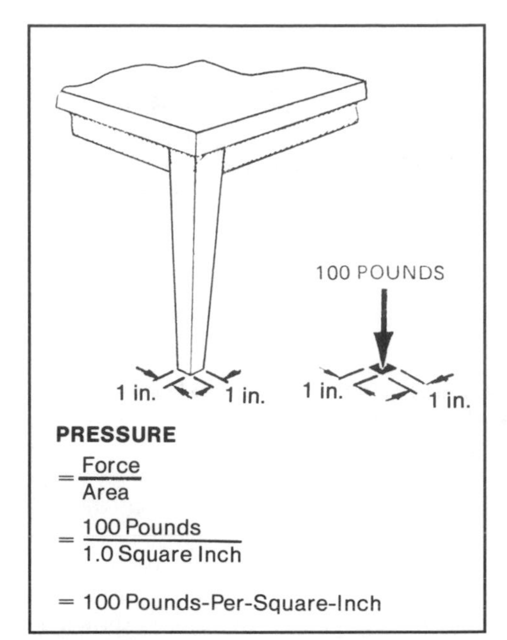

Pressure is force per unit area. A hydraulic piston exerts force on fluid like this table leg exerts pressure on the floor. Pressure is force divided by area on which force is exerted. If leg-to-floor contact area is reduced, pressure on floor is proportionally increased. Drawing by Tom Monroe.

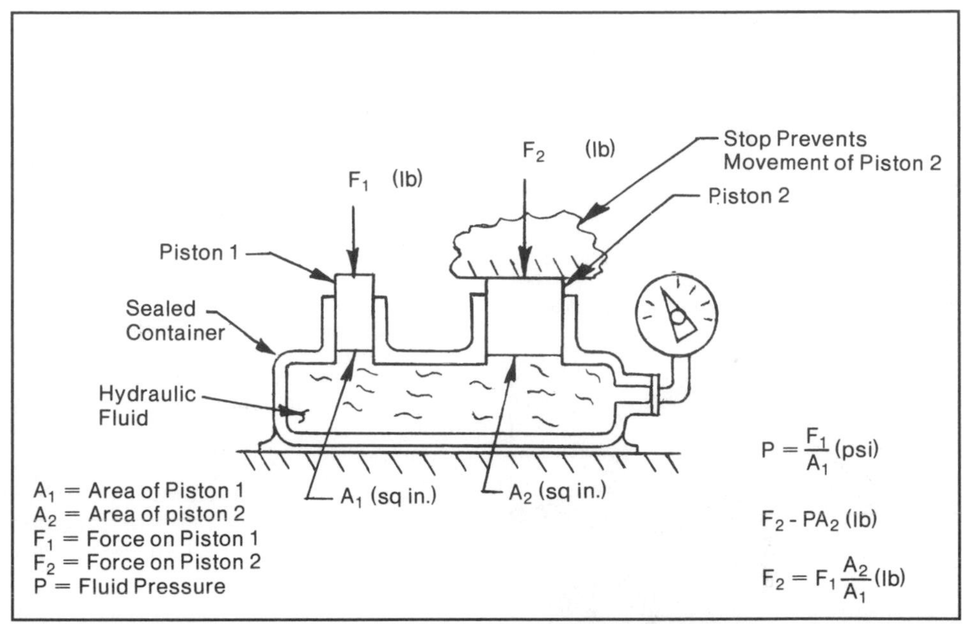

Forces can be multiplied with a simple hydraulic system. In this drawing, force F_1 is applied to a small piston with area A_1. If large piston A_2 is resisted by a stop, force F_2 is created. If this were a hydraulic jack, stop could be the bottom of a car. Force F_2 is greater than F_1 in a ratio that's directly related to piston areas. Because fluid pressure is same throughout, bigger piston has bigger force.

how much force you put on the piston.

If the piston has an area of 2 sq in. and the force is 400 lb, the pressure is calculated as follows:

$$\text{Pressure} = \frac{400\text{lb}}{2\text{ sq in.}} = 200\text{ psi.}$$

A smaller piston gives a higher pressure. If a piston with only 1-sq in. area is substituted for the 2-sq in. piston, a pressure of 400 psi is obtained with the 400-lb force.

Force Multiplication—Pistons can be used to multiply force in a hydraulic system. By choosing pistons of different sizes, any relationship with forces is possible.

A simple hydraulic system with two pistons is shown in the accompanying drawing. The pistons can have different areas. The relationship between the two forces depends on the relationship between the piston areas.

Assume piston 1 has an area of 2 sq in. and a 400-lb force applied to it. I showed in the previous example that this created 200-psi fluid pressure. Piston 2 now is subjected to this 200-psi fluid pressure. Remember that fluid pressure acts equally on all surfaces of the surrounding container. Fluid pressure acts on the contact surface of the piston, causing movement. Force 2 will depend on the area of piston 2.

Force on piston $= P\,A_p$

P = Hydraulic pressure in pounds per square inch

A_p = Area of piston in square inches

This is the same formula as on page 42, in a different format. In the example, assume piston 2 has an area of 1 sq in. The force on it at a fluid pressure of 200 psi is 200 lb. If we change piston 2 to an area of 4 sq in., the force increases to 800 lb. The pressure does not change—only the force on the piston. The larger the piston, the greater the force with the same pressure.

We can write a simple formula for the relationship between the forces on the two pistons:

$$F_2 = \frac{F_1 A_2}{A_1}$$

F_1 = Force on piston 1 in pounds
F_2 = Force on piston 2 in pounds
A_1 = Area of piston 1 in square inches
A_2 = Area of piston 2 in square inches

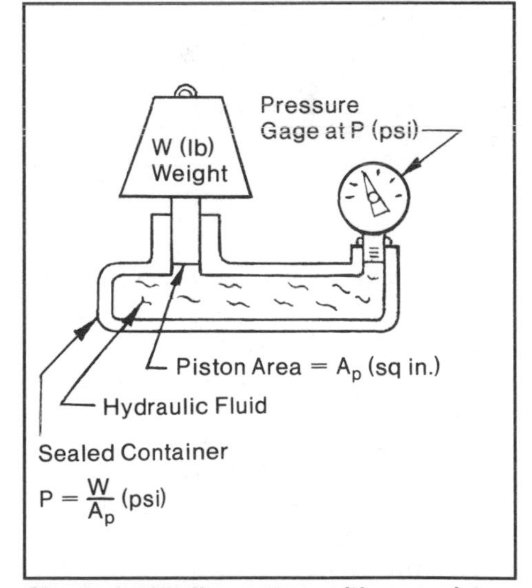

Simple hydraulic system with one piston and pressure gage. By putting force on piston, fluid is pressurized. Pressure is the force—weight in this instance—divided by area of piston. Fluid pressure acts equally on all surfaces of container.

To increase the force on piston 2, you can reduce piston-1 area or increase piston-2 area. Force can only be changed in a brake system by changing piston area(s).

Do not use piston diameter in place of area! Piston area varies as *the*

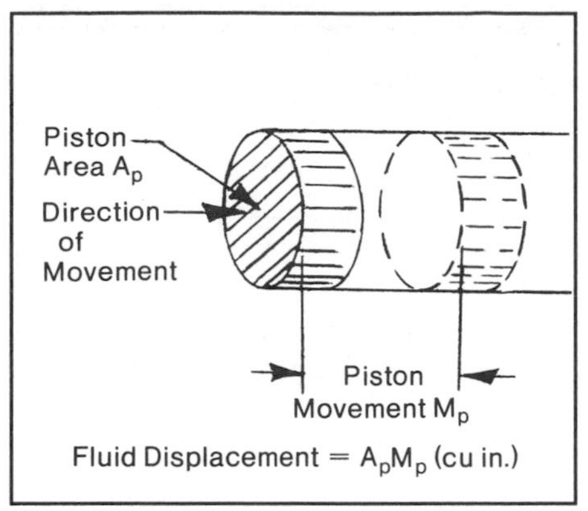

Fluid displaced by piston equals area of piston multiplied by distance it moves. Like a car engine, displacement can be given in cubic inches (cu in.).

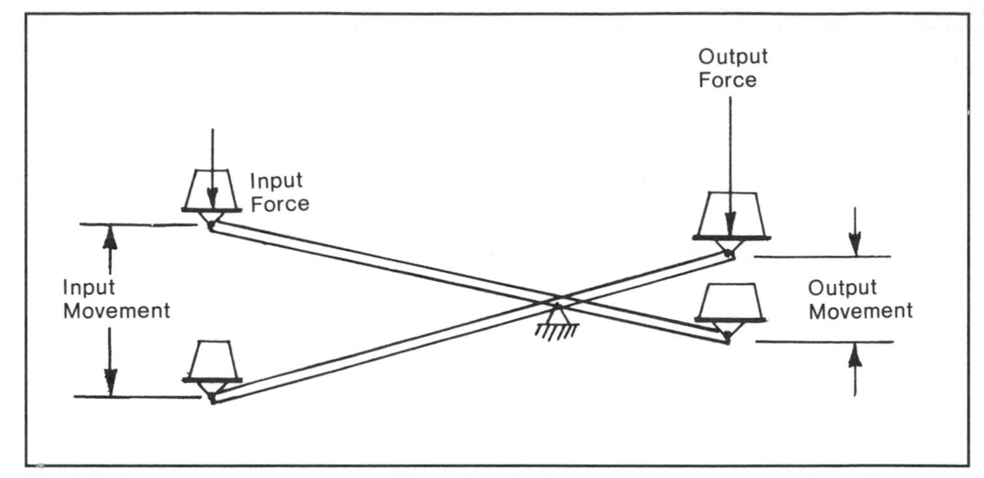

Hyraulic-brake system operates like this simple lever; increases force and reduces movement. Input side is the brake pedal. Input force and movement are supplied by your leg. Output side is the brake. Output force is exerted perpendicular by brake linings against rubbing surfaces. Output movement takes up all clearance in system so brakes can be applied.

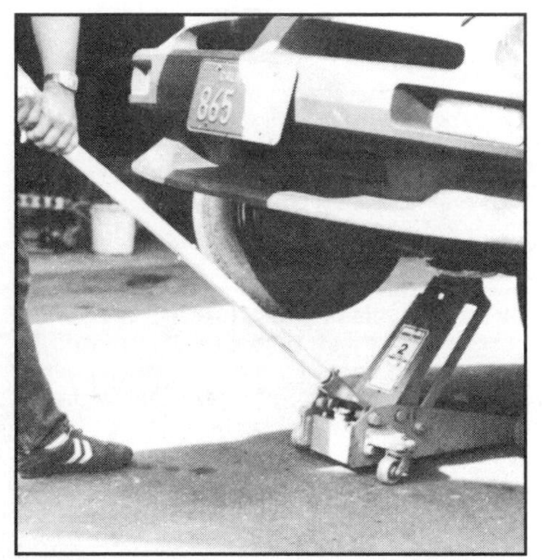

Floor jack illustrates relationship between force and movement in a hydraulic system. Handle has a low force and high movement versus high force and low movement at jack pad.

square of the diameter. Calculate all piston areas first so you won't make costly mistakes.

Piston Movement—Hydraulic systems must provide both force and movement. We just looked at how forces relate. Piston movement also depends on piston areas, but *not* on fluid pressure. Pressure affects force, not movement.

If a system is filled completely with fluid, that fluid cannot change volume, no matter how high the pressure. This is hydraulic-law number 1. This means that any motion inward on a piston must result in an outward motion elsewhere in the system. A good hydraulic system is rigid, so movement should occur only at another piston. If the hoses, lines or cylinders expand when the hydraulic fluid is pressurized, the system will not function as designed.

To calculate movements in a hydraulic system, you need to consider fluid *displacement* when a piston moves. Displacement is the area of the piston times the distance it moves, or the same as cylinder displacement of a piston in an engine. Piston stroke is piston movement; greater piston movement means greater displacement.

Fluid displacement = $A_p M_p$
A_p = Area of piston in square inches
M_p = Movement of that piston in inches

Because hydraulic fluid cannot be compressed, the following relationship exists: *Inward fluid displacement = Outward fluid displacement.*

If the hydraulic system has only two pistons, such as shown in in the accompanying drawing, the inward displacement of one piston must equal outward displacement of the other piston. This basic relationship remains constant regardless of system complexity. *All inward-displacement totals must total all outward-displacement totals.*

Now, let's use a simple example with a two-piston system. Assume piston 1 has an area of 2 sq in. If it moves inward 1 inch, its inward displacement is 2 cu in. The outward displacement of piston 2 must also be 2 cu in. If piston 2 has an area of 1 sq in., its outward movement will be 2 in. To change the movement of piston 2, you must change its area. A larger-area piston will move less; a smaller one more. Assume now that piston 2 is changed to an area of 4 sq in. for 2 cu in. of fluid displacement, piston 2 must move 1/2 in.

The relationship between piston movement in a simple two-piston system is given by the following formula:

$$M_2 = \frac{M_1 A_1}{A_2}$$

M_1 = Movement of piston 1 in inches
M_2 = Movement of piston 2 in inches
A_1 = Area of piston 1 in square inches
A_2 = Area of piston 2 in square inches

To increase the movement of piston 2, increase the area of piston 1 or reduce the area of piston 2.

Remember that fluid movement depends on piston area and displacement. Mistakes in calculation occur when diameter is used in place of area, and movement in place of displacement. Use care with your calculations. Double-check!

Force/Movement Relationships—As seen from previous formulas, there is a relationship between force on a piston and its movement. Both are related to piston area. It is possible to have force without movement, or movement without force. However, a brake system has both forces and movements. The relationships between movement and force is very important.

Formulas for force and movement both depend on piston area. However, piston-area changes cause opposite changes in forces and movements. Changes in area that create more force on a piston result in less movement of that piston. The system works like a simple lever, as shown in the accompanying drawing.

One simple two-piston hydraulic system is used in a hydraulic jack. The jack supplies both force and movement at the same time, such as a brake system. The jack handle is moved with force. This force is applied to the bottom of a car and it moves the car upward. When using a jack, note that the handle force is less than the force on the car. You can easily lift a 2000-lb car using less than 50 lb at the handle. Force is increased by the difference in piston sizes. Note also how great the movement of the handle is compared to the movement of the jack pad or car. Movement is reduced, but force is increased by increased piston size that operates the jack pad.

Input & Output Pistons—A hydraulic system uses force and movement applied to an *input piston* and provides force and movement on an *output piston*. With a hydraulic jack, the handle operates an input piston. The piston is connected to the jack pad, which raises the car. With a brake system, a pedal operates the input piston and output pistons apply the brakes.

BRAKE-SYSTEM HYDRAULICS

These laws of hydraulics explain how a basic brake system works. Although the basics are simple, there are complications with the automotive-brake system. I'll describe each separately.

The input piston is in the master cylinder. Modern brake systems use dual input-piston master cylinders. Each piston operates two brakes. Usually, front brakes are on one system and rear brakes are on the other. But some dual systems operate a *diagonal* pair of brakes with each piston, such as the right-front and left-rear, and vice versa.

Output pistons are located inside the brakes. Disc-brake calipers use one or more output pistons to move the pads, clamping the rotor. Drum brakes use one or more output pistons for moving the shoes to contact the drums. Drum-brake output pistons are about the same diameter as the master-cylinder input pistons. Disc brakes have no servo action, so they require higher force to operate. Consequently, disc-brake-caliper pistons are usually twice the diameter of master-cylinder pistons.

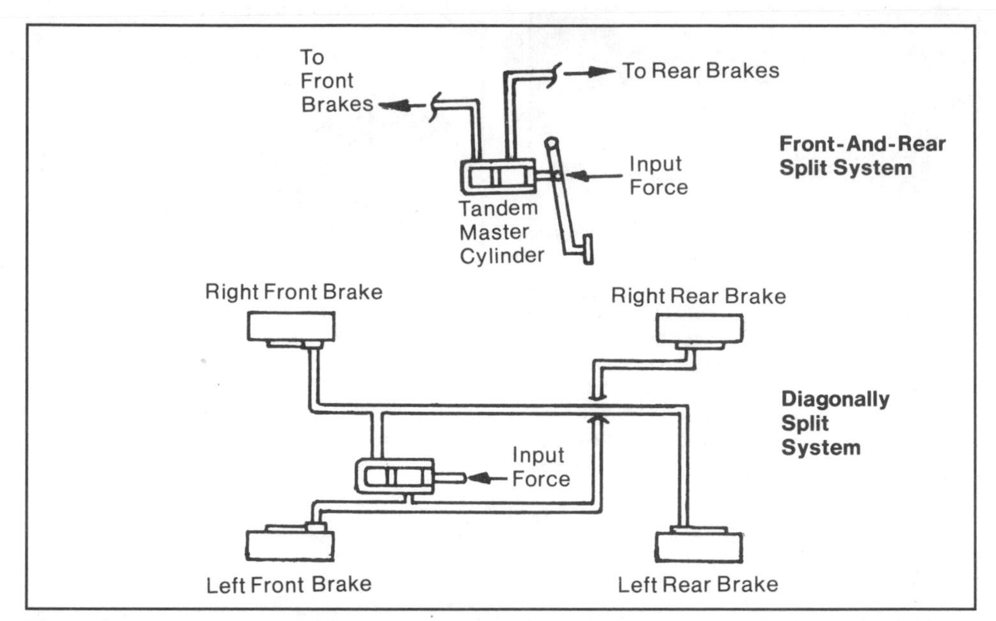

Most modern cars use tandem master cylinders—two pistons in one bore. Each piston operates brakes at two wheels. If one system fails, 50% braking is left. Most tandem master cylinders operate separate front and rear systems. Another one used on road cars has a diagonally split system as shown. I deal only with front-and-rear split systems, because it's the type most high-performance cars use.

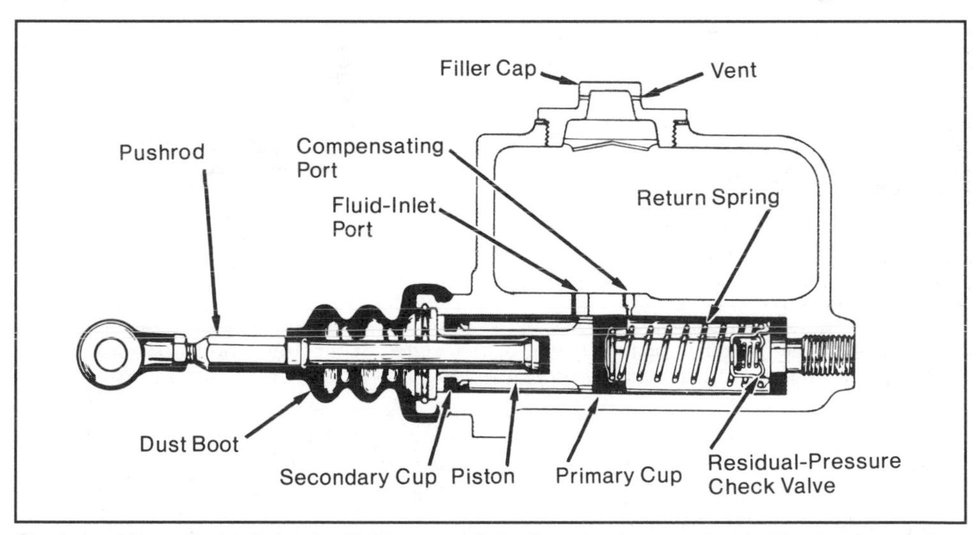

Single master cylinder is basic. Fluid reservoir is directly above cylinder. Fluid enters cylinder through ports in reservoir. Pushrod moves piston to right to pressurize brake fluid. Drawing courtesy Bendix Corp.

A simple approach to hydraulics seems to "say" that large-diameter pistons could be used at the brakes to produce high forces at the brakes. This would give low pedal effort. Therefore, power assist would not be required. Unfortunately, this doesn't work. The problem is input-piston movement, which translates into excess brake-pedal travel. Let's take a closer look at why this is true.

Pedal Travel—To take up clearances in a brake system, pistons must move. Clearance is built into each brake, and is equal to the piston movement required to place the lining in contact with each rubbing surface. The total volume of moving fluid must equal the *displacement* of all pistons that move.

This displacement is provided by master-cylinder-piston movement. Because fluid volume doesn't change, the inward displacement of the input pistons must equal the outward displacement of all output pistons in the system. This rule assumes a hydraulic system so rigid that nothing deflects from forces in the system.

If a small-diameter master-cylinder piston is used with large pistons at the brakes, greater movement is required

Girling single master cylinder is type found on many race cars, typically in dual master-cylinder setup. It is small, simple and lightweight. White nylon cap/reservoir extension increases reservoir capacity. Cylinders are also used to operate hydraulic clutches.

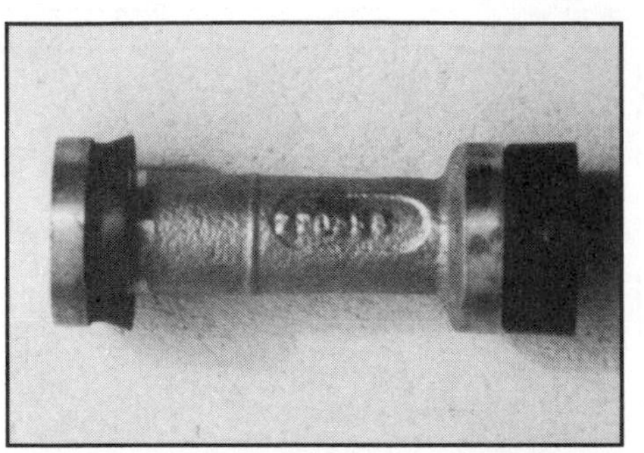

Typical master-cylinder piston with seals; larger seal at right seals high-pressure system. Left seal keeps fluid that's between seals from leaking out open end of master cylinder. Both are cup-type seals.

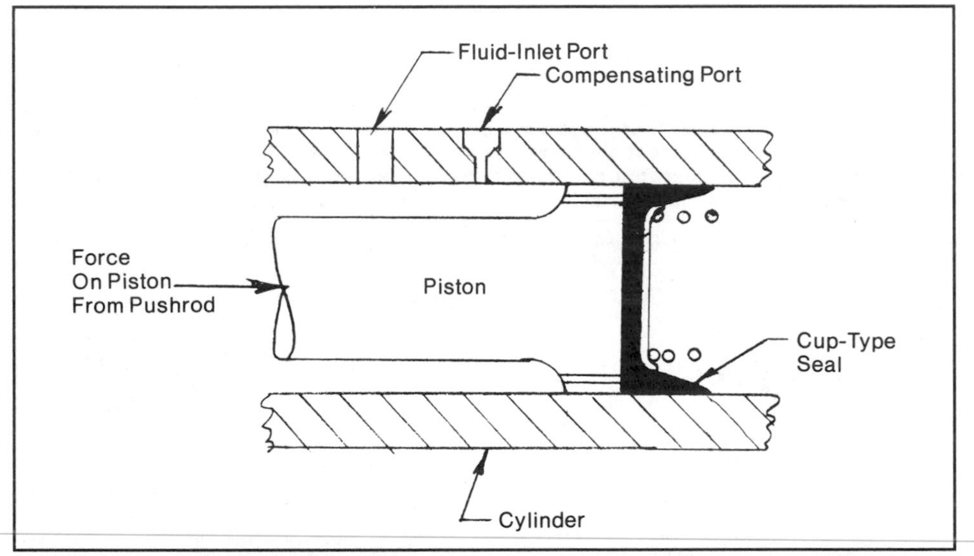

As master-cylinder piston is pushed forward, seal slides past compensating port and traps fluid inside bore. Piston in this position can now pressurize brake fluid to apply brakes.

at the master cylinder to take up clearances. The problem is, only so much pedal travel is available at the master cylinder. This is why you can't design a system for forces without considering movement.

Unfortunately, the characteristics of a real brake system are worse than a theoretically perfect hydraulic system. For example, the forces are as calculated except for a small loss of force at the output pistons from seal friction. However, input-piston movements in a real system *are much larger* than those of the theoretically perfect system due to deflections, or hydraulic hose and line expansion, and bending and twisting of brake components.

Specifically, additional movement comes from places other than clearances in a brake system. Pedal movement comes from the total of the following:

- Brake clearances.
- Designed clearances and wear in linkage.
- Swelling of hydraulic hoses and other parts of the hydraulic system.
- Compression of hydraulic fluid.
- Compression of air bubbles in hydraulic fluid.
- Bending of pedal, linkage or brackets.
- Deflection of calipers, drums or other parts of brake system.

When total movement at the pedal is excessive, the pedal will hit the floor before the brakes are fully applied. Keeping pedal movement within reasonable limits is a major problem in brake-system design.

Brake-system-design problems are discussed in Chapter 9.

MASTER CYLINDERS

Brake-fluid movement and pressure are created by the master-cylinder piston. This piston is connected to the brake pedal with a simple linkage and pushrod.

The typical modern road car has one master cylinder containing two pistons, while race cars are usually equipped with a separate master cylinder for each pair of brakes.

Single Master Cylinder—The simplest master cylinders, such as those used on production cars with hydraulic brakes up to the early '60s, have a single piston to operate all four brakes. See accompanying photo. Today's safety laws have made them obsolete. Regardless, single master-cylinder systems are still worth looking at because they are easy to understand. And, they are used on most race cars.

Note that most hydraulic-fluid reservoirs are integral with their master cylinders and are directly above the piston. The piston is moved forward in the cylinder by a pushrod attached to the brake pedal. The piston is returned by a spring against the piston when pedal force is removed.

There are two holes, or *ports,* in the reservoir that allow fluid to enter the cylinder. These are called the *fluid-inlet port* and *compensating port.* When a piston is at rest, the compensating port is ahead and the fluid inlet port is just behind the lip forming the front face of the piston. In this position, the piston rests against a clip or retaining ring that prevents the return spring from pushing the piston out the end of the cylinder.

As the brakes are applied, a pushrod moves the piston forward in the cylinder. The piston moves forward with the lip of the seal covering the compensating port. Before this port is covered, piston motion forces excess fluid back into the reservoir. Look into a master-cylinder reservoir and you see slight fluid motion just as the piston begins to move. Once the compensating port is covered, fluid trapped in the hydraulic system cannot escape, provided there are no leaks. Beyond this position, the piston moves fluid to apply the brakes.

Initial fluid movement reduces clearances in the brakes. When all clearances in the system are eliminated, fluid stops moving and pressure rises. The driver's foot controls how high fluid pressure gets. As

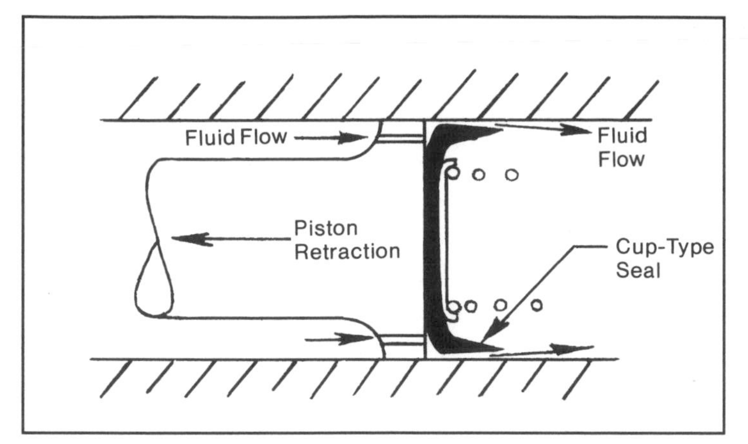

During rapid retraction of master-cylinder piston, fluid flows through small holes in front face of piston and past seal. This prevents air bubbles from forming in front of seal. Fluid-inlet port keeps fluid behind seal at all times.

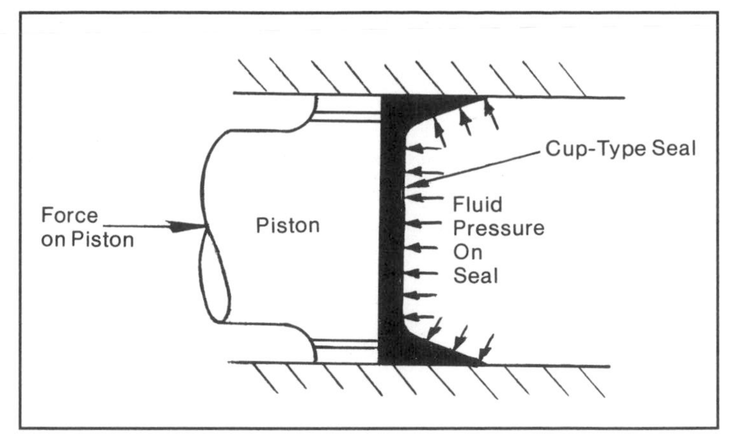

Cup-type seal expands outward against cylinder-bore walls when fluid is pressurized, making a tight seal. Flat end of cup-type seal goes toward piston face; cup end toward high-pressure fluid.

Cup-type seal can withstand high pressure while sliding easily within cylinder bore. Seal lips press tightly against cylinder walls when pressurized. The higher the pressure, the tighter the seal.

Residual-pressure valves are used in special applications. For example, Airheart calipers using mechanically retracted pistons require a residual-pressure valve to prevent pistons from retracting too far.

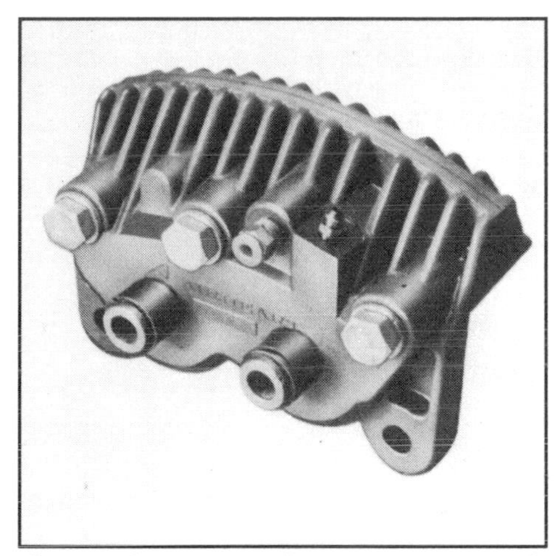

Some Airheart calipers use a pad-retraction system to prevent pads from dragging on rotors. Two cylindrical objects protruding from body of Airheart 175 X 206 caliper are part of mechanical retractors. Residual-pressure valve should be used with these calipers. Photo courtesy Hurst Performance.

force on the pedal is reduced, hydraulic pressure drops, the drum-brake-shoe return springs and/or disc-brake-piston seals retract, and the master-cylinder return spring moves the piston back against the retaining ring.

The fluid-inlet-port function is more complicated. To understand it better, look at the drawings of the piston and seals. Note that the piston has small holes drilled through the front-face lip. The cup-type seal is installed ahead of this face, with the lip extending forward. The fluid-inlet port allows fluid behind the front lip of the piston; the holes in the lip allow this fluid to contact the seal face. This is to help during brake release when the piston returns.

As the brakes are released, the return spring may move the piston faster than the fluid can move. When this occurs, the seal lip is drawn away from the cylinder wall, allowing fluid to flow through the small holes in the seal lip. This keeps fluid ahead of the piston at all times and prevents the formation of a gas bubble in front of the piston. Otherwise, the gas bubble would cause a soft pedal with excessive travel. The secondary seal keeps fluid from leaking out the open end of the cylinder.

Let's look closely at the function of a cup-type seal in the accompanying photo. The cup always faces high-pressure fluid. Fluid pressure acts equally on all surfaces as it contacts the cup face and lip. Pressure against the lip forces it against the cylinder wall, creating a tight seal. The higher the pressure, the tighter the cup lip seals against the cylinder wall. This is a reliable, long-life seal.

Residual-Pressure Valve—A cup-type seal has one problem when used with a drum-brake wheel cylinder. As the seal is retracted, its lip is relaxed and air can be introduced in the system. On drum-type brakes, this motion of the cup-type seals in wheel cylinders will draw in air each time the brakes are released.

There are two devices that prevent this. One, called a *cup expander,* is a thin metal cup installed between the seal and the return spring. This device exerts pressure on the lip and prevents air from passing the seal when the

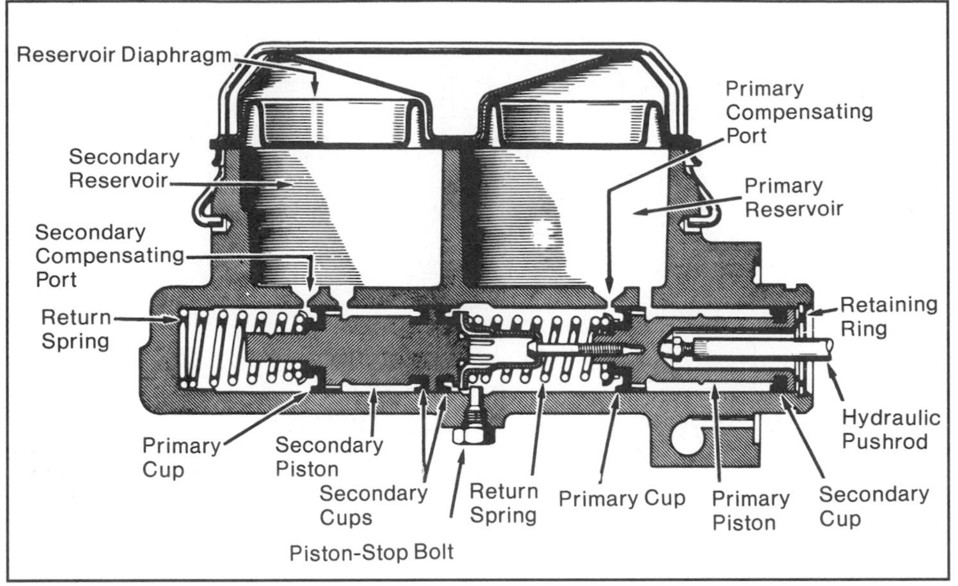

Cross section of typical tandem master cylinder with integral fluid reservoirs. Pushrod is retained on pedal, which is out of view to right, not on master-cylinder piston. Drawing courtesy Bendix Corp.

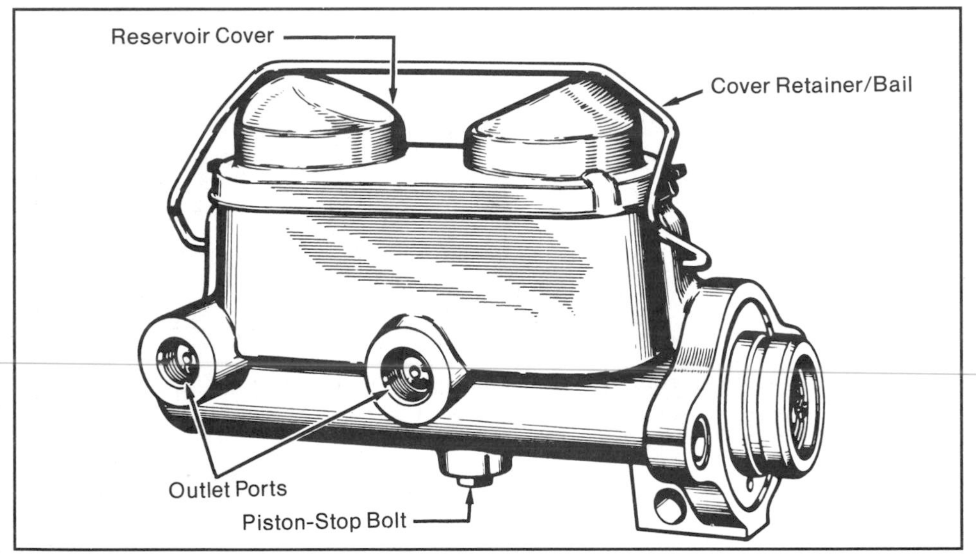

Hydraulic lines from tandem master cylinder to front and rear brakes install in outlet ports. Reservoir cover is retained with spring clip, or *bail*. Drawing courtesy Bendix Corp.

DIAGONALLY SPLIT SYSTEMS

Most front-wheel-drive cars are designed with tandem master cylinders plumbed so they operate diagonally opposite wheels. Each half of the cylinder operates a diagonally opposite pair of brakes. The right front brake and left rear brake comprise one system. Opposite pair of brakes are operated by other half of the master cylinder. Because a front and rear brake always work together, this system is claimed to be safer in the event one system fails.

Brake balance cannot be adjusted in the normal way with a diagonally split brake system. I'm not aware of any race cars or high-performance sports cars using diagonally split brakes, so hopefully you won't have to concern yourself with them. If you have a car with diagonally split brakes, the system would have to be totally redesigned to use conventional front-to-rear balancing techniques discussed in this book.

brakes are released.

The second is a simple spring loaded valve in the master cylinder. This valve, the *residual-pressure valve,* keeps slight fluid pressure on the system at all times—even when the brakes are released. When fluid returns to the master cylinder as the brakes are released, a spring closes the valve when pressure drops to a preset value—usually 6—25 psi. This pressure is enough to keep the cup-type seals in wheel cylinders against their cylinder walls, but not enough to overcome return-spring force. Excess residual pressure would cause brake-shoe drag.

If you change master cylinders, make sure you get one designed for the specific system. It must have the correct residual-pressure valve. Most disc-brake systems and some drum-brake systems *do not* use residual-pressure valves, so master cylinders may not interchange, even if they appear to be identical. Check master-cylinder specifications to be sure.

Residual-pressure valves are used on Airheart disc-brake calipers that have a mechanical-retraction system to pull the pads away from the rotors. A 2-psi residual-pressure valve is also recommended by Airheart for use on brake systems where calipers are mounted higher than the master-cylinder reservoir. This is to prevent fluid flowing back to the reservoir by gravity. Other disc-brake manufacturers do not recommend residual-pressure valves.

This brings up an important point: If you have a special car or a race car with modified brakes, always consult the brake manufacturer before making any changes.

Tandem Master Cylinders—Master cylinders with two pistons in a single bore are used on modern road cars. They are called *tandem* master cylinders. Tandem master cylinders were first introduced on American passenger cars in 1962; they've been universal since 1967. A typical tandem master cylinder is illustrated nearby.

Each section of a tandem master cylinder works as a single cylinder. Each has fluid-inlet and compensating ports. The piston closest to the brake pedal is the *primary* piston and the other is the *secondary* piston. Usually, the primary operates the front brakes and the secondary the rear brakes.

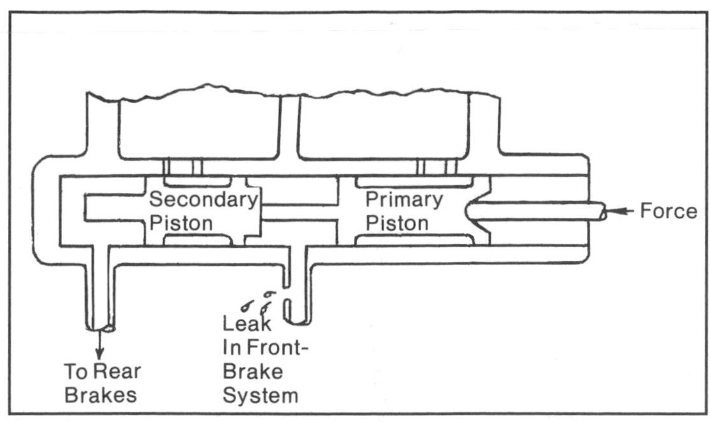

If leak develops in front-brake system, primary piston moves forward until it bottoms against secondary piston. Force from push-rod is transmitted to secondary piston by piston-to-piston contact. This allows secondary piston to pressurize rear-brake system.

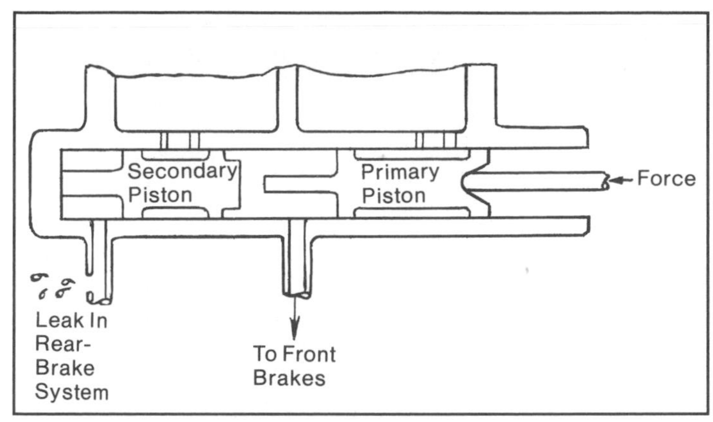

If rear-brake system should leak, secondary piston moves forward until it contacts end of master cylinder. This allows trapped fluid between pistons to be pressurized by primary piston, actuating front brakes.

As the driver operates the brake pedal, the primary piston moves forward, closing the *primary compensating port.* Fluid is trapped between the primary and secondary piston. This trapped fluid cannot compress, so it moves the secondary piston forward to seal the secondary compensating port. With both compensating ports closed, trapped fluid builds pressure in front of the secondary piston, and between primary and secondary pistons. Except for the primary-piston return spring, there is no direct mechanical link moving the secondary piston—only pressure from trapped fluid.

When clearances in the brakes are taken up, the pistons stop moving, fluid pressure increases and force is applied to the brakes. When the driver reduces pedal force, the two piston-return springs plus brake-return springs move the pistons to a retracted position. As mentioned before, holes in the front lip of each piston allow brake fluid to flow to the front of the cup-type seal when the brakes are released. Compensating ports allow excess fluid to return to the reservoir when the brakes are released.

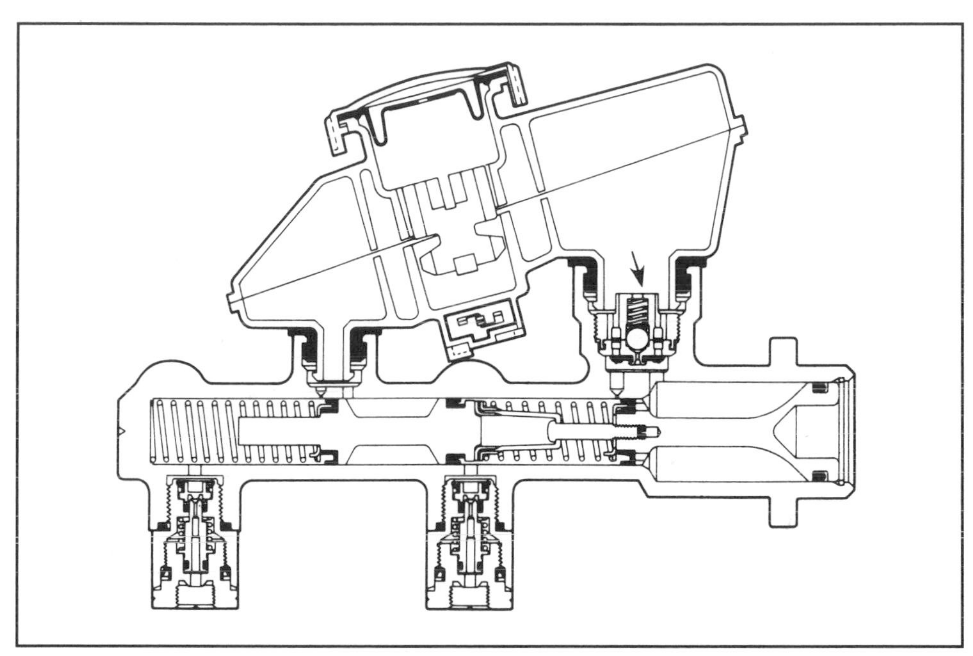

Cross section of fast-fill master cylinder used on front-wheel-drive cars shows fluid-control valve below primary reservoir (arrow). In-line proportioning valves are in rear-brake ports at bottom of cylinder. Front-brake ports are not shown.

The purpose of a tandem master cylinder is to protect against total loss of brakes when failure occurs. With a leak in the front brake system, almost no pressure exists in the front brake system. Note the metal pushrod between the primary and secondary pistons. As fluid leaks out of the front-brake system, the primary piston continues to move forward until this metal pushrod contacts the secondary piston. Then, primary-piston movement is transmitted to the secondary piston, moving it forward and applying the rear brakes. With this condition, a driver will have to exert more pedal force than normal to stop the car.

When a leak occurs in the rear system, fluid is trapped between the primary and secondary pistons, with fluid leakage behind the secondary piston. This allows both pistons to move forward in the cylinder until the pushrod of the secondary piston contacts the end of the master cylinder. When contact occurs, secondary-piston motion stops, allowing pressure to build up between the two pistons. This pressure operates the front brakes. Again, the driver will have to exert more force on the pedal. Also, the pedal moves closer to the floor—travel is higher. During an emergency, with one system malfunctioning, stopping distance is greatly increased because only two wheels are braking.

When designing a brake system using a tandem master cylinder, remember that all force applied to the master cylinder is applied to primary *and* secondary pistons equally. Force it is not split like a dual master-cylinder system using a balance bar.

Some master cylinders are designed to be mounted at an angle to horizontal. If so, the cylinder may have a *bleeder screw.* This screw allows air trapped in the master cylinder to be *bled* from the brake fluid. Bleeder

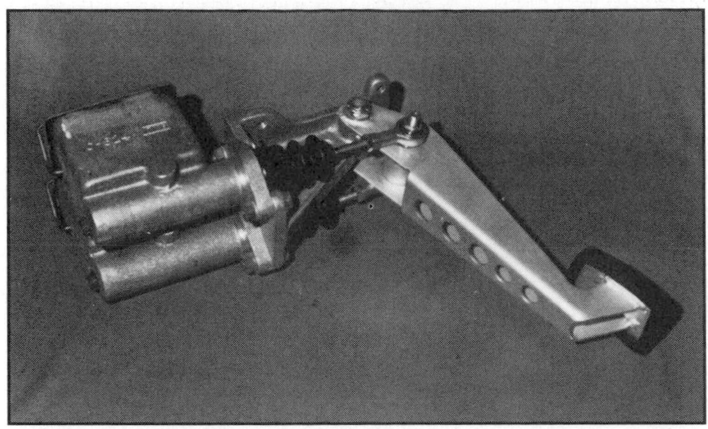

Dual master-cylinder setup manufactured by Neal Products, fitted with Howe cylinders, is popular for use in racing stock cars and other special vehicles. Hurst/Airheart, Girling or Lockheed master cylinders can be used with this setup.

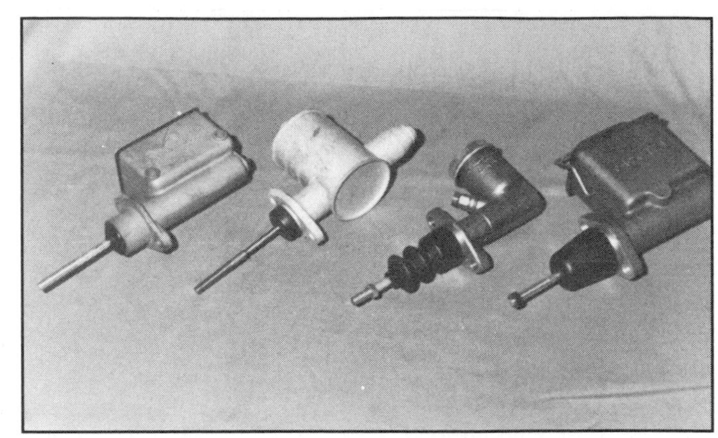

Single master cylinders found on most race cars; from left to right they are Airheart, Lockheed, Girling and Howe.

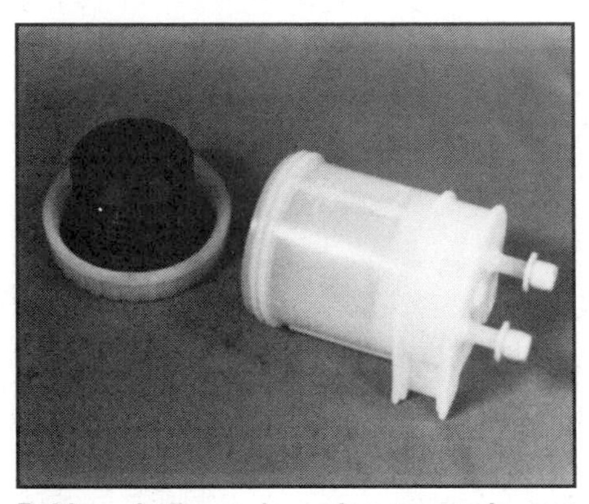

Rubber bellows in nylon-reservoir cap seals air from brake fluid. When reservoir is full, bellows is compressed. Bellows extends as fluid level drops. Vented cap allows air to enter and exit above bellows to compensate for fluid-volume changes.

Large translucent nylon reservoirs used on this Indy Car make it easy to check fluid levels. Caps don't have to be removed. It helps to have good light when checking fluid this way.

screws look similar to grease fittings. I have seen unknowing persons try to shoot chassis grease into the bleeder screws with a grease gun! The internal

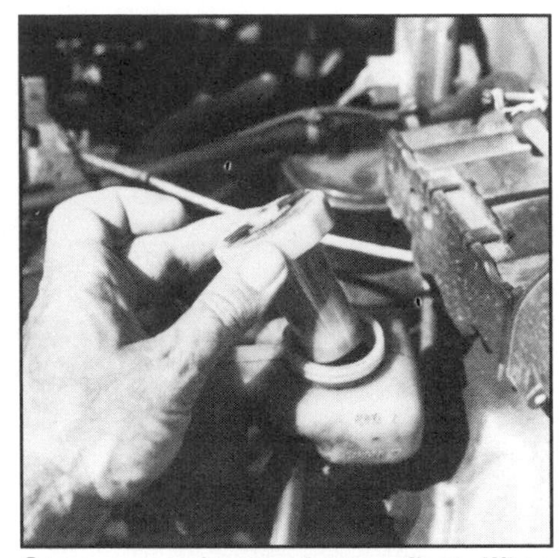

Some reservoir caps have a float. When fluid level falls too low, float activates an electrical switch that turns on a warning light on instrument panel. This light, instead of brake loss, warns of low fluid.

end of the bleeder has a taper that seals when it is tightened down on its seat. When the bleeder is loosened slightly, this seat seal is opened, then brake fluid and trapped air can come out through the hole in the bleeder.

When using this type of cylinder, make sure you mount it where it can be bled. For a master cylinder that is designed to be mounted level, make sure it's level.

Fast-Fill Master Cylinders—Note the master-cylinder drawing, page 49. It differs from a straight-bore master cylinder in two ways: It has a stepped primary, or fast-fill bore, and second, a fluid-control valve.

During initial brake application, the larger fast-fill bore provides fluid for system takeup by forcing fluid past the primary-piston seal into the main bore. After clearances are taken up in the rotors and/or drums, any pressure buildup in the fast-fill bore would result in high pedal effort. The fluid-control valve prevents this by diverting fluid from the fast-fill bore to the primary reservoir. The master cylinder then operates the brakes in a conventional manner using the smaller main bore. During the return stroke, the fluid-control valve allows fluid unrestricted return to the fast-fill bore.

The major advantage of fast-fill master cylinders is to increase fluid displacement during initial pedal travel. This allows smaller master-cylinder main-bore diameter, and reduced pedal effort and travel and/or disc-brake calipers with increased piston retraction distance, or *rollback*, for brake-drag reduction.

Dual Master Cylinders—Most race cars use two single master cylinders mounted side by side with a link between them called a *balance bar*. Balance bars are covered fully in Chapter 6. Dual master cylinders with a balance bar allow the ratio of force between front and rear brakes to be adjusted. This adjustment is necessary due to changing track conditions or variations in car setup. This arrangement is also a fail-safe system similar to the tandem master cylinder. Racing master cylinders are usually paired single master cylinders, often the same as those found on older road cars.

RESERVOIRS

The fluid reservoir is often integral with the master cylinder. However, for service accommodation on some cars the reservoir is separate, or remote. This type of fluid reservoir usually is made of translucent plastic

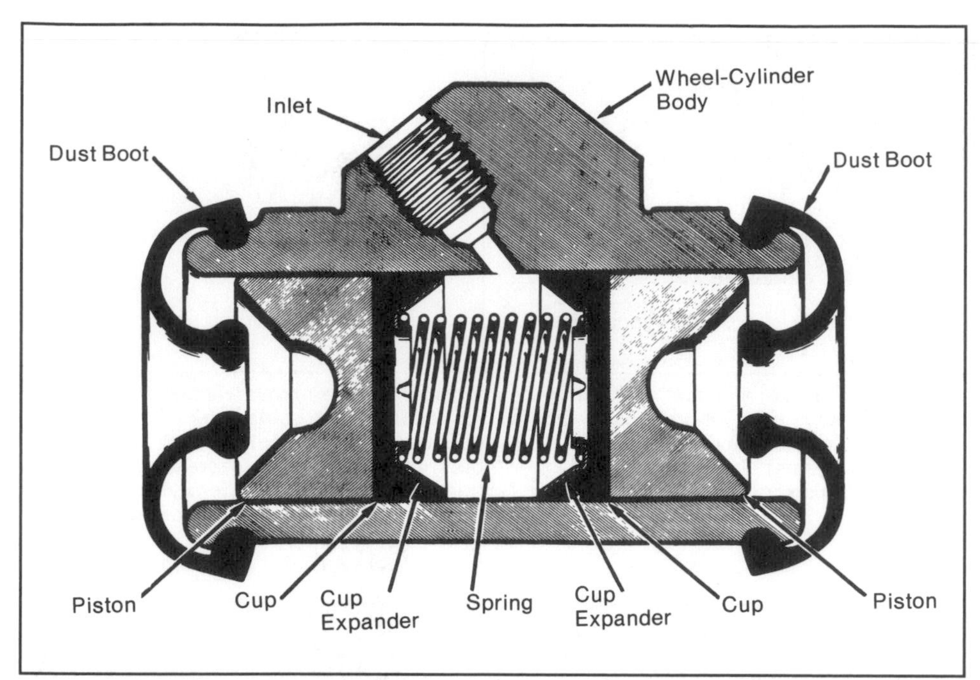

Wheel cylinders with two pistons are used on most modern duo-servo or leading-and-trailing-shoe drum brakes. Spring keeps pistons or pushrod against shoes when brakes are released. Pushrods are used in this cylinder to operate brake shoes. Drawing courtesy Bendix Corp.

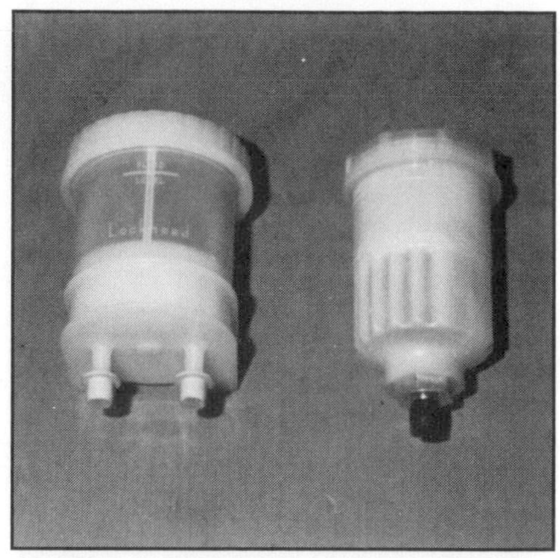

At left is remote reservoir for dual master cylinders; reservoir at right mounts directly on Girling master-cylinder outlet port. Reservoirs are large enough for racing disc-brake systems.

Custom-built remote metal reservoir mounted atop front bulkhead on this Indy Car is stronger than plastic reservoir. Cap must be removed to check fluid level.

so the fluid level can be checked without opening the reservoir. Many of them have graduations on the side indicating full and low levels. Separate fluid reservoirs are mounted above and connect to the master cylinder by flexible hose.

The fluid reservoir must not allow any contamination of the brake fluid. The reservoir cap, which screws on or is fastened by nuts, clamps or bolts, is vented to the atmosphere, allowing the fluid level to rise and fall freely. Water vapor in the air will contaminate brake fluid, so many reservoirs use a rubber diaphragm under the cap that isolates it from the atmosphere. This diaphragm flexes easily, allowing the rise and fall of fluid, while sealing out air and moisture. The diaphragm must be removed on some master cylinders to check or replenish fluid. The best overall design is a translucent reservoir with a rubber diaphragm.

Reservoir capacity must be large enough to allow fluid to fill a system, even when brake linings are worn totally. A brake system must never run short of fluid. Otherwise, total braking loss can occur. Disc-brake pistons displace more fluid, so they usually require more reservoir capacity than drum brakes. Reservoir volume must exceed the displacement of all output pistons, plus allow for lining wear.

With either a tandem or dual master-cylinder system, two separate reservoirs or a single reservoir with two compartments should be used. If this isn't done, a leak in one system will drain the fluid available to both systems, eliminating the fail-safe feature of the tandem and dual master cylinders.

Race-Car Reservoirs—Most race cars use remote reservoirs from road cars. If your race car uses a road-car master cylinder(s) and large racing calipers, a larger reservoir may be installed. The reservoir normally used with road-car master cylinders is too small for large racing calipers. Larger reservoirs are available from Tilton Engineering, Neal Products and other suppliers. Custom-built reservoirs can also be used on race cars.

Remote reservoirs are often used for strength and reliability. Most remote reservoirs on modern road cars are plastic. However, with the risk of heat, fire or physical damage associated with race cars, all-metal reservoirs offer extra protection. Another advantage is that mounting lugs or brackets can be added to a metal reservoir for easy mounting. Obviously, a metal reservoir can be made in any size and shape you need. Remember, air must never enter the reservoir-to-master-cylinder passage during hard cornering, acceleration or braking, so a tall reservoir is better than a short one.

Often, special remote reservoirs are used on off-road race cars. These cars experience incredible up-and-down bouncing and pounding. This results in severe slosh and aeration of brake fluid. To prevent this, tall reservoirs with baffles are used.

DRUM-BRAKE WHEEL CYLINDERS

There are a number of drum-brake wheel-cylinder designs. However, all share common features. For instance, all use cup-type seals. Some have seals with internal springs, or *expanders,* that keep the seal lips in contact with the cylinder walls. With

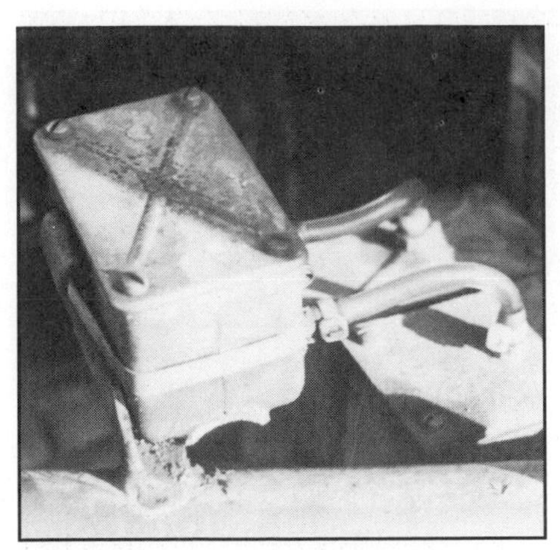

Clever design combats brake fluid sloshing in main reservoirs. Caps to reservoirs were modified to accept plastic tube. Fluid is fed from third reservoir mounted high in car. There's no way air bubbles will develop in the two fully filled reservoirs. Any sloshing and resulting air bubbles will be contained in third reservoir. Modification also increases fluid capacity.

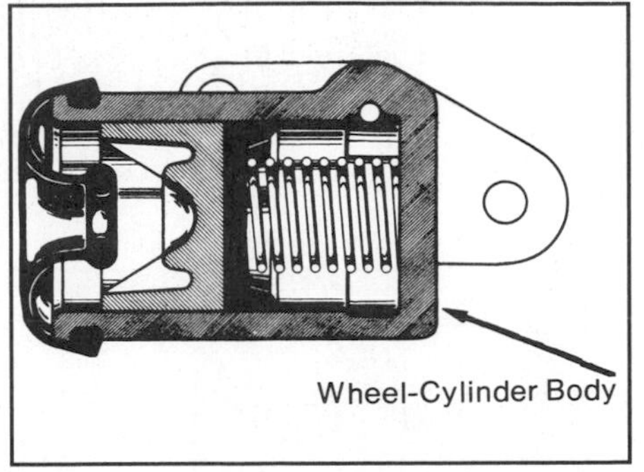

Single-piston wheel cylinder is used on two-leading-shoe drum brakes. Each shoe has its own wheel cylinder. Return spring holds piston against shoe when brakes are released. Drawing courtesy Bendix Corp.

Bleeder screws allows trapped air to be removed from high spots in hydraulic system. One is used at each high spot so air bubbles can be bled off to prevent a spongy pedal. If you mount master cylinders, drum brakes or calipers in a different position from that intended by the manufacturer, bleeder screws may be out of position.

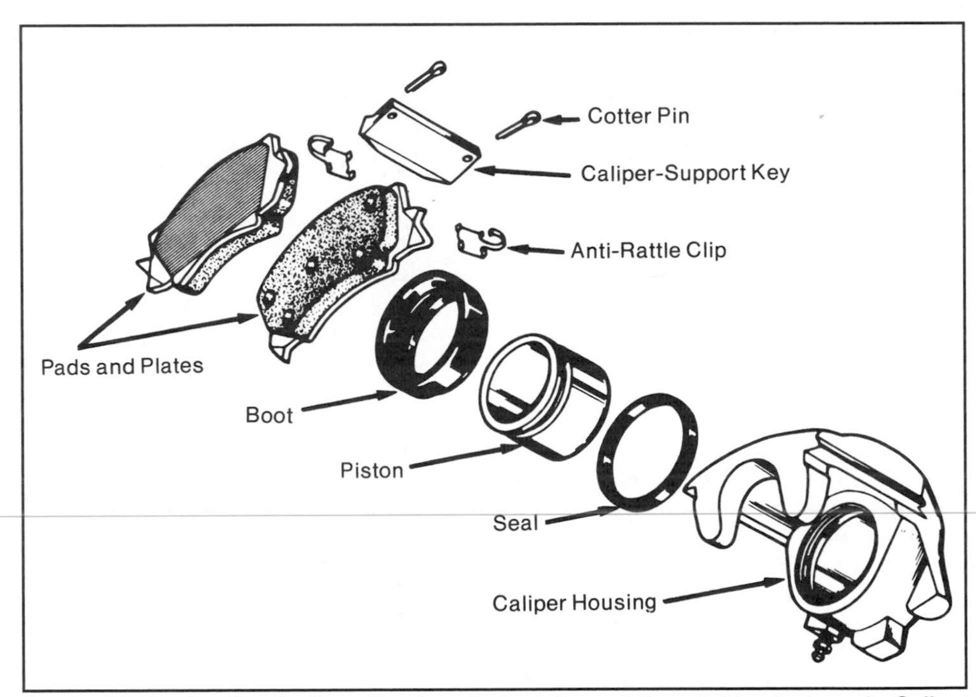

Bendix CA floating disc-brake caliper has single piston on inboard side of rotor only. Caliper housing contains both pads. Not shown is sturdy mounting bracket that bolts to suspension and supports and guides caliper. Drawing courtesy Bendix Corp.

expanders in the cup-type seals, a residual-pressure valve in the master cylinder is not required.

Duo-servo drum brakes, page 17, use a single wheel cylinder with two pistons. The wheel cylinder has both ends open and fluid between the two pistons. A spring between the pistons maintains slight pressure against the shoes. This spring also prevents the pistons from blocking the fluid-inlet port in the wheel cylinder.

There is usually a rubber dust boot on the open end(s) of a wheel cylinder. This prevents lining dust and moisture from contaminating the wheel cylinder, possibly causing a sticking piston. Dirt is a major enemy of any brake system. The rubber boots also hold the piston-to-brake-shoe pushrods in position, on systems that use pushrods.

Some wheel cylinders have a single piston and a closed end, as shown above. They are usually found on double *leading-shoe* brakes—each brake shoe is operated by a separate single-piston wheel cylinder. Similar to the dual-piston type, single-piston wheel cylinders have an internal spring to keep the piston and pushrod in contact with the brake shoe.

A wheel cylinder usually is equipped with a bleeder screw. It must be positioned at the highest point in a wheel cylinder so all trapped air can be bled off. Usually the position of a bleeder is the only difference between a right-hand and left-hand wheel cylinder, so use care not to mix wheel cylinders when installing them.

Some cars have rubber caps over the bleeder screws to keep out dirt and moisture. This is a good idea for road cars because most bleeder-screw breakage is from corrosion caused by moisture. Moisture reaches the bleeder-screw seat and threads through the hole in its center. Brake bleeding is discussed in Chapter 11.

Note: Wheel cylinders are usually made of cast iron or aluminum. Most pistons are aluminum. These metals can rust and corrode from moisture in the system. Wheel cylinders probably fail more often due to these than any other reason. Frequent bleeding of brakes protects against rust and corrosion.

DISC-BRAKE CALIPER

Disc-brake calipers have one or more pistons. They perform the same function as the wheel-cylinder pistons in a drum brake. The caliper piston(s) is moved by hydraulic fluid and pushes the pads against the rotor. As hydraulic pressure increases, the pads are pushed harder against the rotor. The caliper is designed to apply equal

AP Racing/Lockheed caliper has one large piston on each side of rotor. Caliper body is bolted rigidly to suspension. Fixed calipers, such as this, are used on most race cars; road cars typically use floating calipers. Photo courtesy AP Racing.

If brake balance is wrong for wet weather driving, you could be in trouble. Photo taken at Goodyear proving ground near San Angelo, Texas shows how car behaves when wheels lock up during braking. Here they are testing difference between various tires in wet conditions near limit of adhesion. Photo courtesy Goodyear.

pressure to the pads. This is done with one or more pistons on each side of the rotor, or with piston(s) on one side and a moving caliper.

Disc-brake-caliper cylinders differ from drum-brake wheel cylinders in several ways. First, caliper pistons are larger in diameter than drum-brake wheel-cylinder pistons. The reason for this is disc brakes require more force because they have no servo action.

Another difference between disc- and drum-brake cylinders is the type of seals used with the pistons. Cup-type seals are used in drum-brake wheel cylinders; usually, O-ring type seals are used in disc brakes. The O-ring may have a round or square cross-section.

The O-ring seal on most calipers retracts the piston and brake pad from the rotor, just as return springs do in a drum brake. The amount of retraction is controlled by seal and seal-groove design. Movement is slight in a disc brake because the rubber parts only flex a small amount, compared to brake-shoe return springs. However, a little motion is all that is required to release a disc brake. And, it is desired to keep the pad very close to the rotor so fluid movement is slight when the brakes are applied. Otherwise, pedal travel would be excessive due to the larger disc-brake piston.

Some calipers have springs mounted behind the pistons that hold the pads very close to the rotor surface. These springs add to the force supplied by the hydraulic system. However, the piston seal provides a greater force; thus, the piston is pulled back from the rotor.

Like drum-brake wheel cylinders, most disc-brake calipers have bleeder screws. The bleeder screw must also be located on top of the caliper so trapped air can be bled from the system. Right- and left-hand calipers often differ only by location of the bleeder screw because of this, so be careful not to get the calipers mixed up. Otherwise, you won't be able to bleed the brakes properly.

METERING VALVES

In a brake system with discs on the front and drums on the rear, there is often the need for a *metering valve*. This device prevents application of the front brakes below a preset pressure in the hydraulic system. The front brakes are applied at about 75—135 psi, depending on the system.

The reason for a metering valve is the force and fluid movement required to overcome drum-brake return springs. Also, disc-brake pads lightly touch the rotor when not applied, so it takes less pressure and fluid movement to make contact with the rotor. Drum brakes take much more force and movement to bring their friction material into contact with the drums. The metering valve balances these requirements and allows the front and rear brakes to work more evenly.

Without a metering valve, the front brakes do all the braking in an easy stop with a disc/drum-brake setup. This accelerates front-pad wear. Also, rear-wheel-drive cars without metering valves tend to lock the front wheels when stopping on snow or ice.

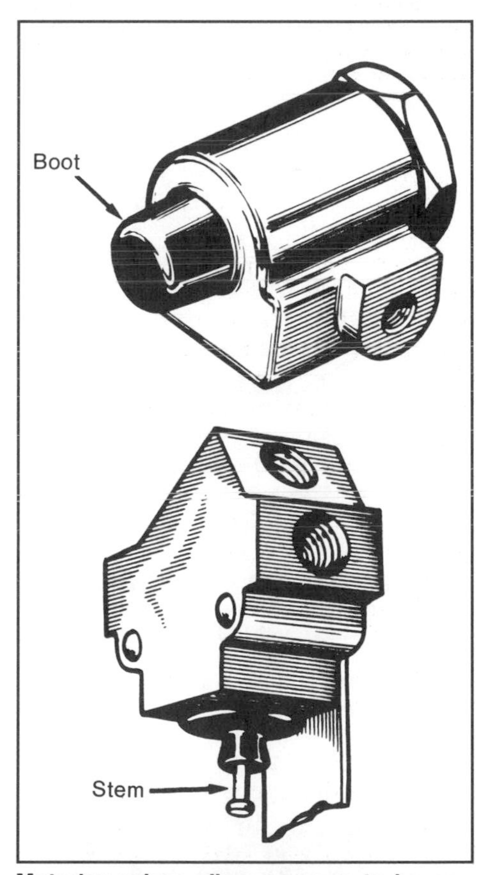

Metering valves allow pressure to be applied to rear brakes before front brakes. Because they may be combined with other brake-system components, you may not recognize them. If your car has a metering valve, you may have to push or pull the stem during bleeding. See your shop manual for specifics. Metering valves are used on road cars with combination drum/disc brakes. Drawing courtesy Bendix Corp.

Not all cars use metering valves with production disc/drum-brake setups. Some have a metering valve

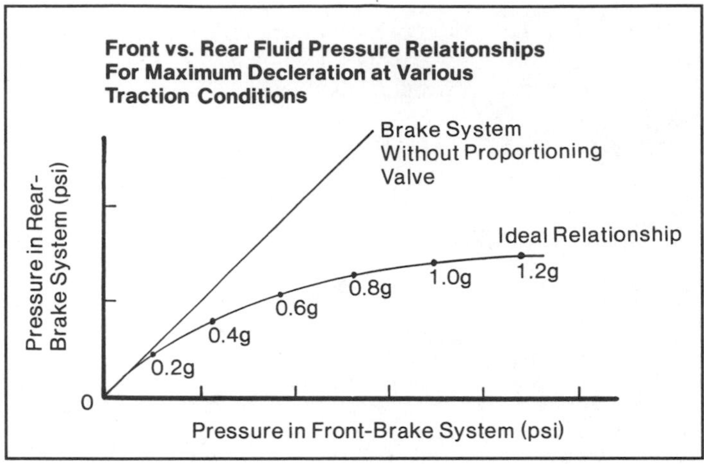

Upper curve represents ratio of front-to-rear brake pressures in system not using a proportioning valve. Pressure ratio is built into master cylinder or balance bar. Ideal pressure relationship for maximum deceleration, regardless of surface, is shown in lower curve. Pressure curve would look like this if brake system could automatically *proportion* pressures to make front and rear tires reach traction-limits simultaneously. The closer a proportioning valve comes to this ideal curve, the faster a car will stop under any condition.

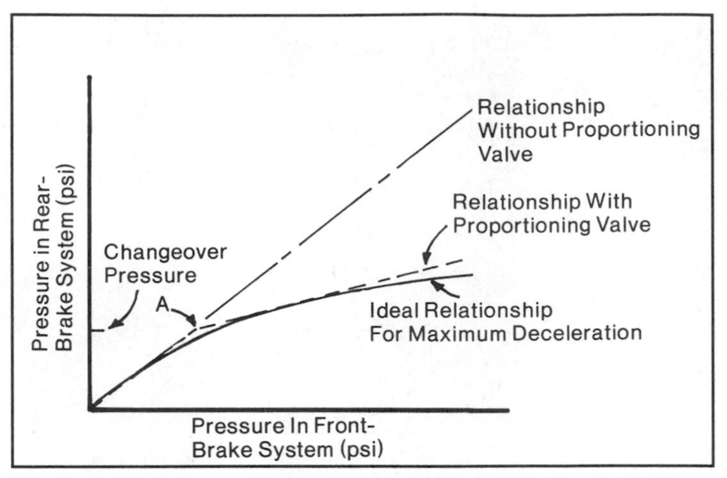

Ideal pressure curve plotted with curve from proportioning valve: At pressures above changeover pressure A, proportioning valve reduces pressure rise to rear brakes. Notice that proportioning valve does not reproduce ideal curve, but it comes close. Without a proportioning valve modifying pressure at point A, pressure to rear brakes would continue to rise in direct proportion to front-brake pressure.

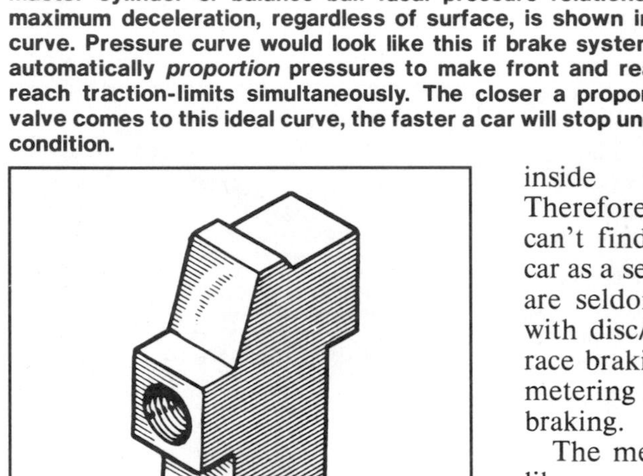

This proportioning valve looks like a brake fitting. It contains a spring-loaded valve that changes pressure-rise rate to rear brakes. Valves such as this are found in many production cars with disc brakes on front and drums on rear. Drawing courtesy Bendix Corp.

Kelsey-Hayes adjustable proportioning valve was used on 1965—'68 Corvettes and some Ford products. It has been used for adjusting brake balance on many race cars. Valve is adjusted by loosening jam nut and turning threaded shaft.

inside the *combination valve.* Therefore, don't be concerned if you can't find a metering valve on your car as a separate unit. Metering valves are seldom used on race cars, even with disc/drum-brake setups because race braking is done very hard, and a metering valve isn't needed in hard braking.

The metering valve operates much like a residual-pressure valve. A spring-loaded poppet valve is seated with a precise amount of spring force. When hydraulic pressure exerts sufficient force on the poppet, the spring force is overcome and the poppet opens to allow fluid flow. Another passage allows fluid return through the valve when the brake(s) is released.

BRAKE BALANCE

As discussed in Chapter 1, a car undergoes weight transfer during braking. Weight is added to the front wheels as it is removed from the rear wheels in the same amount. The problem is that braking force should be applied to each wheel in proportion to the weight on it—more weight, more braking, and vice versa.

Weight transfer depends on vehicle deceleration. Maximum deceleration depends mostly on friction between the tires and road surface. If the friction coefficient of the tires is high, deceleration is high, as is weight transfer. On a slick surface with a low friction coefficient, both deceleration and weight transfer are low.

Correct *brake balance* means that front-to-rear braking forces are proportioned so neither front nor rear wheels lock first. Put another way, the front and rear brakes will lock simultaneously if correctly balanced. If braking was adjusted to be equal on both the front and rear wheels, a hard stop with high weight transfer would cause the rear wheel to lock up first. If the brakes were *proportioned* on this car so more braking occurred on the front, the car could stop quicker. Greatest deceleration occurs when both front and rear tires reach their traction limit at the same time. Tires develop maximum grip just before they slid. Remember, *static friction is higher than sliding friction.*

Correct brake balance is different for high deceleration than for low. Brakes balanced for dry pavement will not be balanced for wet pavement. For example, a hard stop on wet pavement will give less deceleration than for dry pavement; less weight transfer occurs and the front wheels lock up.

A system that corrects the various traction conditions is needed. *Proportioning valves* are designed to do this.

A hydraulic-brake system has a certain ratio between front and rear brake pressures. A tandem master-cylinder brake system with equal piston diameters has the same pressures front and rear. This means that if you push on the pedal with enough force to produce 100 psi in the front brake lines, the rear brake lines will also have 100 psi in them. If you step down harder and raise the pressure to 400 psi, both front and rear brakes will have 400-psi pressure. This is not an ideal situation for maximum stopping on all surfaces.

Let's assume that we have a car with four-wheel disc brakes. Disc brakes have no servo action, so that simplifies this example. Let's also assume that all four brakes are the same—rotor diameter and caliper-piston sizes are the same. This means that the torque output of the brakes is the same front and rear if front and rear hydraulic pressures are the same.

We will also assume that the car has a 50/50 weight distribution—50% of the weight is on the front wheels and 50% is on the rear. If we make a panic stop on a slick surface, weight transfer is nearly zero. Equal braking torque is desired from both front and rear brakes. If we are trying to stop in a hurry on ice, the condition of hard braking with low weight transfer might actually exist.

Now let's try to design a brake system that gives maximum deceleration, regardless of the coefficient of friction between the tires and the road. This brake system would have to adjust the ratio of pressures between the front and rear brakes to account for the difference in weight transfer. The ideal brake system would be designed to lock all four wheels at once, regardless of weight transfer.

In the graph on page 54, I plotted the relationship between front- and rear-brake pressures for our example car. If master-cylinder-piston sizes are equal, there is always equal pressure in the front- and rear-brake systems. This is the straight line in the graph. When the pressure in the front-brake system is doubled, the rear also doubles. This is not ideal for maximum deceleration under various traction conditions.

With a high-traction surface, weight transfer increases. This requires less pressure at the rear brakes and more at the front because the vertical force on the front tires is greater than on the rears. A brake system with equal brake pressure and equal brake torque will lock the rear wheels before the fronts. This is not desirable for maximum deceleration or stability. A car with locked rear wheels and rolling front wheels will spin easily.

The lower curve shows the ratio between front- and rear-brake pressures automatically changed to account for weight transfer. It shows the need to gradually reduce the percentage of brake pressure going to the rear brakes as traction increased. This reflects high deceleration rates obtained with high tire grip and the resulting weight transfer.

PROPORTIONING VALVE

Now that we have an ideal front-to-rear brake-pressure curve for this car, how can we modify the straight-line pressure relationship to get close to the ideal curve? Answer: It is done with a *proportioning valve.*

A proportioning valve *reduces the pressure increase* to the rear brakes. This does not mean that pressure to the rear brakes is prevented from rising—it merely rises at a lower rate than front-brake pressure when a specified pressure—*changeover pressure*—is reached. The curve of front-to-rear-brake pressure with a proportioning valve in the rear brake line is plotted in the graph, page 54.

The accompanying pressure curves do not have numbered axes because they vary with each car. Exact pressures in a brake system vary with hydraulic-cylinder dimensions, brake design, CG location, weight and wheelbase of the car. The curve always has the shapes shown in the two figures, but the numbers vary from car to car.

Most proportioning valves are preset at the factory. They cannot be adjusted or serviced later. Each valve

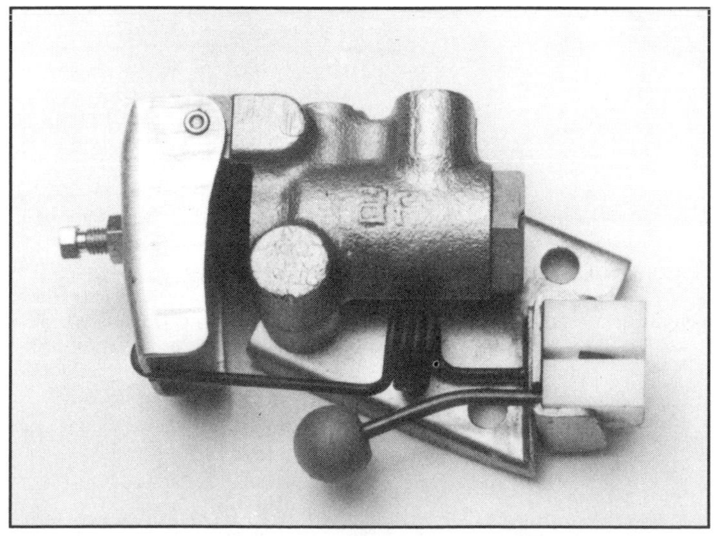

Tilton Engineering offers this AP Racing adjustable proportioning valve. Lever has five positions, each with a different changeover pressure. Valve is ideal for road car occasionally raced on racing tires. Improved grip and weight transfer with racing tires means less pressure should go to brakes. Photo courtesy AP Racing.

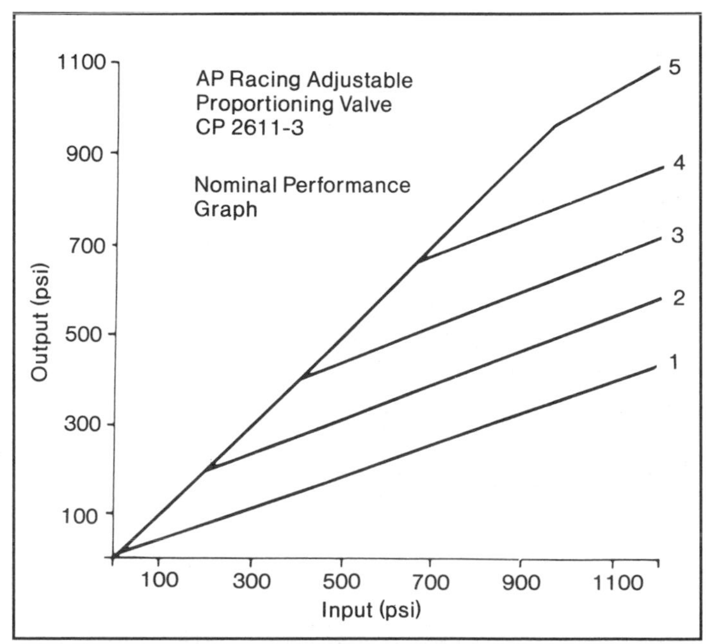

Graph shows AP Racing adjustable proportioning-valve pressure changes to rear brakes. With valve at setting 2, pressure rises as if there were no valve, until 200-psi changeover point is reached. Pressure then rises at a lesser rate. This effectively reduces pressure-rise rate to rear brakes and delays or prevents rear-wheel lockup. Graph courtesy Tilton Engineering, Inc.

is suitable for use only on the car for which it was adjusted. This is unfortunate because you may like an adjustable proportioning valve as a means of balancing your high-performance or race-car brakes. Fortunately, some proportioning valves allow an external adjustment. Some of these are shown in the accompanying photos.

Porsche, Chrysler, Chevrolet and Ford have adjustable proportioning valves for high-performance use. The Porsche 914 uses an adjustable valve that can be changed by loosening a lock nut and turning a screw. Turning the screw inward on this valve puts more pressure on the rear brakes during a hard stop. Adjusting the valve is a trial-and-error process, as described in Chapter 10.

The Chevrolet Corvette adjustable proportioning valve, 3878944, is available both from Chevrolet and from racing shops specializing in brake components. Chrysler's valve, P4120999, is available through their performance-parts division, Direct Connection. Ford's proportioning valve, M-2328-A is available through Ford Motorsport dealers. These valves work much the same as the Porsche valve. Proportioning begins between 100 and 1000 psi, depending where adjustment is set.

Another proportioning valve, which operates like the Porsche and Corvette valves, is available from Alston Industries of Sacramento, California. This one was designed by Frank Airheart to go with his high-performance dual, tandem master cylinder.

A proportioning valve is installed in line between the master cylinder and the rear brakes. It will work with either a single or a dual master cylinder in the system, because all it does is affect the maximum pressure at the rear brakes.

An adjustable proportioning valve, CP2611-3, is sold by Tilton Engineering of Buellton, California. This device can be mounted in the cockpit of a race car so the driver can adjust it during a race. He can move a lever into any one of five different positions as brake or track conditions, or the weight distribution of the car changes during a race.

Curves of inlet pressure versus outlet pressure for this valve are shown on page 55. Note that pressure rises in a one-to-one ratio until the changeover pressure is reached. For position 5, pressure rises as if the proportioning valve is not in the system until about 950 psi is reached. At that point, the spring-loaded poppet valve inside the proportioning valve opens, and output-pressure increase is reduced with increasing input pressure. Meanwhile, front-brake pressure is not affected. It increases as it did without a proportioning valve.

Most road cars have proportioning valves that give approximate correct brake balance with an average load using street tires. If the car is set up for racing, its total weight, weight distribution and tire grip are likely to change from the average values the car was designed for. Thus, brake balance may be incorrect for racing, even though it has a proportioning valve. Remember, a proportioning valve gives the approximate correct brake balance over a limited range of traction conditions.

If you drive a car on the street and race it on occasion, you may want to use an adjustable proportioning valve

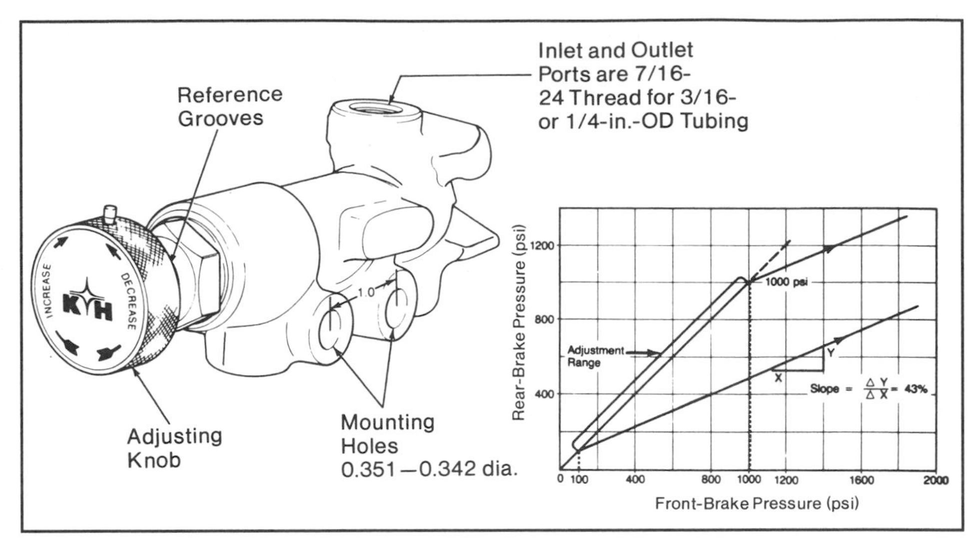

Kelsey-Hayes adjustable proportioning valve is available from Kelsey Products Division and many U.S. automobile dealers through their high-performance catalogs. Valve functions the same as AP Racing valve except it's infinitely adjustable between 100 and 1000 psi. Adjustments are made with thumbscrew. Drawing courtesy Kelsey-Hayes Corp.

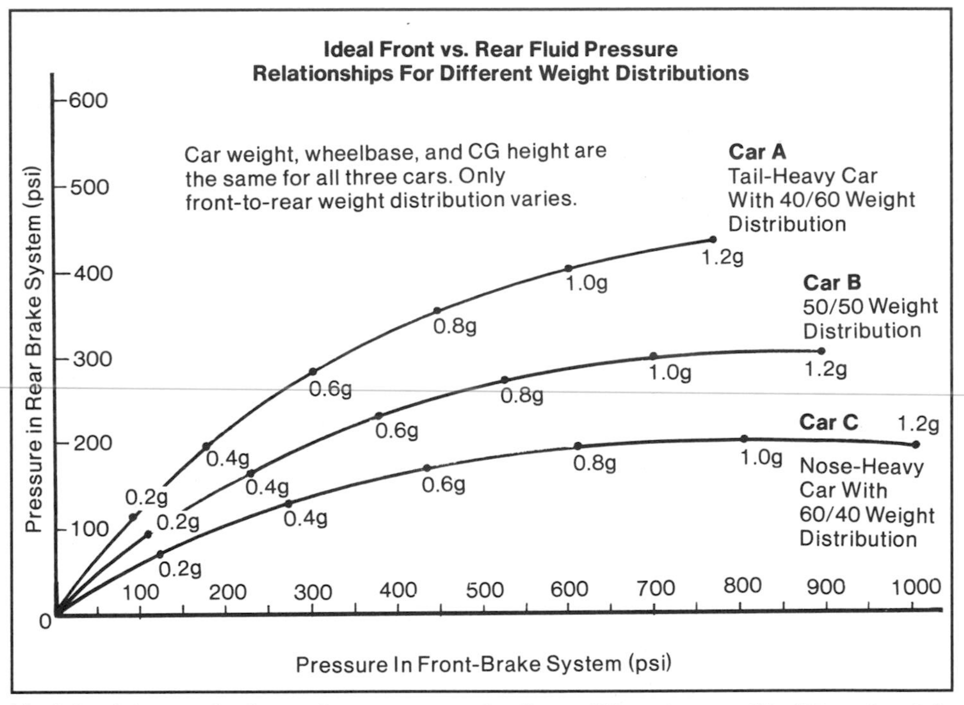

Ideal front-to-rear brake-system pressures for three different cars with different weight distributions: Nose-heavy car has less pressure to rear wheels; tail-heavy car has more.

to change brake balance. For street driving, the car should be correct as is. For racing, brake balance probably will have to be readjusted to provide more braking force on the front—less on the rear. Racing tires have more grip than street tires; thus, weight transfer is greater under maximum braking.

Let's say you have a sports car that you drive to work and slalom on the weekends. With the quick-adjustable Tilton valve, you don't even have to get your hands dirty. In the example given, you might run the valve at position 5 for street use and drop to position 4 for racing. Position 4 reduces rear-wheel braking, starting at a lower pressure than position 5. This compensates for the greater weight transfer when braking on racing tires. More about setting up brake balance in Chapter 10.

One disadvantage of proportioning valves is that they increase pedal effort without reducing displacement during high-deceleration stops.

If you wish to add a proportioning valve to a car that doesn't have one, make sure the car does *not* have a diagonally split brake system. Some rear-wheel-drive cars and most front-wheel-drive cars are designed so the left-front and right-rear brakes are operated by the same half of a tandem master cylinder; right-front and left-rear brakes are operated by the other half of the master cylinder. With a diagonally split brake system, a single adjustable proportioning valve will not work. It only works where both rear brakes are fed by a single hydraulic line. I cover adding a proportioning valve to an existing car in Chapter 12.

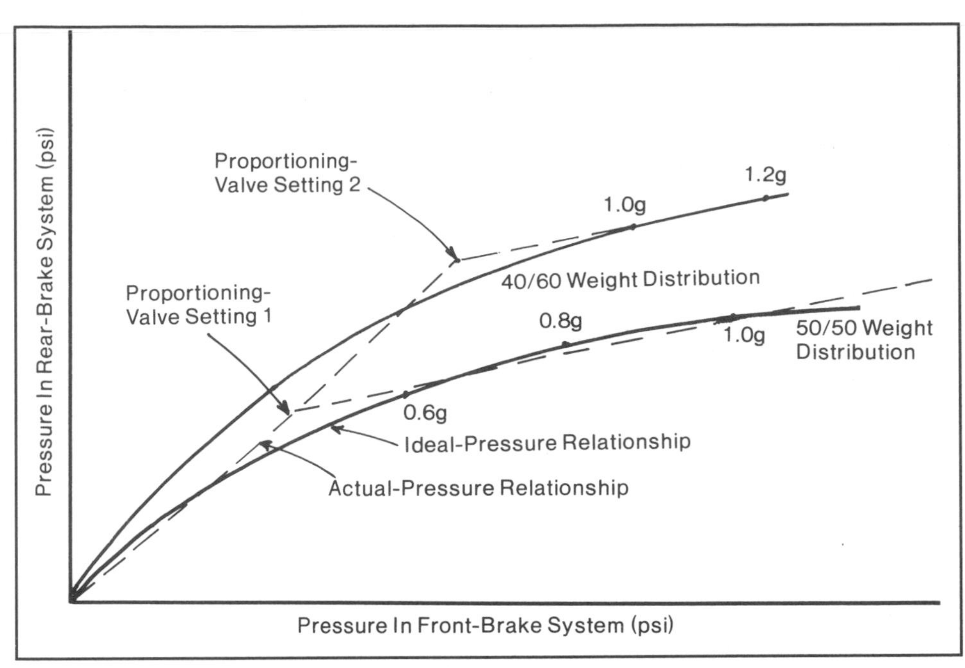

Two ideal pressure curves for cars of different weight distributions show how one proportioning valve can be adjusted to work with either.

PRESSURE-LIMITING VALVES

A *pressure-limiting valve* is used on some cars with disc brakes on the front and drums on the rear. It performs a similar function as a proportioning valve. The pressure-limiting valve prevents the pressure at the rear brakes from exceeding a preset value. The difference between a proportioning valve and a pressure-limiting valve can be seen by comparing the accompanying graphs.

The relationship between pressure in the front-brake system and rear-brake system is shown in the graph, below. Note how the pressure-limiting valve differs from a proportioning valve. Remember, a proportioning valve allows the pressure in the rear system to rise if the driver pushes harder on the pedal—but at a reduced rate. However, a pressure-limiting valve limits pressure to the rear brakes.

When this preset value is reached,

Honda Civic has diagonally split braking system with two proportioning valves. Valves are shown at bottom. One adjustable proportioning valve cannot be used with diagonally split braking system. Two proportioning valves will work, but adjustment must be synchronized.

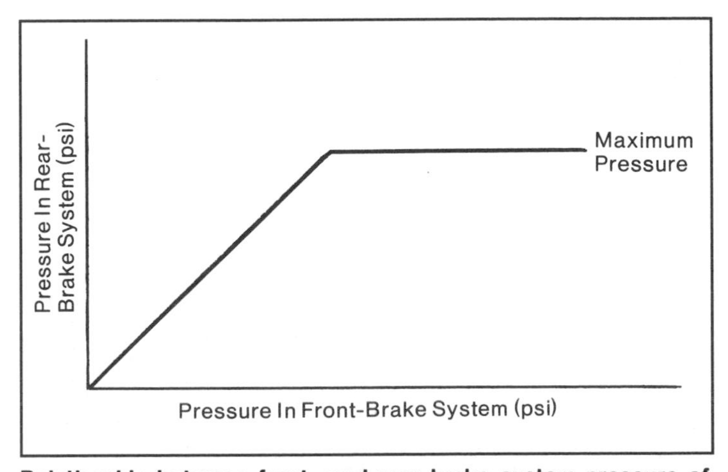

Relationship between front- and rear-brake system pressure of pressure-limiting valve: Compare curve to that of ideal pressure curve on page 54. With a proportioning valve, pressure continues to rise in rear-brake system beyond changeover pressure. With pressure-limiting valve, once changeover pressure is reached, that's as high as it gets. For a high-performance car, use a proportioning valve, not a pressure-limiting valve—unless car is extremely nose-heavy.

Both proportioning valves are adjusted by spring between rear axle and proportioning-valve lever. Valve adjustment changes with rear-suspension travel from weight on rear wheels. Rear-axle movement moves lever and changes valve setting. As weight on rear wheels increases, so does rear-wheel-brake pressure. Cars that have such large weight-distribution changes, such as this Plymouth Voyager/Dodge Caravan (left) or Peugeot station wagon (right), will benefit from a height-sensing proportioning valve.

pressure to the rear brakes cannot increase, regardless of how hard the driver pushes on the pedal. A pressure-limiting valve works well only on an extremely nose-heavy car, such as a front-wheel-drive sedan. Also, pressure-limiting valves usually are not useful for race cars because most are not nose heavy.

ANTI-SKID SYSTEM

Anti-skid systems are designed to prevent wheel lockup, no matter how hard the brakes are applied.

Many manufacturers claim their systems have anti-skid properties. What they really do is prevent one pair of wheels from locking before the other. In a sense, a good proportioning valve is an anti-skid device; it keeps the rear wheels from locking before the fronts. In this section, I discuss systems other than *fixed* proportioning or pressure-limiting valves.

Some anti-skid devices amount to nothing more than adjustable proportioning valves controlled by some external means. The valve is adjusted by rear-suspension movement or by sensing the actual deceleration of the car. The type that operates from rear-suspension movement compensates for both deceleration and vehicle weight. This is particularly important in a truck or a station wagon, where the CG can move a great deal with changes in load in the cargo area.

The anti-skid valve that senses deceleration uses the *ball-and-ramp principle*. The valve is positioned so that the ramp angles up, toward the front of the vehicle. The valve contains a conventional proportioning valve and a bypass. At low deceleration, brake fluid passes the ball and bypasses the proportioning valve, regardless of pressure. At a predetermined *transition deceleration*—about 0.3 g—the ball rolls up the ramp, closes the bypass, and proportioning starts.

This valve can compensate for a changing CG location. As weight at the rear of the vehicle increases, the rear suspension compresses and the ramp angle increases. This increases the transition deceleration and, therefore, rear-brake pressure at transition. Thus, the rear brakes share a larger portion of the braking.

A true anti-skid system senses when a wheel starts to lock. It closes a valve in the line to that wheel to reduce the pressure. It then senses the wheel speeding up and reapplies pressure. A typical system uses magnetic sensors to sense wheel speed.

Wheel-speed signals are sent to an electronic module, or computer, which computes wheel deceleration and slip values for each wheel. Deceleration and slip values for that wheel are compared with values for the other wheels and with fixed *thresholds* (preset values). When thresholds are exceeded, the module sends a signal to the proper controller. The valve is closed and pressure to that wheel is reduced. When wheel deceleration and slip go below threshold levels, the module signals the controller to increase pressure. All this shuttling happens rapidly during hard applications, so the wheel(s) never completely lock.

ATE skid-control front-wheel sensor, introduced on '85 Lincoln Mark VII, is illustrated on display model with sectioned front-wheel rotor/hub. As wheel speed changes rapidly when brake lockup begins, brakes are rapidly cycled on and off to prevent lockup. Photo by Tom Monroe.

Note: Controllers require an external power source, which can be hydraulic, electric or vacuum.

Skid-control systems have been used on large aircraft since the '50s. The first vacuum-powered road-car systems developed by Kelsey-Hayes and Delco were introduced in the late '60s by Ford, GM and Chrysler. The Ford and GM systems operated on the rear wheels only; Chrysler had a four-wheel system. Porsche experimented with more sophisticated systems in the early '70s on 911 and 917 race cars.

Today, most major car manufacturers are developing four-wheel systems. 1984 Porsche and Mercedes models use Bosch anti-skid systems. Honda and Mitsubishi have systems in production. Ford introduced a

Safety Braker is fancy cylinder mounted in brake line from nearest master cylinder. It damps pressure pulses in brake fluid, giving more consistent braking and less pedal pulsation. It is *slightly* more difficult to lock one wheel with a Safety Braker in the system. Because device is an accumulator, it displaces fluid and increases pedal travel slightly. Safety Brakers are no longer being made.

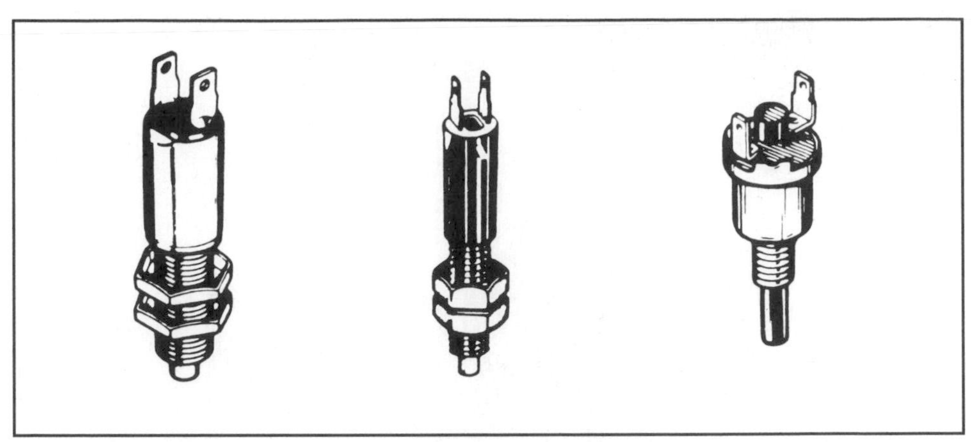

Three different types of mechanical stoplight switches each have plunger to operate electrical switch. Usually, switch is operated by movement of brake pedal. Drawing courtesy Bendix Corp.

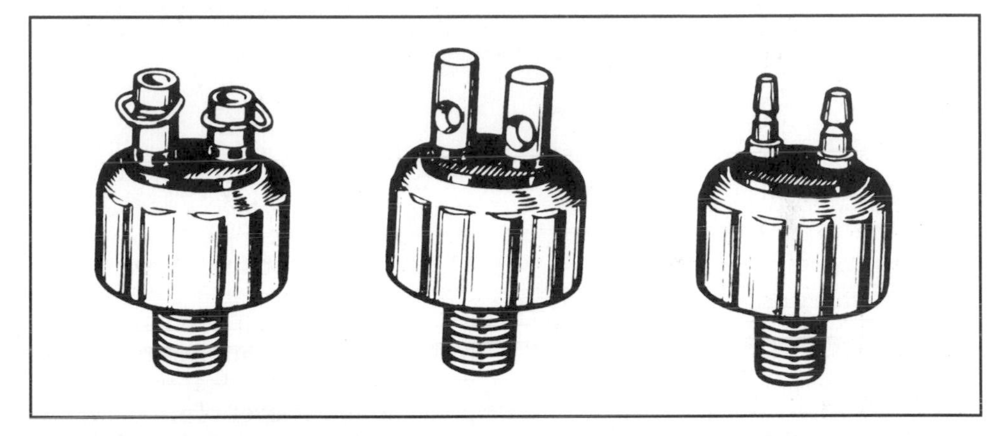

Hydraulic stoplight switches screw into a fitting in hydraulic system. Fluid pressure operates switch. Although easier to install, mechanical switches are better for racing. Hydraulic switch takes a small amount of fluid movement to operate it, and a mechanical switch does not. Also, hydraulic switches can leak. Drawing courtesy Bendix Corp.

system developed by ATE for the 1985 Mark VII. GM released systems developed by Bosch and ATE for 1986 models. These systems are expensive, but they provide much improved directional stability, steerability and deceleration in critical braking maneuvers on all road surfaces.

Anti-skid systems can provide these same advantages in racing, plus they can eliminate tire flat spotting. To date, anti-skid systems have been used little because of their weight and questionable reliability. However, recent advancements in technology should change that. Many high-powered, exotic road-racing cars will probably be using anti-skid devices by 1990. But for now, a skilled racing-driver's feel is considered to be the most effective anti-skid device.

STOP-LIGHT SWITCHES

Stop-light switches are either mechanical or hydraulic. Both are connected to the electric circuit and turn on the brakelights when activated. The hydraulic switch is designed as a hydraulic fitting and is operated by brake-system pressure. The mechanical type is operated directly by brake-pedal-linkage motion. Most newer cars use the mechanical switch.

With both switch types, the electrical circuit is open until the brakes are applied. When the switch closes, the stop lights illuminate. Usually, the mechanical switch is designed to close when the brake pedal has been moved about 1/2 in. The hydraulic switch is designed to close with a small pressure buildup in the system. The exact point at which the switch closes is not critical because any modest brake application should close the electrical contacts.

If you are selecting a stop-light switch for a race car, use a mechanical type. Although more difficult to install, a mechanical switch will not affect the brake system if it fails. A hydraulic switch is just another possible leak. Also, slight fluid movement is necessary for operation. Every possible failure must be eliminated from a race car.

BRAKE WARNING-LIGHT SWITCHES

Brake warning-light switches have been in use since tandem master cylinders were introduced in the '60s. This pressure switch illuminates a warning light on the instrument panel, which indicates a problem with the brake system. The switch senses any hydraulic-pressure imbalance between the front and rear brake systems. And, the light illuminates when a preset pressure imbalance occurs. This will warn the driver of problems such as fluid leaks, master-cylinder-fluid bypass or air in the system.

A typical brake warning-light switch is shown on the following page. At the center of the switch, a piston is located between two springs. Pressure from the front brakes is on one side of the piston and rear-brake pressure is on the other. Differences in the force of the two springs compensate for the designed-in pressure differential of the front and rear systems. When severe pressure differential occurs, the piston is moved toward the low-pressure side. This closes the switch and turns on the warning light.

There are many variations of this simple switch. Some have centering

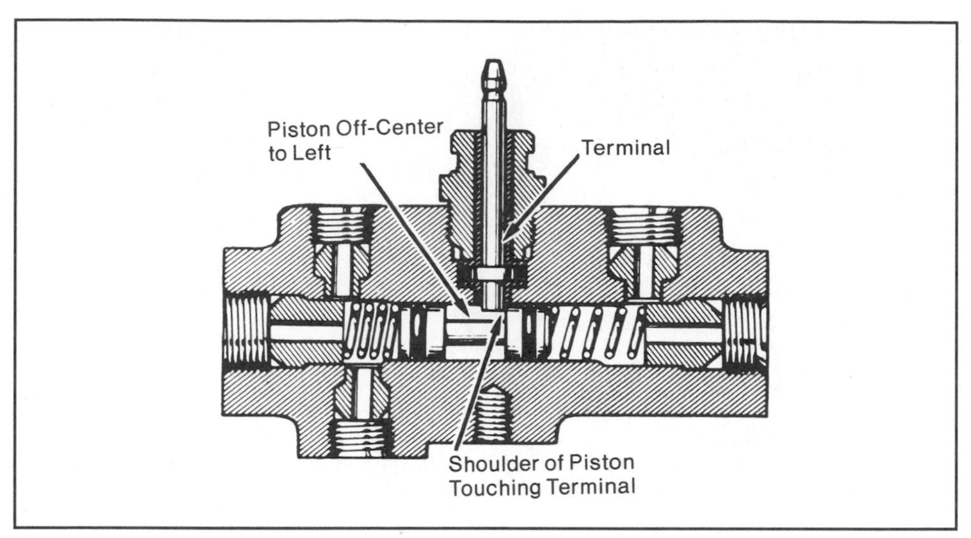

Brake warning-light switch is indicating hydraulic-system malfunction. Piston between two centering springs is pushed to left by higher pressure in right chamber. Piston contact with electrical terminal turns on instrument-panel warning light. Drawing courtesy Bendix Corp.

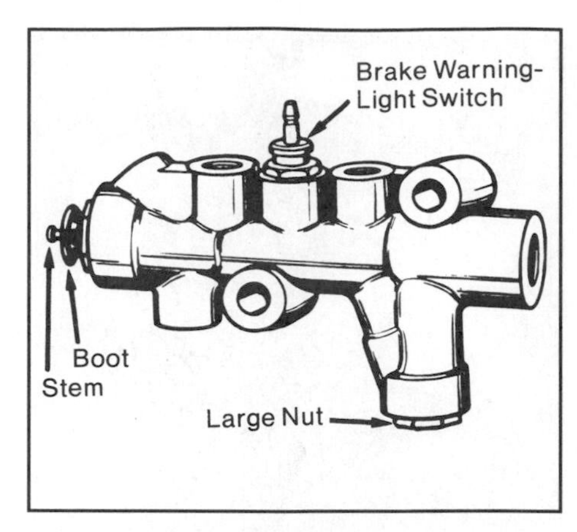

Three-function combination valve combines metering valve, proportioning valve and brake warning-light switch into a single unit. Metering valve is at left end and proportioning valve is under large nut on bottom. Drawing courtesy Bendix Corp.

springs; some even have two pistons. The switch type used varies with the design of car and brake system.

COMBINATION VALVE

In recent years, car manufacturers have combined the function of the metering valve, proportioning valve and brake warning-light switch into a single *combination valve.* Some manufacturers combine the brake warning-light switch with either the proportioning valve or the metering valve. Some front-wheel-drive cars with diagonally-split systems combine two proportioning valves with the brake warning-light switch. Because external appearances of these valves are similar, be careful when moving brake-system components from one car to another. Check part numbers and car shop manuals to be sure.

Combination valves are found only in brake systems using a tandem master cylinder. Usually, they are found on cars with front disc and rear drum brakes. Most combination valves cannot be disassembled for service or adjustment. If defective, a valve must be replaced.

Combination valves work similar to the individual valves they replace. By combining in a common housing, fewer fittings are used. Consequently, the valve is simpler, costs less and leak possibilities are reduced.

Pictured is a typical three-function combination valve. The metering valve operates the front system and the proportioning valve the rear. A combination valve must be mounted near the master cylinder so lines to the front and rear brakes can be easily connected to the valve.

FLUID-LEVEL INDICATOR

A *fluid-level indicator* consists of a device in the fluid reservoir and a brake warning light. It switches on when reservoir fluid drops below a predetermined level. This indicating device consists of a float and a mechanically or magnetically actuated switch.

Like the brake warning-light switch, low fluid-level indicator warns the driver of fluid leaks or master-cylinder bypass. It can also signal worn disc-brake linings, but not air in the system. Usually, two are used with tandem master cylinders. Because fluid-level indicators perform many of the same functions as brake warning-light switches, they are rarely used together.

HOSES

Hydraulic fluid is carried to the calipers or wheel cylinders by flexible hoses. These hoses allow suspension and steering motion.

In the early days of the automobile, high-pressure flexible hoses did not exist. Consequently, it was difficult to get hydraulic fluid to brakes at the wheels. It was amazing that early hydraulic-brake systems worked at all. Fluid was routed through axle and suspension links to the brakes. This was particularly difficult on the front wheels. Sliding seals were used to allow steering and suspension motion. To contrast this, the modern brake hose is simple and reliable.

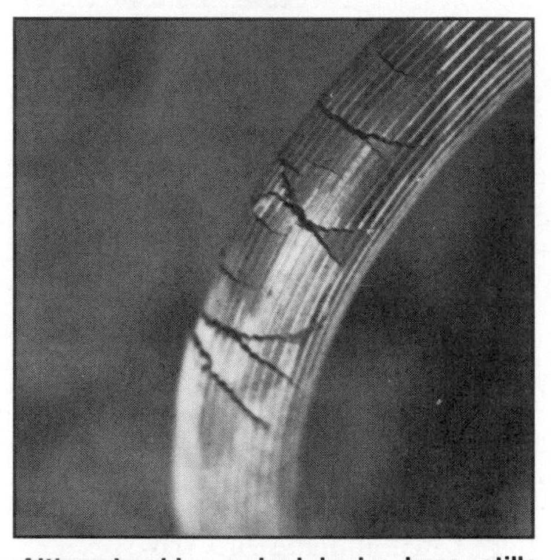

Although old, cracked brake hose stills holds pressure; it should be replaced. Condition of outside rubber is the only way to judge condition of a brake hose. It's better to replace a cracked hose than to risk a sudden brake failure.

Brake hoses are made from multiple layers of fabric impregnated with synthetic rubber. They are tough and last many years, but are still the weakest link in the hydraulic system. It is easy to damage a hose through contact with a wheel, tire, suspension member, exhaust or road debris. Or an overzealous mechanic can damage a brake hose during installation. A brake hose can fail if twisted or improperly installed.

Brake hoses expand slightly under high internal pressure. This *swelling* requires additional fluid movement to compensate for the increased hose volume. The driver senses hose swelling or expansion as a soft or spongy brake pedal. Racing and high-

"Ultimate" brake hoses are steel-braided teflon. Because exterior is braided stainless steel, hoses are more resistant to mechanical damage than ordinary rubber hoses. Steel-braided hoses are also stiffer, giving a firmer pedal.

Don't use high-pressure plastic tubing for a brake hose! After repeated flexing, tubing cracks at fittings. I've seen this failure on several race cars. Plastic tubing is also easily damaged by heat or mechanical contact. Although plastic tubing can be used if rigidly mounted to frame, weight saved isn't worth risking brake failure.

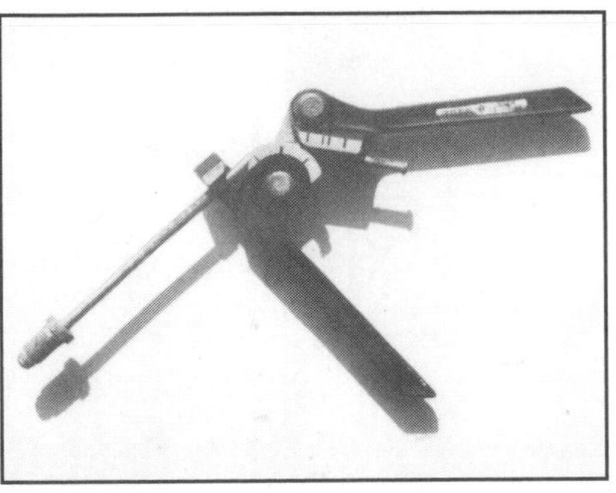

Imperial tube bender is available from good tool stores. Neats bends can be made without risk of crimping tube. Before you bend brake lines, invest in a good tube bender.

performance brake systems require that rubber brake hoses be replaced with something better.

High-pressure stainless-steel-braided brake hoses, developed for aircraft and military use, are available from specialty shops. Suppliers are listed at the back of the book. Steel-braided hoses are available in custom lengths with adapter fittings to suit special applications. These hoses are Teflon lined and externally reinforced with stainless-steel braid. They are almost "bulletproof" and resist expansion under pressure.

Aeroquip 6600-series Teflon-lined hose with Super Gem fittings is a type commonly used on race cars. I highly recommend this flexible brake hose to make the brake pedal firmer, and for overall safety. Remember, use flexible hose only where there is motion between components being connected. Even the best flexible hose expands more under pressure than steel hydraulic tubing.

Some race cars use plastic hose for both the brake tubing and the flexible brake hoses. This tubing may work for some low-pressure applications, but it is dangerous to use as flexible brake hose. Plastic tends to crack where the tube enters the fitting. After repeated flexing, even the special fittings developed for this tubing will not prevent failure. If you have a race car with plastic brake hoses, replace them with steel-braided Teflon hoses. Failure of the plastic hose will occur without warning and cause sudden braking loss.

HYDRAULIC TUBING

Most cars use tubing manufactured from steel for the brake system. Commonly called *brake tubing,* it is available in diameters from 1/8 in. through 3/8 in. Most of today's cars use 3/16-in. tubing; 1/4-in. tubing is found on most older cars. Smaller tubing is stiffer and lighter and easier to bend. Larger tubing has less resistance to fluid movement and is less prone to damage during handling. The 3/16-in.-diameter tube is a compromise arrived at after years of experience. I recommend using 3/16-in. hydraulic tubing for all brake systems.

Copper tubing or any tubing of an unknown material should *never* be used for brake tubing. Copper is soft and prone to cracking. Ordinary steel tubing may be strong enough, but it doesn't meet the rigid quality and design standards of brake tubing. Tubing manufactured for brake systems by Bendix is made of soft-steel strips sheathed with copper. These strips are then rolled into a double-wall tube and bonded at high temperatures in a furnace. The tube is then tin-coated for corrosion resistance. Ordinary steel tube is only single wall and not copper- or tin-plated. It could rust out after a few years and suddenly

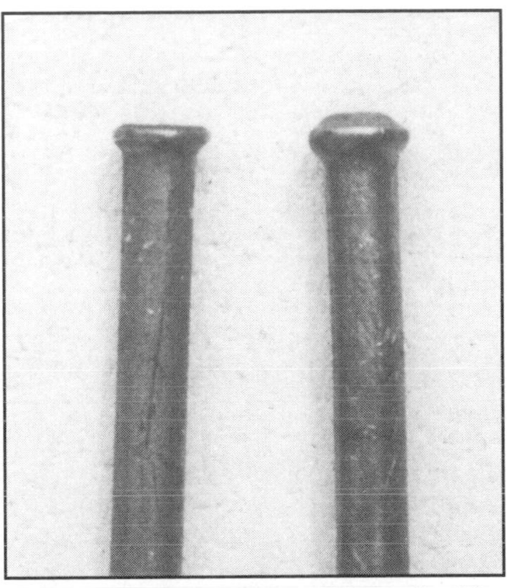

Automotive brake-tubing flares: Double-flare used on most American cars is at left; at right is *iso-flare* used on some foreign cars. Flares must match fittings to which they attach, otherwise a leak or failure will result.

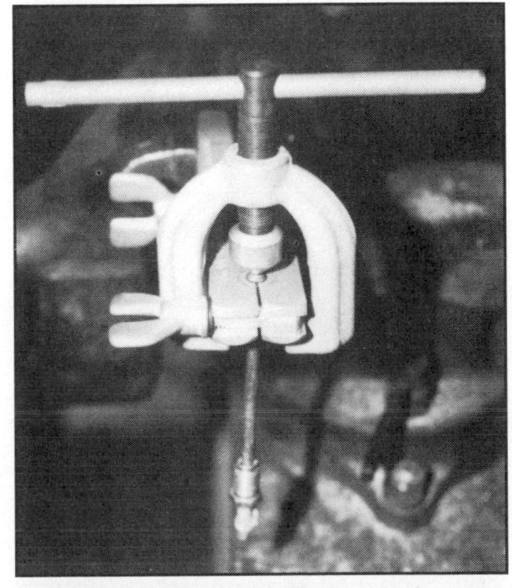

Making double flare: Flaring tool comes with dies needed to do both forming operations. Tool and tubing are set up to do second forming operation.

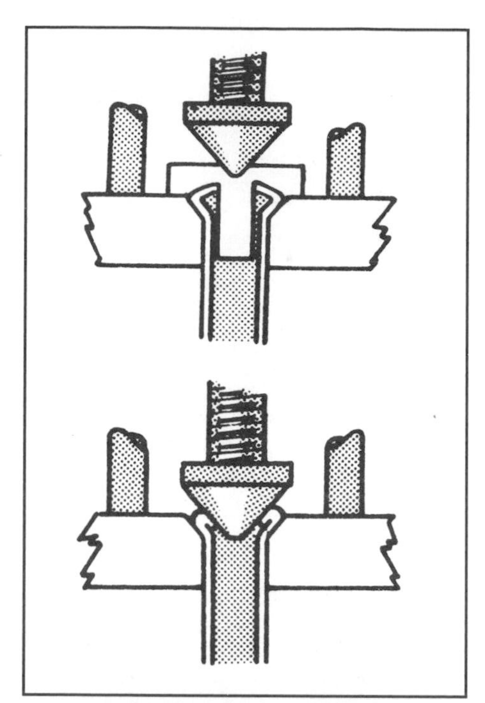

Flaring tool folds tube end back on itself to finish double-flaring operation. Drawing courtesy Chrysler Corporation.

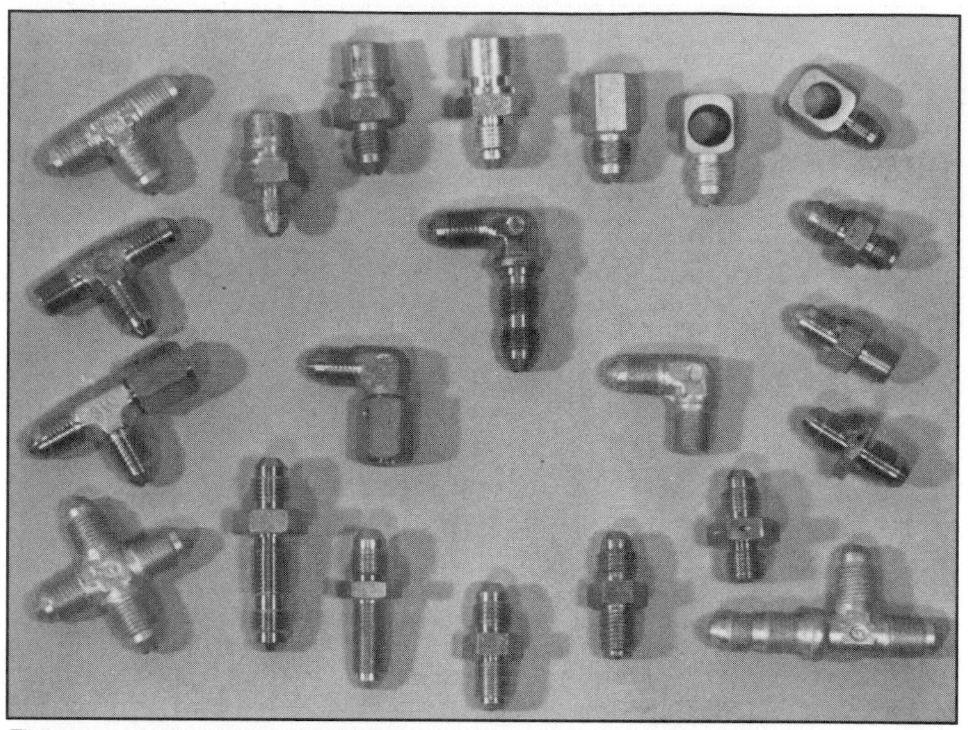

Fittings used with steel-braided hose are different than standard automotive brake fittings. High-pressure steel-braided-hose fittings meet aircraft standards, not automotive standards. Special adapter fittings attach aircraft-type fittings to automotive brakes. Photo courtesy C & D Engineering Supply Ltd.

Race cars often require fittings not normally found in auto-parts stores. *Bulkhead* fitting is handy for mounting brake lines where they pass through a panel such as this fire wall. Two jam nuts support fitting at hole in panel.

burst. Always use proper-quality brake-tubing material, available from auto-parts stores.

Double Flare—Brake tubing is available in several standard lengths with male fittings at each end. Common lengths are 8, 12, 20, 30, 40, 50 and 60 in. Commonly available diameters are 3/16 or 1/4 in. for American cars and 3/16 in. or 4.762mm for imported cars. Each end is *double-flared*—tube is doubled over on itself—in a convex or concave shape. American and imported cars use different flares and threads. Accompanying photos show some of these differences.

Double flares are important for the strength and safety of a brake system. Anyone with a cheap flaring tool can make *single flares.* **NEVER** single-flare brake lines. If you don't want to invest in a special double-flaring tool, buy preflared brake tubes or pay someone to custom-make double flares. Single flares are weak, will eventually fatigue, crack and cause a leak in the hydraulic system.

Armor—Brake tubing is exposed to harsh conditions under a car. It gets blasted by dirt and debris, is immersed in water and salt, but should continue to function for years without failure. Anytime you are under your car, look for brake-line damage. A rusty, kinked, nicked or leaking line must be replaced. If you want extra protection with new brake tubing, it is available wrapped with a coil spring. This *armor* protects the tube from damage by rocks or other sharp objects. Armored brake lines can be found in major auto-parts stores.

FITTINGS

Brake lines are connected with steel or brass fittings of various types, such as tees, junction blocks and unions. Be careful when selecting brake-line fittings. They must have the correct seat for the flare they mate to. Otherwise, they will leak. Many fittings that look OK will not mate properly with the tubing. Never use tapered pipe threads in a brake system. They will leak under high pressure, particularly if installed without sealant. They will also split when overtorqued.

Some stop-light switches have tapered pipe threads. They will seal if Teflon tape is used. However, you must be very careful not to allow any Teflon tape to extend past the first thread. A shred of Teflon tape floating around in the brake fluid can plug a critical port in the system. The best solution is avoiding the use of tapered pipe threads. Take a tip from the aircraft industry: Leave Teflon-tape sealant to plumbers and pipe fitters. It can cause many problems in a brake system.

On race cars, steel-braided hoses are used to improve reliability and provide firmer brakes. Fittings used with steel-braided hoses are different than those used with conventional hoses. They have entirely different specifications for threads, flares and other dimensions. Adapter fittings must be used when installing steel-braided hoses on vehicles with automotive-type brake fittings.

Adapter fittings are available from

Brake tube in fitting fractured during a race, causing complete brake failure. Driver was without brakes, but luckily it happened where he could recover without crashing. Failure occurred because fitting-mounting bolt was not installed. Vibrating fitting and hose was supported by tube, which eventually failed from fatigue. Brake tubes and fittings must be securely mounted to prevent such failure.

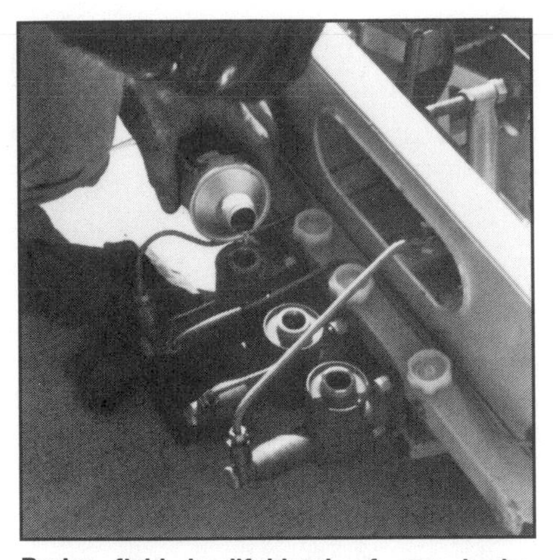

Brake fluid is lifeblood of any brake system. For race cars, you must keep fresh new fluid in brake system at all times. Fluid with highest possible boiling point is a must for racing.

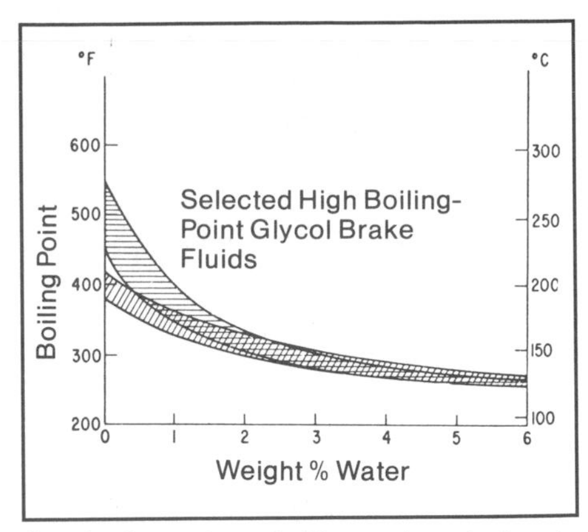

Graph shows how boiling point of glycol-based brake fluid drops with exposure to moisture. Notice how brake fluid with highest boiling point when new is affected most by water content. Graph courtesy Yankee Silicones, Inc.

the sources that supply the hoses, such as C & D Engineering Supply Ltd., or WREP Industries. Some shops offer brake-hose kits that include adapter fittings and correct-length hoses. Don't mix automotive and aircraft fittings. This will result in damaged threads, leaks or cracked fittings. If you can't recognize the difference, ask the advice of a shop that supplies these fittings.

Adapter-fitting usage will be minimized if a race car is constructed using all aircraft fittings in the brake system. This can be done by getting the proper fittings from an aircraft hardware supplier. You may still have to adapt the brake line(s) to your master cylinder(s) and brakes using automotive flares or fittings.

Aircraft flares are not the same as automotive flares. They use *ferrules*—sleeves—over the tubes to reinforce them. Aircraft hoses and fittings are identified as AN (Army-Navy) or MS (Military Standards). AN and MS specifications, written by the U.S. government, are standards recognized throughout the world. Automotive specifications are written by the Society of Automotive Engineers (SAE). SAE standards also have the same worldwide recognition.

BRAKE FLUID

Once a brake system is installed, it must be filled with fluid so it can function. Hydraulic fluid for an automotive brake system is simply called *brake fluid.* This distinguishes it from fluids used for other hydraulic applications and systems. Automotive brake fluid is designed to work the brake system of a car. Other hydraulic fluids have totally different chemical and physical properties. *Always* use automotive brake fluid in a car. Aircraft brake fluid is totally different and *will not* work in a car.

Brake fluid is a chemical mixture, which varies with manufacturer. Each manufacturer has a secret formula designed to perform a "better job." All manufacturers must formulate their brake fluid so it meets rigid SAE, industry and government standards. In the U.S., the Department of Transportation (DOT) sets the standards.

Hydraulic brake fluid used in racing and road cars is mostly a Polyalkylene Glycol Ether mixture—commonly called *glycol.* This type of fluid can be found in any auto-parts store and is compatible with the fluid originally installed in your car. A newer type of brake fluid has a silicone base. Its properties differ from glycol-based fluids. Silicone-based brake fluid is discussed later in this chapter.

Glycol-based brake fluid must have the following properties to work correctly in an automobile:

- Must not boil at temperatures encountered during severe braking conditions.
- Must not freeze or thicken in cold temperatures.
- Must not compress.
- Must flow through system passages with minimal no resistance.
- Must not corrode or react with brake-system materials.
- Must lubricate internal sliding parts of brake system.
- Must not alter its properties after being in system for extended periods.
- Must be compatible with other glycol-based brake fluids.
- Must not decompose, or form gum or sludge at any temperature.

Fluid Boiling—The most critical property of brake fluid is its resistance to boiling at high temperature. When brake fluid boils, small bubbles form. These bubbles collect and are trapped in the system. Because gas bubbles are compressible, this results in a spongy brake pedal.

Boiling generally is caused by water vapor being absorbed by the brake fluid. The more water brake fluid contains, the lower its boiling point.

New, uncontaminated, high-quality brake fluid boils at 550F (288C); pure water boils at 212F (100C). A mixture of brake fluid and water boils somewhere between these temperatures. Two boiling points are referred to in brake-fluid specifications: *dry boiling point* and *wet boiling point.* Without getting into details of the specifications, the meaning should be clear. Dry boiling point refers to fresh fluid out of a

BRAKE-FLUID SPECIFICATIONS

The print on brake-fluid cans is confusing. Even worse, specifications change constantly, so an old can of fluid may say something different than a new can. The following is an explanation of current specifications.

All brake fluids must meet federal standard 116. Under this standard are three Department of Transportation (DOT) specifications for brake-fluid performance: DOT 3, DOT 4 and DOT 5. All brake fluids are suitable for use in either drum or disc brakes. Boiling point for the three types of fluids are as follows:

		DOT 3	DOT 4	DOT 5
Dry Boiling Point	Degrees F	401	446	500
	Degrees C	205	230	260
Wet Boiling Point	Degrees F	284	311	356
	Degrees C	140	155	180

AP 550 Racing Brake Fluid is used in most-severe racing applications. Its dry boiling point of 550F (288C) is highest for a glycol-based fluid. Photo courtesy AP Racing.

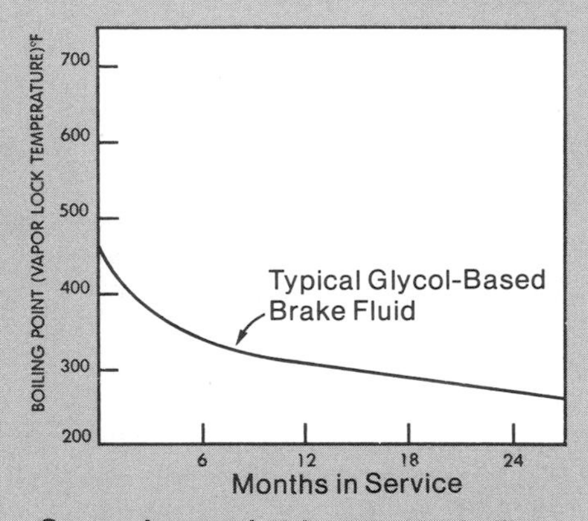

Curve shows what happens to boiling point of glycol-based brake fluid in typical road-car use. Weather changes curve—boiling point gradually drops in extremely damp climates. Silicone brake fluid is not susceptible to moisture as it doesn't pick up moisture. Consequently, silicone brake-fluid boiling point remains high after long use. Graph courtesy Yankee Silicones, Inc.

Old brake fluid meets DOT 2 or SAE J1703 or 70R3 specifications. These obsolete requirements refer to fluid with a dry boiling point of 374F (190C). If you have any of this old fluid, discard it. Don't use anything less than DOT 3 fluid in a brake system. Prior to the DOT 2 specifications, there was a 70R1-specification brake fluid meant for drum brakes only. Throw this away too if you find any. It has a dry boiling point of only 302F (150C).

Specified boiling points are minimums. Many brake fluids have higher boiling points, as indicated on their containers. As for what to use, a dry boiling point over 500F (260C) is best for hard driving.

DOT 3 and DOT 4 specifications are for glycol-based brake fluid and DOT 5 is for silicone. You can tell immediately what the basic chemistry of a fluid is by its DOT number. A glycol-based brake fluid with a 500F (260C) dry boiling point is tested to DOT 3 or DOT 4, not DOT 5.

Sometimes boiling point is referred to as ERBP on the brake-fluid can. This means *equilibrium reflux boiling point,* which is another way of saying wet boiling point. ERBP refers to a method in the specification for how fluid is exposed to moisture and tested.

Note that fluid boiling point drops after it is exposed to water. Glycol-based brake fluids absorb moisture. The resulting mixture of brake fluid and water results in a boiling-point drop. Fluids that absorb water easily are called *hydroscopic.* Water in brake-fluid also causes corrosion and impurities to accumulate in the hydraulic system.

sealed can; wet boiling point refers to fluid that has been exposed to moisture under specific conditions. Both boiling points should be of concern to you, but the most important one will depend on how you use your car.

Dry boiling point is important for racing. It is assumed that a properly maintained race car will always have fresh brake fluid. For best performance, the fluid must be changed many times during a racing season to avoid a lower boiling point. Changing the fluid is an inexpensive way to ensure the brakes work best at temperatures encountered in racing.

Wet boiling point is most important for street use. Brake fluid in street-driven cars is not changed often—if ever. If you could bleed your brakes every week or so, the brake fluid would have properties similar to a race car's. But, who can take the time to bleed their brakes weekly? Normally, the brakes must function with moisture in the brake fluid over long periods of driving during all types of weather conditions.

Brake fluids with a high wet boiling point do not necessarily have the highest dry boiling point. Most racers use a glycol-based brake fluid with a dry boiling point of 550F (288C) for racing. A common high-performance brake fluid used in road racing is AP Racing 550. This fluid exceeds DOT 4 specifications for dry boiling point, and meets DOT 3 specifications for wet boiling point. To maintain your racing-brake system properly, it is a good choice.

For street use, you should use brake fluid that meets DOT 4 specifications for wet boiling point. If the fluid meets DOT 4 specifications, it also meets *all* parts of DOT 4, including wet boiling point. Your choice of a brake-fluid brand should be based on the following requirements:

- Use a fluid that your car manufacturer recommends.
- Don't mix brands of fluids.

Violate either of these rules, and long-term harm may occur to your brake system. For instance, some brake fluids may react with rubber seals in a brake system. Over a long period of time, this may cause swelling or leakage. This occurs commonly when U.S.-manufactured brake fluid is used in an older foreign car with natural-rubber seals. The seals swell and have to be replaced eventually.

You wouldn't dare do this with ordinary glycol-based brake fluid; it would remove the paint. There's no danger to paint with silicone fluid. Silicone fluid is ideal for cars that have problems with brake-system contamination and corrosion. Photo courtesy Yankee Silicones, Inc.

Consult the factory shop manual or your dealer for brake-fluid recommendations.

Brake-fluid brands contain blends of chemicals, so their properties change when mixed. If you must change brands, flush out the brake system and replace *all* brake fluid. See Chapter 11 for tips on changing brake fluid.

If you want the highest wet boiling point and a different combination of properties, you might consider changing to silicone-based brake fluid. Silicone fluid is required to meet DOT 5 specifications. This fluid does not absorb moisture like a glycol-based fluid. Therefore, its wet boiling is much higher. Also, silicone fluid will not damage the paint on your car if spilled. But glycol-based brake fluid makes good paint remover on some finishes.

A good application for silicone brake fluid is in antique and collector cars. Brakes in these types of cars are never subjected to high temperatures encountered in racing—high temperatures that can cause silicone fluid to compress! And protection of the internal parts from corrosion is most important. Silicone brake fluid serves this purpose beautifully, because the absence of moisture in the system practically eliminates the chance of dreaded internal corrosion. However, for racing, use only a glycol-based racing brake fluid.

Silicone fluid has been tried in racing, but it has a tendency to give a spongy pedal after exposure to high temperatures. This is due to the slight compressibility of silicone brake fluid at high temperature. For ordinary street driving, this is not critical, but a racer needs all the brake-system stiffness he can get.

Switching to Silicone—If you change to silicone brake fluid, you must first clean *all* of the old fluid from the brake system to get the maximum benefit of the silicone fluid's properties. The best way to accomplish this is during brake-system overhaul. Totally drain and clean out the old fluid. If you merely bleed out and install new fluid, you will have a mixture of the two fluids. This will work, but it won't be as good as it could be.

Keep Out Contaminants—If you don't use silicone brake fluid, do everything possible to keep water out of the brake fluid. Although it's more expensive, always buy brake fluid in small cans. Keep the cans closed before using. Once opened, don't keep an old can of brake fluid around long. Use it immediately or throw it away. Periodically bleed some of the old fluid. This removes contaminated fluid from the system. *Never reuse drained fluid.*

Howe master cylinder has rubber diaphragm built into lid. This keeps air from contacting and contaminating brake fluid. Unfortunately, cap must be removed to check fluid level, partially defeating advantage of rubber diaphragm.

If you have a choice, always use a master cylinder with a rubber diaphragm in the cap to seal the fluid. This keeps the fluid from exposure to air. Most moisture in brake fluid is picked up from air on damp days. All master cylinders and reservoirs must be vented to atmosphere, so don't plug vent holes.

In addition to moisture, brake fluid can be contaminated by other materials. Dirt can enter the system when fluid is added to the reservoir. It is critical to keep dirt from the system, so use extreme care when working on your brakes. Usual contamination is by petroleum products such as oil, grease or solvents. Petroleum solvents mix with brake fluid and can damage rubber seals in the system. *Never clean brake-system components with solvent.* Use alcohol, brake fluid or brake-system cleaners. Solvents are too difficult to wash off, and will contaminate brake fluid.

Brake Pedals & Linkages 6

Bird's-eye view of foot-box area of formula car shows pedals, linkages and brake and clutch master cylinders. Dual brake master cylinders with remotely adjustable balance bar is used. Note bumper protecting master cylinders from head-on impact.

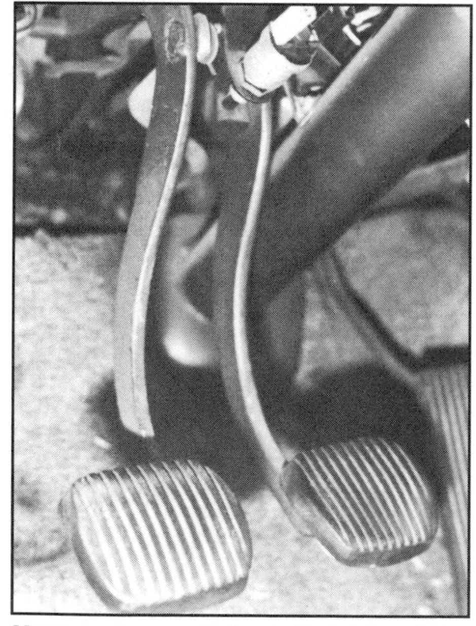

Most road-car pedals are hanging—pivots are above pedal pad—and pedal bracket doubles as a steering-column bracket. Device near top of brake pedal is a mechanical brakelight switch.

The brake pedal and linkage transmit force and movement from the driver's foot to the master cylinder(s). Design and construction of these parts greatly affect how a brake system operates and how it feels to the driver. The brake pedal must be the most reliable part of a brake system because failure can mean a complete loss of braking. The result can be catastrophic.

When building a car, you may want to design and fabricate the brake pedal and linkage. These parts are some of the components that can be fabricated by the home craftsman. This chapter describes how pedals and linkages are designed and fabricated for safe and reliable brake operation. The pedal and linkage are normally designed as a part of a complete brake system. This is discussed in Chapter 9.

BRAKE PEDAL

A brake pedal is familiar to everyone who has driven a motor vehicle. Take a good look at the pedal in a car. It has important design features. Good brake pedals have the following characteristics:

- Must not break or permanently bend under the greatest load a driver can apply.
- Must be stiff and not flex during hard braking.
- Must be free of excessive friction.
- Must provide proper leverage at master cylinder.
- Must match requirements of driver and master cylinder, linkage and system.

A brake pedal consists of the *arm, pad* and pivot *attachments.* The pedal is connected to a linkage. This linkage transmits force and movement to the master cylinder. The linkage can be as simple as a straight pushrod, or more complex.

The pedal arm is subjected to bending loads from the driver's foot. It is usually constructed of steel plate. The arm may have holes in it for the pivots. Often pivot bushings are contained in steel tubes—*pivot housings*—that are inserted in the arm and welded in place. A good pedal design will not have welds completely across the arm, particularly at the pivots. A weld is a source of weakness—*stress riser*—at a critical area of the arm.

A pedal arm is usually straight when viewed from the driver's seat. If it has bends or curves in it, the arm tends to twist when the pedal is pushed. Twisting in the pedal arm can create movement at the pad, giving the driver feel of sponginess. Also, stresses in a pedal will be higher due to twisting forces. To overcome these problems, pedals with bends, offsets or curves must be fabricated of heavier material.

A good pedal arm should be designed to take side loads and forward loads. Side loads on a pedal arm are most likely to cause failure, because most arms are weaker laterally. Pedals should be installed so they won't be loaded from the side. A brake pedal should be in line with the driver's leg.

Pedals in most road cars are made of steel. This gives a pedal the highest

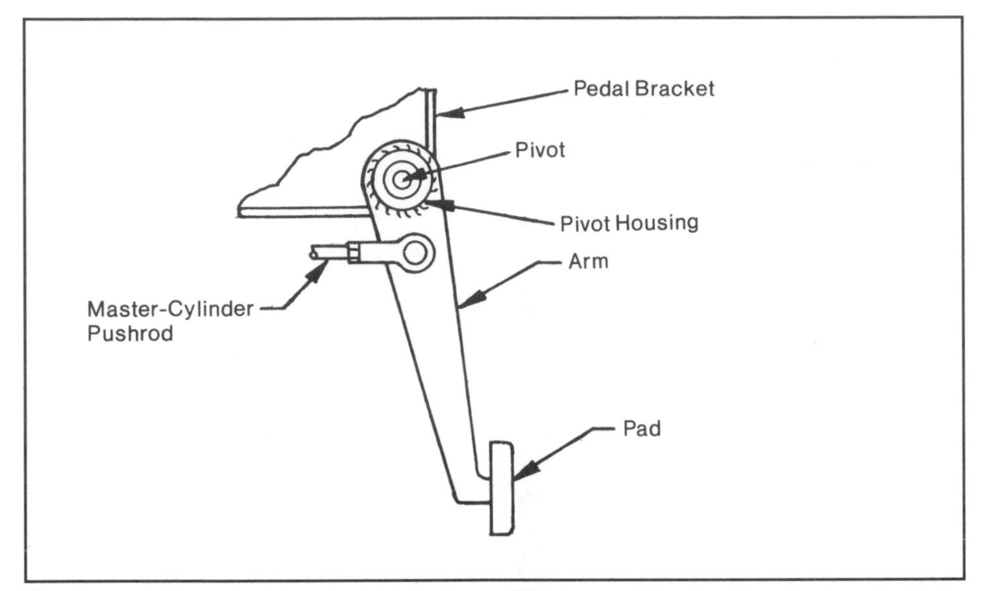

Hanging pedal has pivot housing at top of arm. On many pedals, pivot housing is a steel tube welded into a hole in pedal arm. Pivot bushing is inside pivot housing. For rear-mounted master cylinder, master-cylinder-pushrod pivot installs above pivot, page 70.

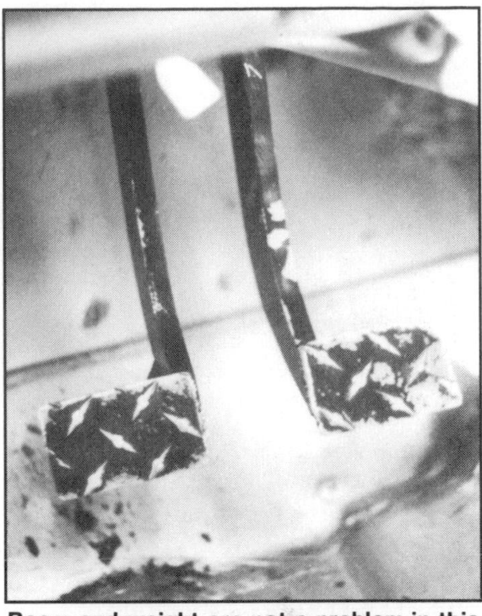
Room and weight are not a problem in this racing stock car, allowing big and sturdy pedal pads. Such pedals would not fit in a cramped Grand Prix-car cockpit.

stiffness and strength of any low-cost material. Older cars often used steel forgings for pedals, but modern cars use parts cut or stamped from steel plate. Race cars sometimes use lightweight materials such as aluminum or titanium. Consequently, race-car pedals are expensive when compared to those for road cars. However, steel will offer the best results. Its stiffness for a given weight is as high as any other material, and it is easier to fabricate. Designing an aluminum pedal can be difficult because of its low stiffness.

A pedal arm must be rigid. Therefore, it usually has a deep cross section. A road car usually has a flat-plate or an open-channel cross section, but a race cars often use a tube. Tubular pedals are lighter for a given stiffness. However, they are expensive and difficult to mass produce. With a tubular pedal, a thin-wall tube can be used and still have enough stiffness.

A pedal pad should be shaped to fit the foot. A tiny pad is hard to push with comfort and your foot can slip off easily. A good pedal pad is about shoe width; wider if space allows. A rubber pad is used on road cars for aesthetics, comfort and grip. For race cars, a textured metal pad is used for maximum grip. This is especially important in open race cars that run in the rain.

Small pedal pads are in a Formula Ford road-racing car. These little single-seat racers have barely enough room for the driver's feet, making small pedal pads a must. Nonskid surfaces are made from expanded metal welded to pads.

Some race-car pedals have an adjustable pad, adjustable to suit the driver. Seats in many race cars are fixed, so adjustable pedal pads compensate.

With adjustable pads, you can vary the distance between the steering wheel and brake pedal. This will affect arm position when the driver is seated in the car. A race car with adjustable pedal pads and seat can be made to suit the driver nearly perfectly.

Pedal pad is adjusted by releasing jam nut and screwing pad in and out. Pad is welded to a bolt, which threads into a tapped tube at top of pedal arm. More than 1-in. of adjustment requires moving driver's seat.

If *heel-and-toe* brake- and accelerator-pedal operation is desired, the two pedals should be positioned no more than 1-1/2-in. apart laterally. Also, the pedals should be adjusted so that during hard braking, the brake pedal travels to about 1/4-in. above the accelerator pedal.

PEDAL FORCES

The two forces a pedal designer must consider are:

- Highest force that can ever be ap-

Once wheels are locked, driver's strength is the only thing limiting brake-pedal force. As fear increases, the harder he pushes. This condition represents the maximum force that can be applied to a brake system. Brake-pedal force during a maximum-deceleration, controlled stop is much lower.

Bird's-eye view of formula-car pedal box: Note that pedal pad is not centered on pedal arm. This twists pedal arm when driver pushes on pad. Although tubular arm can withstand this twisting load, pedal assembly would be stronger if pad and pedal arm were centered.

plied to a pedal.

- Force normally applied in a hard stop at maximum deceleration.

These two forces are *not the same.* Maximum force is considerably higher than the force normally applied during a hard stop. To keep the terms separate, the force normally applied during a hard stop is called *pedal effort.* This is the force applied to a pedal by the driver when he tries to stop at maximum deceleration. In a race car, pedal effort is applied by the driver almost every time he brakes the car. This is rarely the case with a road car.

During an emergency, the driver should stop the car at maximum deceleration with the tires on the verge of lockup. Often, though, a driver in trouble forgets about controlling the car. Instead, he panics and jams on the brakes as hard as he can and locks the wheels. Often he loses control and skids. This type of stop is where *maximum force* on a pedal occurs. The pedal and linkage must be designed not to break or permanently bend under this condition.

Pedal Effort—A high-performance brake system usually is designed for about 75-lb pedal effort at maximum deceleration. This feels OK to most people. Any one person can easily apply 75 lb with his leg. If you are designing a truck or race car, you may wish to use a higher pedal effort. A 100-lb pedal effort is still manageable, but it feels like a hard push. I don't recommend using more than 100 lb. Then there would not be any reserve force left for brake fade or other system failures.

Most passenger cars have power-assisted brakes. With power assist, pedal effort is usually less than 50 lb.

Maximum Pedal Force—A heavy person can easily push with a force greater than his weight—about 300 lb for a large man. If he stomps quickly on the pedal, the effect of the sudden force further increases stress on the pedal. Engineers know that if a load is applied quickly, stress on a part is doubled. Most brake pedals are designed to take *twice* the maximum force that a strong leg can apply. I use 600 lb for designing pedals.

Safety Factor—The pedal should be designed to withstand a force much higher than the maximum force. The ratio of the force required to break a part to the maximum applied load is called the *safety factor.* I use a safety factor of 3 on critical parts such as a brake pedal. This means the part is designed to break at a load at least three times the maximum force applied to the part. Therefore, the breaking strength for a brake pedal should be at least 1800 lb, or 3 X 600 lb.

If you wanted to test your brake-pedal design, apply an 1800-lb force on the pad in a direction in line with the force that would be applied by a person's foot. If the pedal doesn't break, the safety factor is adequate. You may bend the pedal doing such a test—never use a pedal in a car after testing its breaking strength.

The master-cylinder pushrod must withstand the maximum pedal force *times* the *pedal ratio.* Pedal ratio can be obtained by dividing pedal travel at the foot pad by pushrod travel. Pedal ratio is discussed in more detail on the following page. Also, a pushrod should always be straight. One that is straight is many times stronger than a bent or offset rod.

In addition to normal load, the pedal must be able to resist some side load. A driver's foot is not always aligned with the pedal pad. The worst case occurs where the driver's foot hangs halfway off the pad. This puts a side force into the pedal arm. I recommend using a side load of 200 lb on the pedal pad in addition to the 600-lb forward load. If you apply the safety factor of of 3, the pedal should not break if tested with a 600-lb side load on the pad. Note: The side load should be applied at the same time as the forward load.

All this suggests that perhaps you should leave brake-pedal design to an engineer. If you are not trained in stress analysis, a mistake here could be fatal. An option is to use a pedal that fits from another car. All pedals are designed using similar techniques. Safety factors may vary, but the maximum force a pedal can withstand is about the same for all cars. Be aware that a pedal from a road car will probably be stronger than a race-car pedal. Many race-car designers try to save

Bad pedal design on homemade race car has pedal arm off-center to pad and pivot housing. When pushed on, twist of arm causes excessive flexibility and higher stresses. If arm were moved to right, pedal would be stiffer and stronger.

Spherical bearings on master-cylinder pushrods minimize friction and wear. Neal Products pedal uses a balance bar with special spherical bearing at each end. Balls in spherical bearings are pinned to allow motion in only one direction. Ordinary spherical bearings would allow rod ends to cock, with potentially disastrous results.

weight at the expense of strength.

If you insist on designing your own brake pedal, follow these rules:

- Do not weld across a pedal arm.
- Make the pedal arm straight when viewed from the driver's seat.
- Use steel, not aluminum or other light alloys. Steel gives maximum stiffness and ease of fabrication. Welding drastically weakens aluminum.
- Position the pedal arm in the center of the pad—not off to one side.

PEDAL PIVOTS

Pivots in brake pedals should have plastic or bronze bushings to reduce friction. Plastic bushings are often used on road cars to reduce cost; bronze bushings are found on some race cars. The best pedal pivots have provision for lubrication. Most plastic bushings are designed to operate without lubrication, but this does not always give maximum durability.

Avoid pedal pivots that use unlubricated clevis pins or bolts. Some brake pedals have been so designed. They work for a while, but the lack of adequate bearing area and lubrication usually causes the pivot to wear out rapidly. Then the pedal gets loose and sloppy, requiring extra pedal movement to take up the slop. Pedal movement is a valuable item. Don't waste it with potentially sloppy pivots.

If you have a car with sloppy or unbushed pedal pivots, consider modifying them. If you have to enlarge a hole to install a bushing, make sure you don't weaken a critical area. If you are uncertain, leave it alone. Or, have an engineer do a stress analysis of the modification. If you don't bush the pivots, at least lubricate them at regular intervals to reduce wear.

PEDAL RATIO

The force required on the master-cylinder piston is usually much higher than the pedal effort supplied by the driver. Thus, the pedal must act as a lever to increase the force supplied by the driver. The leverage of the pedal—*pedal ratio*—is also known as *mechanical advantage.* Pedal ratio for manual brakes is about 5 to 1; power pedal ratios are about 3 to 1.

If you plan to design or modify a brake system, you need to determine the pedal ratio. There are three ways to do this. Choose the method that works best for you. They are:

- Measure forces.
- Measure movements.
- Measure brake pedal and linkage geometry.

Measure Forces—If you can measure the force on the pedal pad and the simultaneous force on the master-cylinder piston, the pedal ratio can be

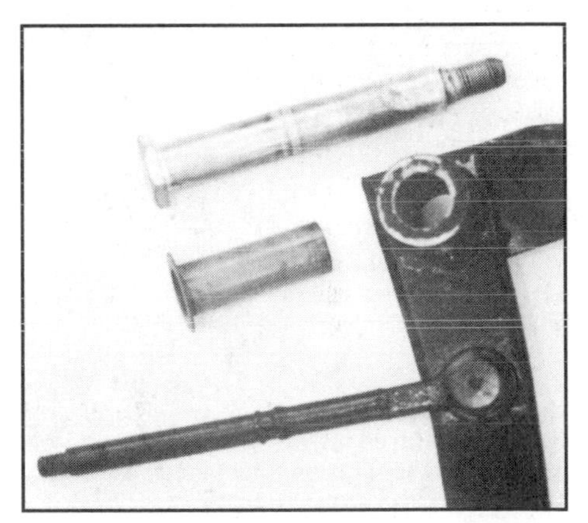

Typical road-car pedal design; pivot bolt and bushing are removed. Plastic bushings last longer if greased occasionally.

calculated from this simple formula:

$$\text{Pedal ratio} = \frac{F_{MC}}{F_{PP}}$$

F_{MC} = Force on master cylinder in pounds
F_{PP} = Force on pedal pad in pounds

Forces are measured in pounds; the pedal ratio is just a number. If you could push on the pedal with a scale or instrument to measure force, you could simultaneously measure hydraulic-system pressure. If you know the master-cylinder-piston area, pressure can be converted into force on the piston.

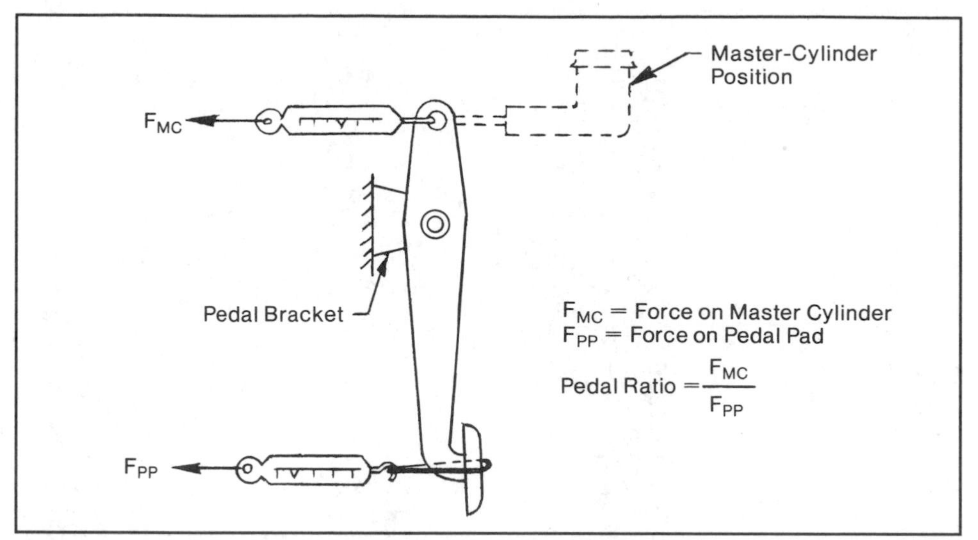

Pedal ratio can be found by simultaneously measuring forces at pedal pad and master cylinder. Small spring scales may be used for this. Disconnect master-cylinder pushrod so it doesn't affect readings. Make sure one scale lines up with pushrod and other scale with line-of-action of driver's leg. See drawing at right. Pushrod force divided by pedal force is pedal ratio.

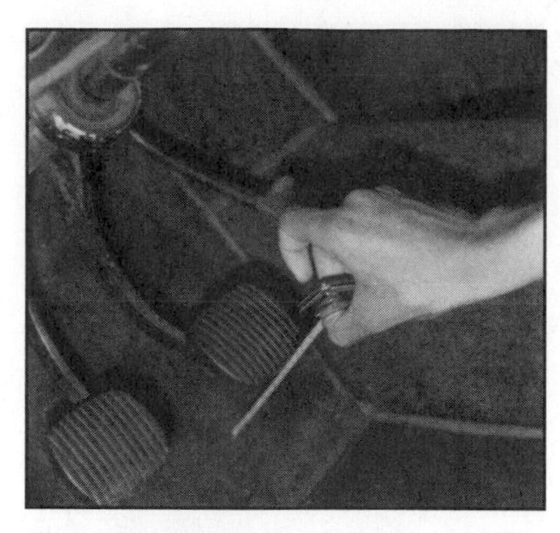

Steel tape is used to measure pedal-pad movement. Measure movement of master-cylinder piston at same time. Pedal travel divided by pushrod travel is pedal ratio.

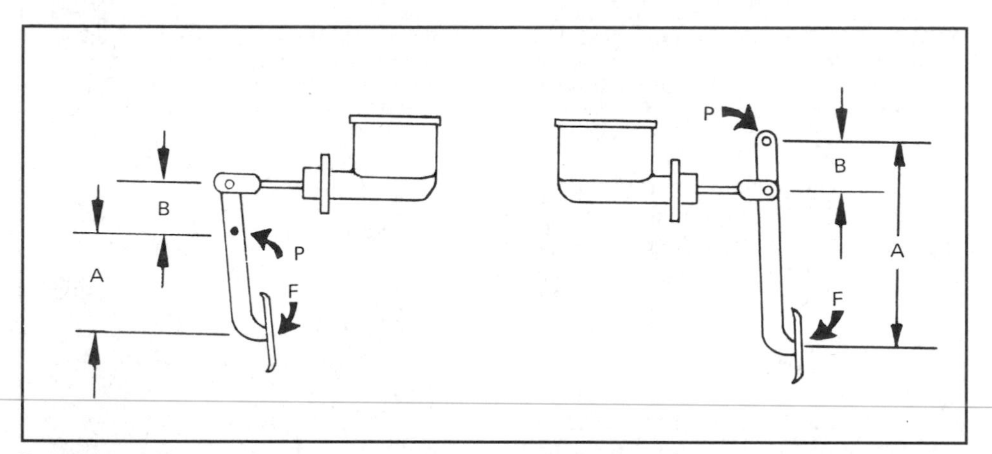

Pedal ratio is equal to A/B. Pedal pivot is at point P. Force F, applied by driver, is multiplied by pedal ratio to produce a larger force at master cylinder. If driver applies 50 lb and pedal ratio is 4, force at master cylinder is 200 lb. Trick in measuring distances A and B is measuring them perpendicular to line-of-action of forces.

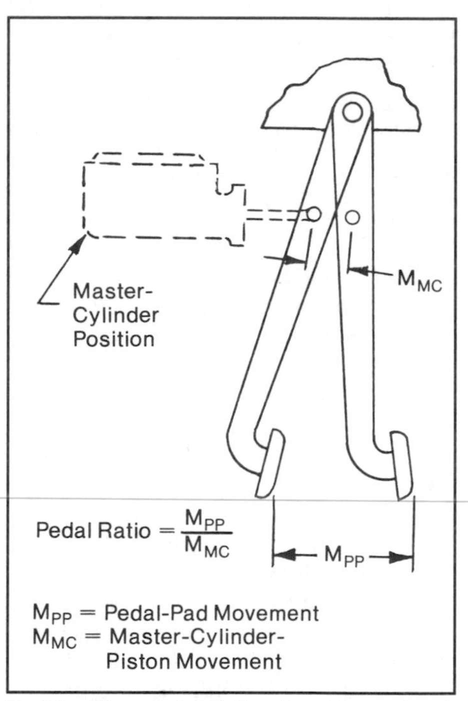

Pedal ratio can be determined from simultaneous pedal-pad and master-cylinder-piston movements: Measure large movements to keep measuring errors small. Measure movements as shown.

The force method requires special measuring equipment. So skip this method, unless you have the equipment that will measure force and pressure.

Measure Movements—Pedal ratio can be found by measuring movement at the pedal pad and master-cylinder piston. Don't include movement needed to take up clearance or free play. The formula for pedal ratio using movements is:

$$\text{Pedal ratio} = \frac{M_{PP}}{M_{MC}}$$

M_{PP} = Movement of pedal pad in inches

M_{MC} = Movement in inches of master-cylinder piston for a single or a tandem master cylinder, or movement of center pivot of the balance bar for a car with dual master cylinders.

To use this formula, you must be careful to get the direction of movement right. Pedal-pad movement should be measured through the full stroke of the pedal. You will have to open a bleeder or drain the fluid from the hydraulic system to do this. Start the measurement from the point where all clearances are taken out of the linkage and the pushrod *just starts* to move the master-cylinder piston. To hold the pedal in this position, put a spacer between the pedal stop and pedal.

Starting from this point, move the pedal all the way to the floor. Measure piston movement through full travel of the pedal. Indicate the starting and stopping points of the pedal pad by placing a piece of plywood or cardboard alongside the pad and making marks on it. Total movement of the pedal pad between the starting and stopping points is M_{PP}, or the number to use in the formula. Master-cylinder-piston movement is how far the piston moved down the bore. How to measure this movement is illustrated above.

Finding the pedal ratio by measuring movements gives an *average* pedal ratio for the full stroke. It is not a very accurate number. The problem is pedal ratio changes as the pedal moves. To get a more accurate pedal ratio, use the method described below. This uses the geometry of the pedal and linkage. It can give instantaneous pedal ratio at any point along the pedal movement.

Pedal Geometry—Dimensions of a pedal determine its ratio. Formulas to use for various pedal mountings are given in the drawing at left. Use the formula that matches your pedal setup. In each formula is a distance A and B. Measuring these distances accurately is the only problem. To do it right, you need to know the *line-of-action* of the forces on the pedal.

The line-of-action is drawn from the driver's hip to the ball of his foot. This line changes, depending on where the pedal pad is in its travel. To simplify matters, pick a spot in the movement of the brake pedal to measure lines-of-action—and stick to it.

The line-of-action for a pushrod is always a straight line joining the ends of the pushrod. Brake pushrods are usually straight, so it is easy to determine. If your pushrod makes an angle with the pedal arm, be sure to measure distance B at a 90° angle to the pushrod. This is illustrated nearby.

If the brake-pedal linkage has an intermediate mechanism, such as a bellcrank, determining pedal ratio becomes more complicated. The pedal ratio for the pedal itself can be measured using the methods already discussed. Then you must find the leverage of the rest of the linkage. Call that leverage the ratio of the linkage. Pretend the force into the linkage is the force from a driver's leg, and treat the linkage the same as you would a pedal. When you find the "pedal" ratio of the linkage, multiply it by the ratio for the pedal to get the total brake-pedal ratio. The accompanying drawing shows how a complex system is analyzed for overall pedal ratio.

Using pedal and linkage geometry, pedal ratio can be found for a particular position of the pedal. You should measure pedal ratio at several positions of the pedal to see how it changes. It is useful to draw a graph of pedal ratio versus pedal-pad position. This gives you a picture of how your brake system works. You want a pedal that avoids a radical pedal-ratio change, particularly as the pedal gets close to the floor.

Choosing the correct pedal ratio on a special car is a compromise because of conflicts between high and low pedal ratios. A low pedal ratio increases pedal-linkage stiffness and reduces forces on the pivots. A high pedal ratio reduces pedal effort and allows a large-diameter master cylinder to be used. I detail these trade-offs in Chapter 9. Total travel at the pedal pad is limited by the driver's leg. His anatomy determines the distance he can move his foot back from the floor or accelerator pedal to the brake pedal. If he has to stop quickly, he doesn't want to move his foot very far.

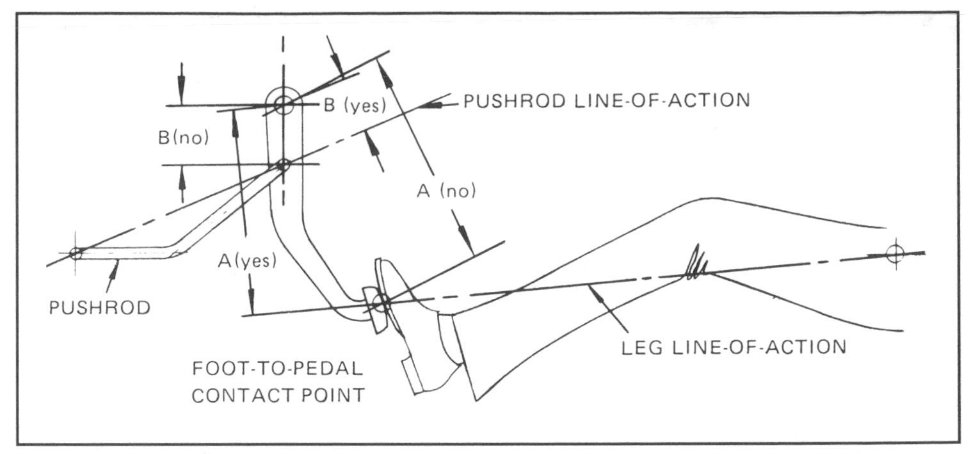

Line-of-action of leg is between pedal pad and driver's hip. Measure distance A perpendicular to this line. Distance B is measured perpendicular to line-of-action of master-cylinder pushrod. Pushrods should not be bent as shown, but if one is, its line-of-action would be through pushrod ends. Drawing by Tom Monroe.

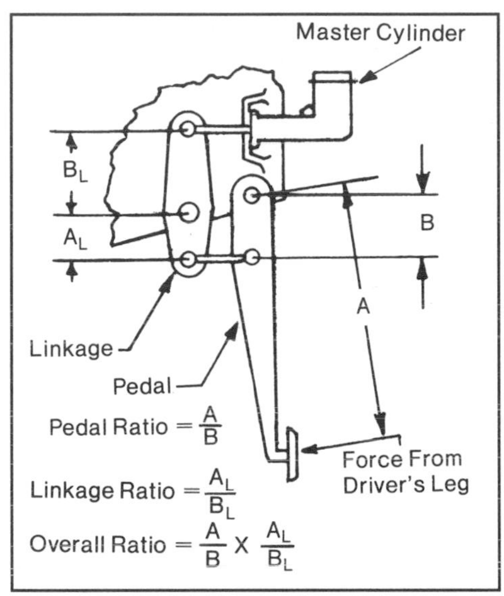

If pedal operates through a bellcrank, rocker or other linkage, pedal ratio alone is not the final answer. *Overall ratio* between driver's foot and master cylinder is required for brake-system calculations. Once pedal ratio is determined, ratio of intermediate linkage is determined. When determining linkage ratio, force from pedal is the same as force from driver's foot. Note that linkage pictured has a ratio less than 1.0. Force from pedal is reduced by linkage because A_L is smaller than B_L.

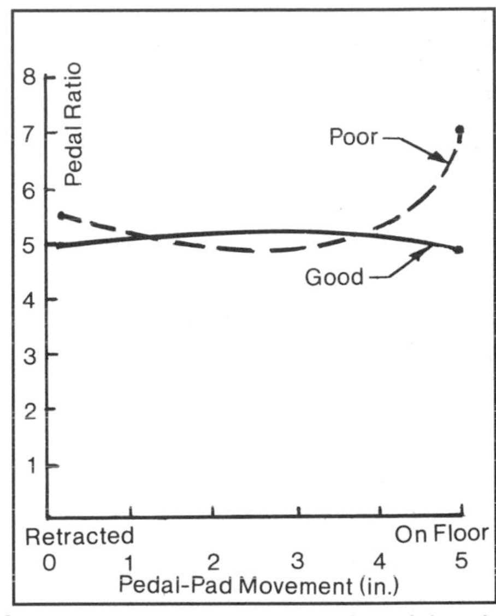

Graph shows two curves of pedal-pad movement versus pedal ratio. Solid line is for a good design—pedal ratio changes little with pedal movement; dotted curve is poor—pedal ratio changes considerably. Angle changes with pedal or linkage movement determines curve shape.

Pedal Movement—Another consideration is how far the brake pedal moves during brake application. Most brake pedals move from the retracted position to the floor in less than 6 in. measured at the pad. Pedal movement must be less than this for safety and comfort. When designing a special brake system, keep pedal movement during normal braking as short as possible.

Let's review some basics: The brake pedal must supply movement as well as force. Pedal movement includes all slop and clearance takeup in the brake system, movement of friction material, and combined deflection of all parts in the system. In

Tilton Engineering pedal-and-bracket assembly mounts dual master cylinders to rear. Assembly is used with hanging-pedal mount. Stock cars with clearance problems in engine compartment benefit from this setup. Master cylinders install in driver's compartment, away from engine heat.

Neal Products offers a complete line of pedals, brackets and other brake-system hardware for special cars. Neal pedals and brackets accept various master cylinders. Dual units shown include balance bar and are set up for floor mounting. Pedal and cylinders can be changed for hanging-pedal installation.

addition, extra movement must be available at the pedal pad in case there is fade, non-flat lining wear, hot-lining compression, or air in the brake fluid. Normal brake-pedal movement must be as short as possible to allow extra movement for these emergencies or failures in a brake system. There's more if you're working with a dual-brake system or drum brakes.

In a tandem or dual master-cylinder system, extra pedal movement must be provided in case one brake system fails. It must have enough pedal travel available to lock up two wheels when the brakes are hot and one system has failed.

In addition, overheating drum brakes can cause extra pedal movement. When the drums get hot, they expand away from the shoes. Consequently, the brake shoes must move farther. If the pedal hits the floor, you may not be able to stop the car. This is not a problem with disc brakes.

If you cannot achieve a pedal travel less than the 6-in. maximum, possible cures are bigger brakes, power assists or other design changes. Chapter 9 discusses this problem. Before you panic, though, minimize each contribution to total pedal movement. Start with some careful measuring and testing to find what is causing most of the pedal movement.

SPECIAL PEDALS

If you'd rather not design the brake pedal, excellent aftermarket pedal assemblies are available from several sources. Neal Products of San Diego, California offers a complete line of pedals, linkages and other brake-system hardware. A Neal pedal assembly can be purchased with its own integral bracket, including master cylinders. This assembly is easily mounted on a special car. Neal pedals come in various sizes and mounting configurations to suit most applications. Winters Performance Products of York, Pennsylvania also offers similar pedal-and-master-cylinder assemblies with integral brackets.

A special brake-pedal assembly is offered by Tilton Engineering. They offer a racing brake-pedal-and-bracket assembly that positions the master cylinders toward the rear of the car rather than the front. This installation is designed for stock-car and GT racing, where having the master cylinders in the engine compartment can cause problems. With rear-facing master cylinders, the master cylinders are not subject to heat from the engine, and they do not interfere with parts such as carburetors, valve covers or headers, or with sparkplug removal. Instead, the master cylinders end up under the instrument panel or front cowl. The one disadvantage with this setup may be servicing the master cylinders.

PEDAL BRACKETS

The strength of a brake-pedal bracket should match that of a properly designed brake pedal. It is essential to mount a brake pedal on a strong and rigid bracket. In addition, the master cylinder should be part of the brake-pedal mount to reduce the relative deflection of the two assemblies. Forces on the brake-pedal bracket can exceed 2000 lb, so it must be rigid. Otherwise, the bracket may be dangerous and can cause a spongy feel to the brake pedal.

Neal, Tilton and Winters brake pedals come with their own bracket, which links the pedal directly to the master cylinder. This is important. The bracket is mounted to the car, but forces on the bracket mounting are lower than forces on the brake-pedal pivot and master cylinder or the pushrod. Still, the mounting of the pedal bracket to the car is critical. This mounting should also be rigid so it will support the force of the driver's foot. Therefore, don't mount a brake pedal on a flat sheet-metal fire wall or unreinforced floorboard. Mount the pedal in a corner or on a boxed beam so there won't be excessive deflection.

Many road cars have brake-pedal brackets that can be adapted to other cars. Some have small compact brackets that can be purchased with the pedals for a reasonable price. It is much better to adapt an existing bracket-and-pedal assembly to a car

Pedal bracket on homemade sports car is built into frame structure. It pays to work in pedal-bracket design with frame or roll-cage design. Save time, weight and cost by planning ahead.

Eagle Indy Car has large hole in pedal box for easy pedal access. Balance bar has remote brake-balance adjuster, so frequent access to the brake pedal is not required. Cable balance-bar adjuster (arrow) goes to knob on instrument panel.

than to design and build one yourself. It's a lot less expensive, too, to buy an existing unit.

Many race-car brake-pedal brackets are integrated into the frame structure. This approach is excellent if you are designing the entire car. Most race-car frames have tubes or rigid bulkheads that make excellent structures for mounting pedals. With a large sheet-metal structure like a passenger-car fire wall, the mounting requires a special bracket for the pedals.

Pedal Location—There are two types of brake-pedal mounts: *floor-mounted* and *hanging*. Both are used and both have advantages. Most modern road cars use hanging pedals. There is a lot of room under the dash for brackets, and often the steering column is supported by the same structure.

Hanging pedals also can be long, are out of sight and the pivots are less susceptible to corrosion than floor-mounted pedals. Additionally, the master cylinder ends up high in the engine compartment, where it is easy to service. Another advantage of this is air bubbles rise to the master cylinder, then into the fluid reservoir. Thus, the system is self-bleeding, except for a few spots at the wheels.

On low-profile race cars, floor-mounted pedals are more popular. Race-car design demands a low CG, so it's best to keep components mounted as low as possible. Also, the height of the body nose of a rear- or midship-engine car is often determined by the height of the driver's feet inside the cockpit. If hanging pedals are used, the added height required above the driver's feet raises the height of the nose. Therefore, a racing sports car or formula car usually has floor-mounted pedals.

Master cylinders with floor-mounted pedals are difficult to service, unless they are in front of the front suspension. The balance bar is also difficult to adjust. Worst of all, air in the system tends to end up at the wheels rather than the master cylinder(s) and reservoirs(s). Con-

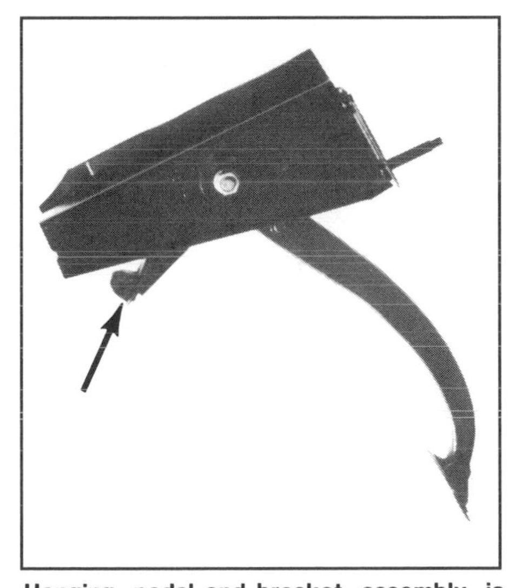

Hanging pedal-and-bracket assembly is ready to mount in car. Unit includes pivot bushings, return spring and pedal stop (arrow). Such bracket-and-pedal assemblies can be found in junkyards for a modest price.

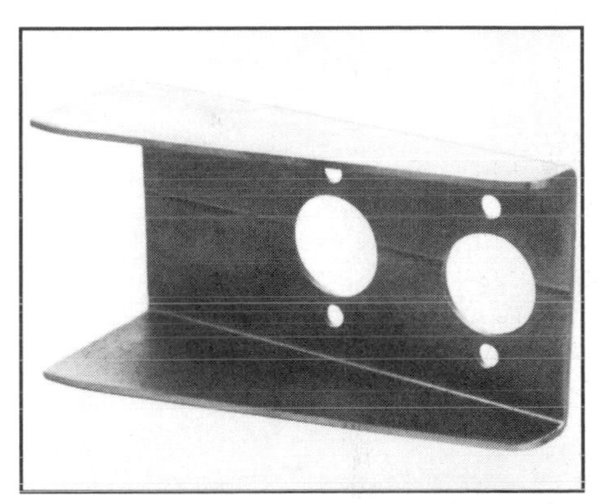

Sturdy universal bracket mounts dual master-cylinder-and-bracket assembly to frame or roll cage. Photo courtesy Tilton Engineering.

Hanging pedals are neat and package well in most large cars. Such a setup with dual master cylinders can be difficult to adjust, particularly when you're in a hurry—just the place for a remote adjuster.

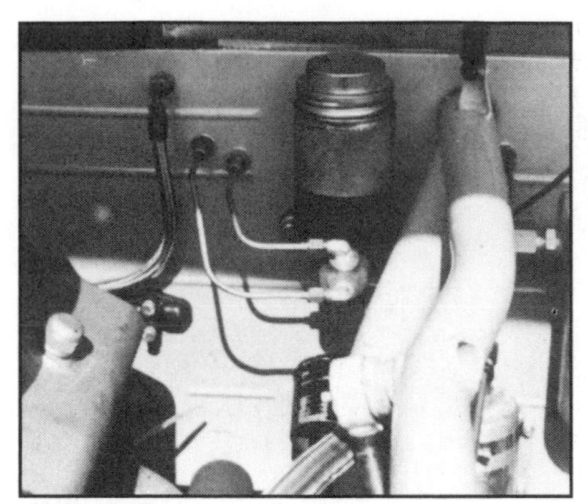

Stock car uses tandem master cylinder and adjustable proportioning valve mounted in cockpit. The driver simply moves lever on valve to adjust brake balance. It would've been difficult to use dual master cylinders with a balance bar because of frame-tube interference.

sequently, floor-mounted pedals usually are more sensitive to air buildup in the system. As a result they require more frequent bleeding.

Brake-fluid reservoirs for a floor-mounted pedal often must be remote from the master cylinders. This adds complexity to the brake system. In addition, dirt from the driver's shoes can get into the pedal linkage and cause problems; hanging pedal linkages stay relatively clean.

To summarize, use a hanging pedal mount if possible. If not, a floor-mounted setup will work even though it has disadvantages.

BALANCE BARS

A balance bar is used to adjust the brake balance for only one value of tire grip. It does not compensate for changing traction conditions the way a proportioning valve does. For this reason, balance bars usually are not used on a road car.

Balance bars are used on most race cars with dual master cylinders. The balance bar proportions the force from the brake pedal to the two master cylinders. Most balance bars are adjustable to vary the braking force going to the front and rear cylinders, thus the front and rear brakes. The most-exotic balance bars can be adjusted by the driver from the cockpit.

The simplest balance bar is a cross link, or *beam*. The brake pedal pushes on a pivot near the center of the cross link, which then pushes on the two master-cylinder pushrods that are mounted to the ends of the beam. Where the brake-pedal pivot is in relation to the two master-cylinder pushrods determines front-to-rear brake balance. Some balance bars can be adjusted; some cannot. A simple balance bar is shown above. This type is not adjustable as is.

There are other simple balance bars that are adjustable, but not easily. Shims or washers must be moved to change brake balance. Balance can be determined through testing, and once set, it never gets changed again.

For racing, a balance bar must be adjusted according to track conditions. For instance, if the track is slippery, the balance bar should be adjusted to give less force on the front brakes. Less forward weight transfer with a slick track means the front wheels must do less braking. If the track has good traction, the front brakes should do more braking. Balance-bar adjustment would then be made to give more front-braking force. Normally, the balance bar would be adjusted during a prerace practice session. Once the race starts, the bar is left alone.

The ultimate in sophisticated equipment is a balance bar that can be adjusted by the driver *during* a race. If the driver desires more or less front brakes as track conditions change, he

BALANCE BARS VS. ADJUSTABLE PROPORTIONING VALVES

Both a balance bar and an adjustable proportioning valve can be used to change the relationship between front- and rear-brake hydraulic pressure and, consequently, brake balance. However, they work in different ways.

A balance bar divides the pedal force between two separate master cylinders in a preset ratio. This force ratio can be adjusted by moving the center pivot of the balance bar closer to one master-cylinder pushrod or the other, thus, increasing pressure in one system and reducing it in another. On most road cars, a tandem master cylinder is used, which cannot be changed to adjust brake balance.

An adjustable proportioning valve varies the ratio of front-to-rear brake pressure in the direction of the front brakes above a preset *changeover* pressure. The valve is usually placed in the rear brake system to compensate for weight transfer to prevent rear-wheel lockup under hard braking.

A proportioning valve gives a front-to-rear brake hydraulic-pressure relationship that approximates the ideal relationship for various values of tire grip. A balance bar does not automatically compensate for variation in tire grip.

Nonadjustable proportioning valves are used on most modern road cars. An adjustable proportioning valve can be easily added to any brake system to make the balance adjustable within a limited range.

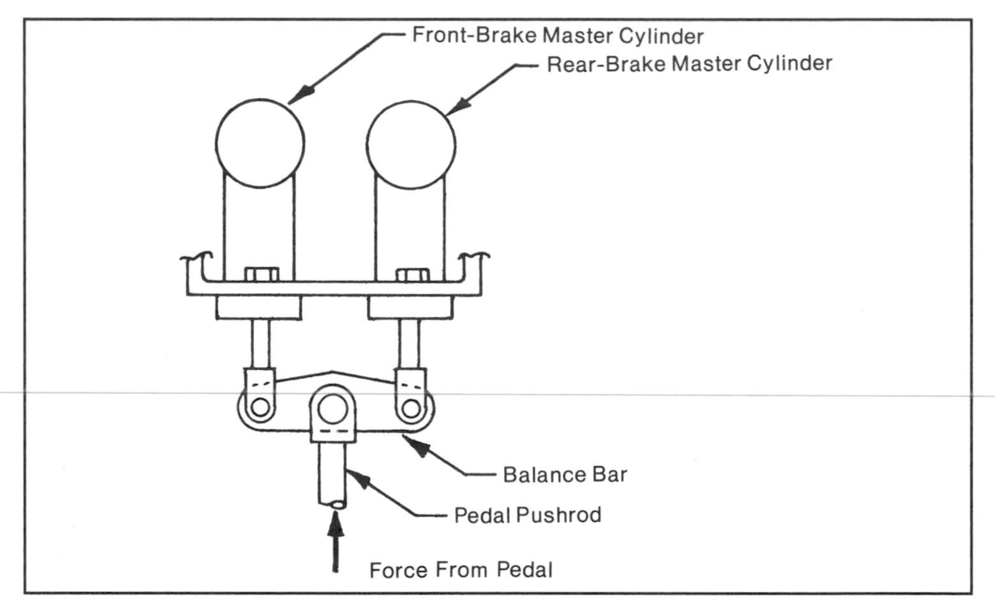

Simplest balance bar is a link with three holes. Ratio between front and rear brake pressure is determined by proximity of center hole to outer holes. To adjust this type of balance bar, it must be replaced with a new bar with center hole in a different place.

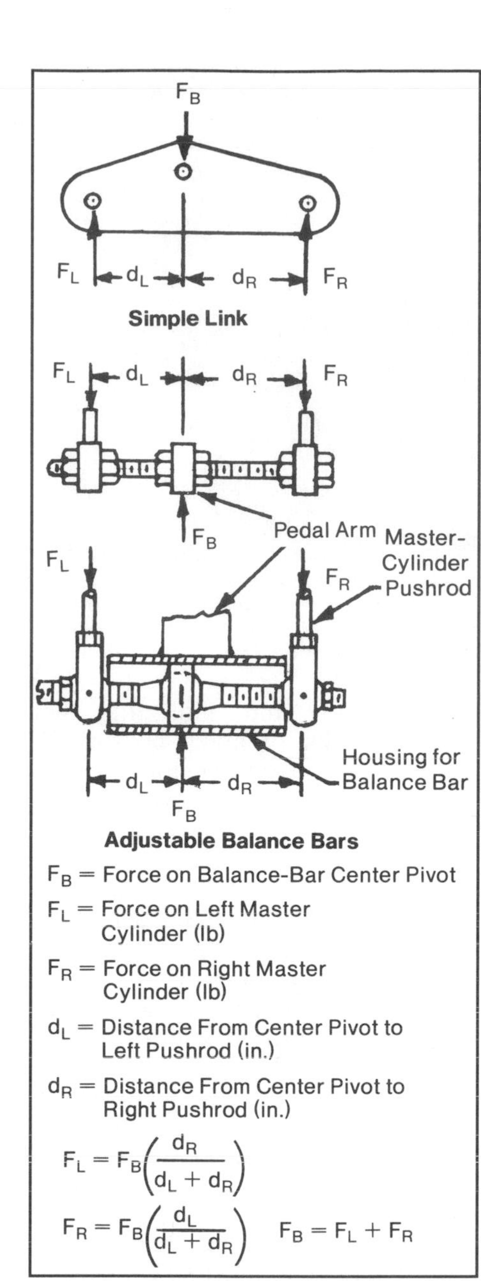

Compute forces on master cylinders with formulas given above. If center bearing (F_L) is moved toward front-brake pushrod, front brakes get more hydraulic pressure and vice versa. Note that spherical bearing in bottom drawing is off-center in pedal-arm housing. It slides sideways when adjusting screw is turned. Distances from spherical-bearing center to pushrod bearings determines balance—not distances from center of pedal arm.

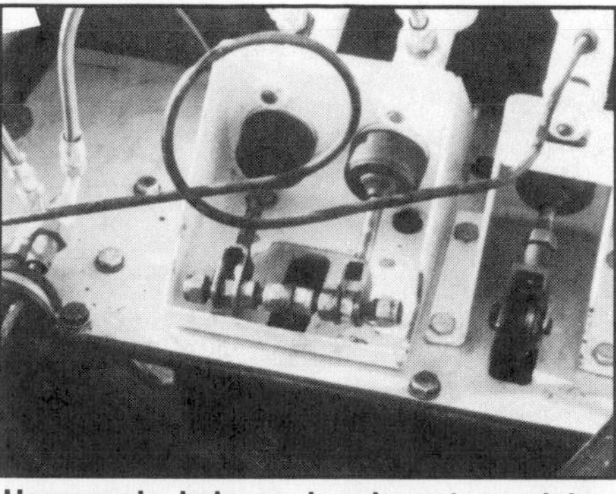

Homemade balance bar is not good because it is weak and hard to adjust. Threaded rods are not strong in bending—a failure could be disastrous. Correctly engineered balance bars are large at the center where stresses are highest. Also, balance bar probably would bind if one master-cylinder pushrod moved much more than other pushrod.

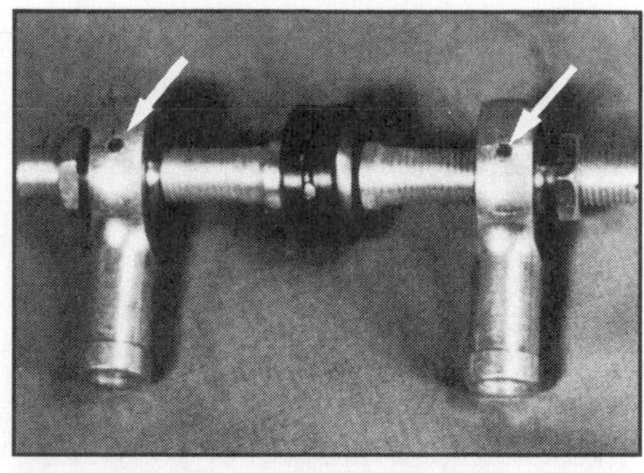

Neal Products' adjustable balance bar: Center ball slides in tubular housing welded into pedal arm. Turning threaded bar moves center bearing right or left inside pedal housing. Special spherical bearings have threads inside bore and pins (arrows) to restrict ball movement to only pivoting in one direction.

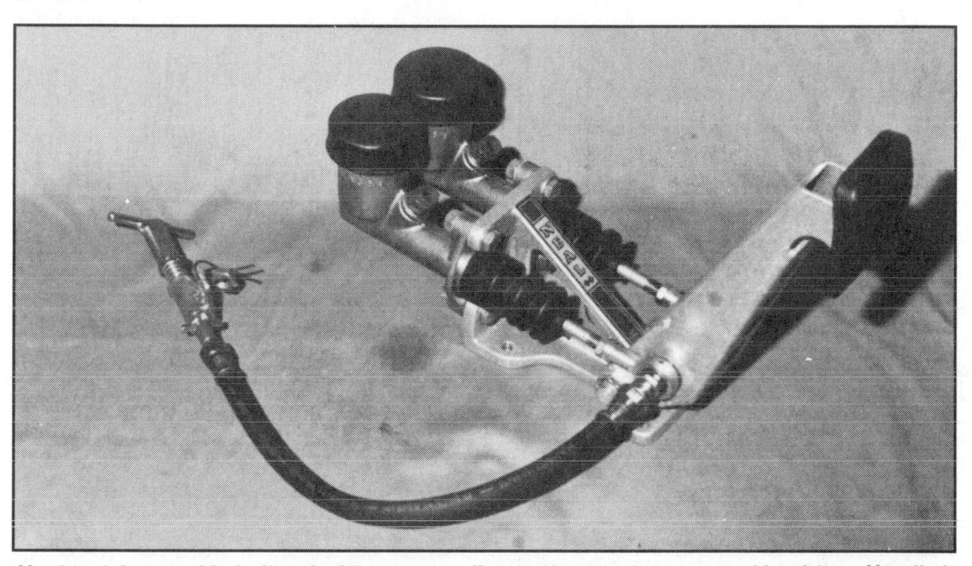

Neal pedal assembly is fitted with remote adjuster that can be operated by driver. Handle is mounted so adjustments can be made while racing. Driver must be able to sense need for brake-balance changes before he can use adjuster effectively.

can make the change. For example, if it rains, and traction suddenly drops, a rearward brake-balance adjustment is required. It takes a skilled race-car driver to know when a brake-balance change is needed.

In-cockpit adjustable balance bars are used in Grand Prix cars and similar high-priced machines, where the drivers are sufficiently skilled to make use of the feature.

The most popular balance-bar design is one where the bar is mounted in the pedal arm. A spherical bearing rides inside a tube welded into a hole in the brake-pedal arm. The balance bar is a high-strength rod mounted firmly inside the bore of the spherical bearing. A clevis to each brake pushrod is mounted at each end of the rod. Each end fitting is threaded so a turn of the rod will move the spherical bearing right or left inside the pedal arm. The force on the bearing is then closer to one end of the rod, making that master cylinder share more of the braking force and the other less.

Most road-racing cars use adjustable balance bars. You can adapt one of these to your car. You can purchase a balance bar from Neal Products, Tilton Engineering, Winters Performance Products or other racing-component outlets.

Setting up a balance bar can be difficult. First, choose master-cylinder size(s) so the balance bar can be set in the center for average track conditions. A balance bar should not be used to compensate for the wrong master-cylinder diameter(s). If you need to vary the balance bar a great deal from the center position to obtain proper brake balance, a change in a master-cylinder diameter would be better. The balance-bar adjustment range should be sufficient to compensate for conditions encountered at the track.

The motion of a balance bar must allow one master cylinder to operate when the other one bottoms. The system should be designed so the pedal does not run out of travel before both master cylinders bottom. Sometimes people overlook this, so difficul-

Handle in upper left corner of Indy Car instrument panel adjusts balance bar. Slots in handle mount provide positive setting and lock handle in position.

Hanging pedals operate reverse-mounted master cylinders installed above pedal box. Setup gets master cylinders out of engine compartment. As shown, balance bar must be able to operate at extreme angles if one hydraulic system fails.

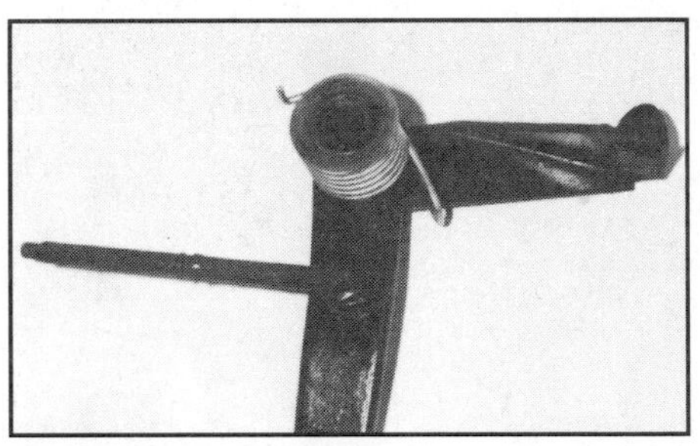

Upper end of production road-car hanging pedal is equipped with return spring, pedal stop and pushrod. Pedal stop is rubber pad at end of arm to right—it's not adjustable. Hairpin-type return spring wraps around pivot housing.

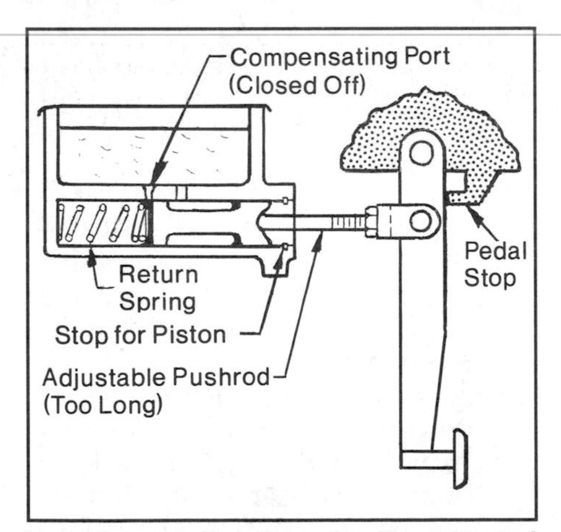

If piston can't return to its retracted position, seal covers compensating port and brakes won't fully release. Adjust pedal stop or pushrod to move piston back in its bore. If pedal stop or pushrod can't be adjusted, shim master cylinder forward.

ties arise in emergency situations.

A balance bar must not bind when operating at extreme angles possible when one master cylinder strokes much more than the other. For example, boiling fluid or a slow leak may require one master cylinder to stroke a great deal. This potential movement should be checked using accurate drawings during design. Actual balance-bar action should be checked on the car by opening a bleeder on one brake system and stroking the pedal. This forces the balance bar to move into extreme angles.

RETURN SPRINGS & PEDAL STOPS

There must be a return spring to push a brake pedal back to its retracted position when the brakes are released. The return spring that pushes the master-cylinder piston back can also return the pedal, if the pedal is light, the pedal is mounted vertically, and the pivots have low friction. On some road cars, a separate return spring is mounted on the pedal to make sure it returns. On race cars, the master-cylinder return spring(s) usually does the job.

A brake pedal must have a positive stop in the retracted position. This position must always be the same so the driver can find the pedal immediately. The positive stop or the pushrod usually is adjustable to make sure the master-cylinder piston returns fully. If the piston is prevented from reaching its retracted position, the ports in the master cylinder may not open as they should and hydraulic pressure will build, causing severe brake drag.

With the pedal in the retracted position against its stop, there should be a small amount of free play in the master-cylinder pushrod. This ensures that the master-cylinder piston is fully retracted. Most road cars have positive-return stops built into the master cylinder or brake booster. If your car has an external stop, see the shop manual for adjustment instructions.

If you are designing your own system, be careful not to use a separate return spring on the pedal if the pushrod can be pulled from its socket in the master-cylinder piston. Most pushrods are retained in the piston, but some are loose. If a return spring forces the pedal back faster than the master-cylinder piston, the pushrod can fall out unless it is retained in the piston. Therefore, use *only* the master-cylinder-piston return spring to retract the pedal if the master-cylinder pushrod is not retained to the piston.

All tandem and most single master cylinders are strong enough so that the closed end of the master-cylinder bore can be used as the forward pedal stop. If you are not sure about this or don't want to bottom the master-cylinder piston for fear of damage, install a forward pedal stop. Many cars use the floorboard as the stop. If you use a stop, allow some stroke in the master cylinder with the pedal hard against the floorboard.

Test the forward pedal stop when you bleed the brakes. With the bleeder screws open, push the pedal gently all the way down. Make sure it is hard against the stop or the floor before the master-cylinder piston bottoms. The piston should have about 1/16 in. of travel left before it bottoms. If the pedal hits the floor, make sure it cannot get jammed in the carpet or any other object.

Power Assist 7

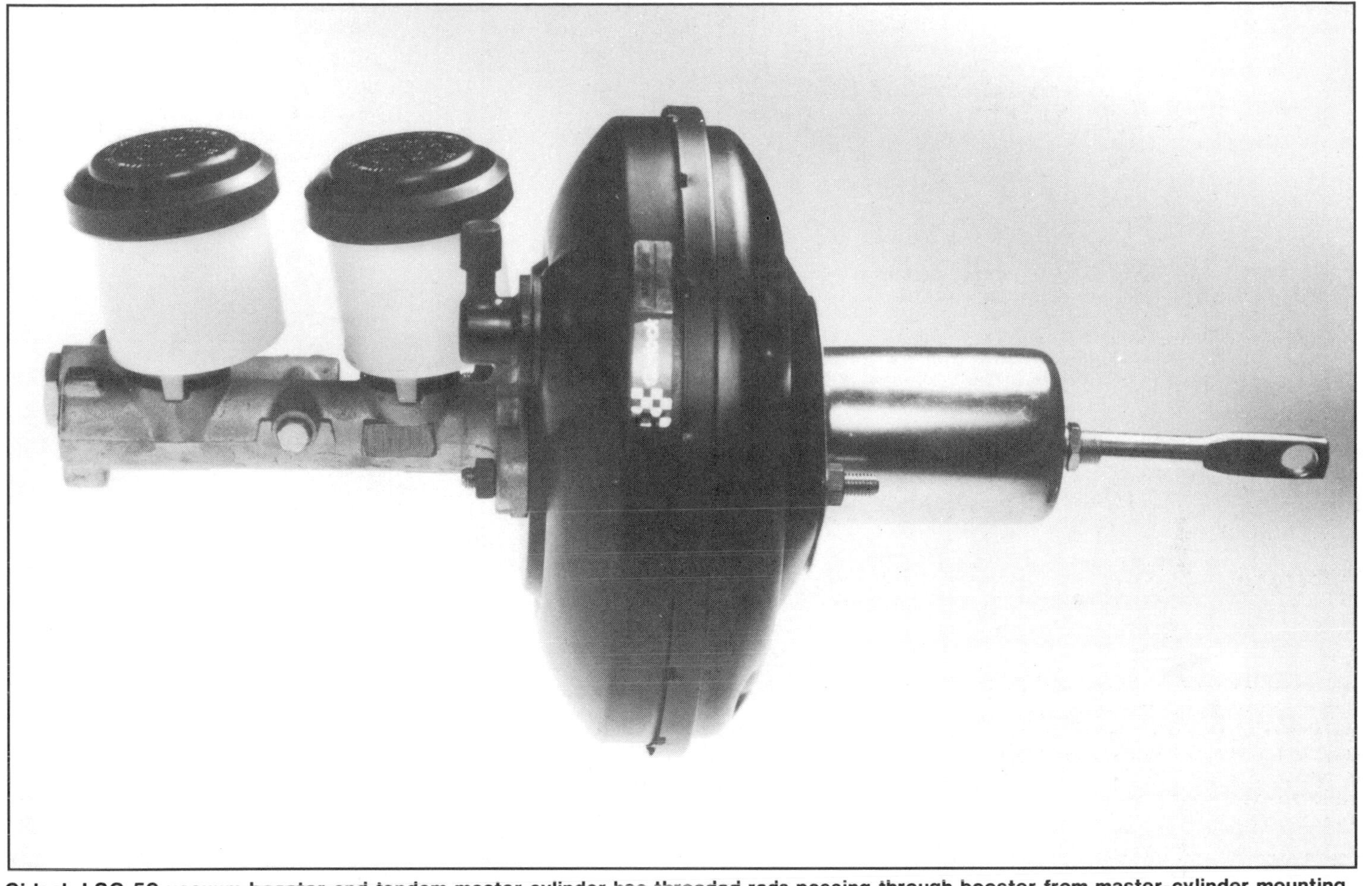

Girlock LSC 50 vacuum booster and tandem master cylinder has threaded rods passing through booster from master-cylinder mounting flange. Rods transmit forces on master-cylinder mounting flange directly to booster mounting bracket. Because booster chamber is not loaded by forces from master cylinder, booster is lighter than it would otherwise be. Booster weighs only 4 lb. Photo courtesy Girlock Ltd.

Most road cars and a few large race cars use *power assist—boosters*—to reduce braking effort. A car with a power assist is said to have *power brakes.* To the driver, the brakes feel like they have more stopping power; however, it's just lower pedal effort. Most power-brake systems are designed with a lower pedal ratio so the driver also senses less pedal travel. Power brakes can fade the same as standard brakes. Luckily, power brakes are reliable and usually perform well for the life of the car.

The *power* in power brakes comes from a unit called a *brake booster.* It is usually mounted on the fire wall between the brake-pedal linkage and the master cylinder. The brake booster looks like a large tank, if it is the type operated by engine vacuum. Some cars use a hydraulic-powered booster that looks like a complex valve mounted behind the master cylinder.

Whatever the design of the booster, they all do the same thing—add extra force to the master cylinder when the driver pushes on the brake pedal. The harder the driver pushes, the more force the booster adds. The effect to the driver is reduced pedal effort.

Power brakes are used on heavy cars and on cars with disc brakes. Power assist was used before the switch was made from drums to discs. Now, most large cars use power brakes. Many small cars using disc brakes also have power assist. Power assist makes the job of designing brakes easier. For example, the disc brake's lack of servo action requires more master-cylinder-piston force than a drum brake. If a drum brake is designed with a large servo action, it can stop a heavy car without power assist. With power assist, drum brakes can be designed with less servo action. That makes them less affected by changes in friction between shoes and drum. Thus, a car can have drum brakes that are more fade resistant by using power assist.

Many early power-brake systems were too sensitive. The power assist was so great that the slightest push on the brake pedal locked the wheels. Nothing was gained from this design, and the brakes were dangerous. In a panic situation with overassisted brakes, it is nearly impossible to avoid locking the wheels and losing control. Fortunately, most modern power-brake systems are designed to feel more like an unboosted system. Therefore, a driver should have few problems with a later power-brake system. Power assist combined with the consistency of disc brakes gives the modern car excellent brakes.

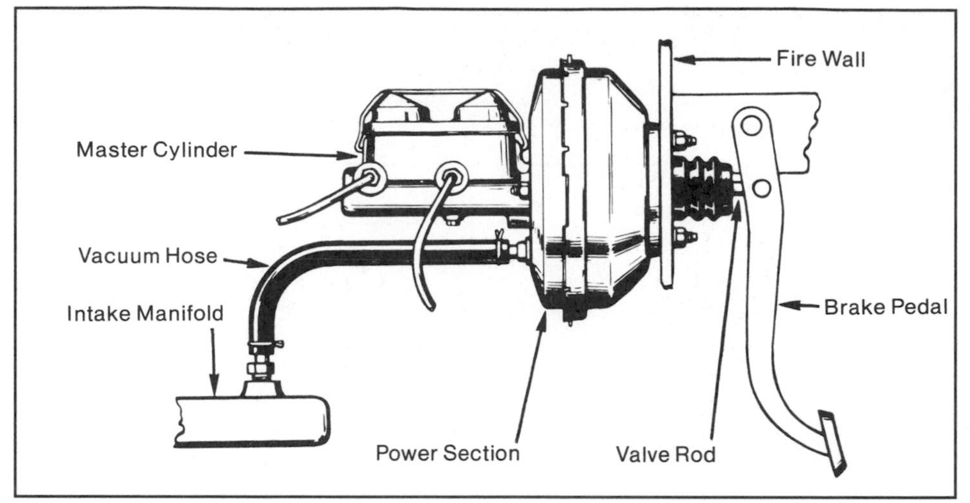

Typical vacuum-booster setup: Intake manifold supplies vacuum to booster. Brake pedal pushes directly on booster-valve rod, which controls force booster applies to master cylinder. If vacuum supply drops to zero or booster fails, a mechanical linkage operates master-cylinder piston by pushing harder on pedal. Effort increases significantly without boost. Drawing courtesy Bendix Corporation.

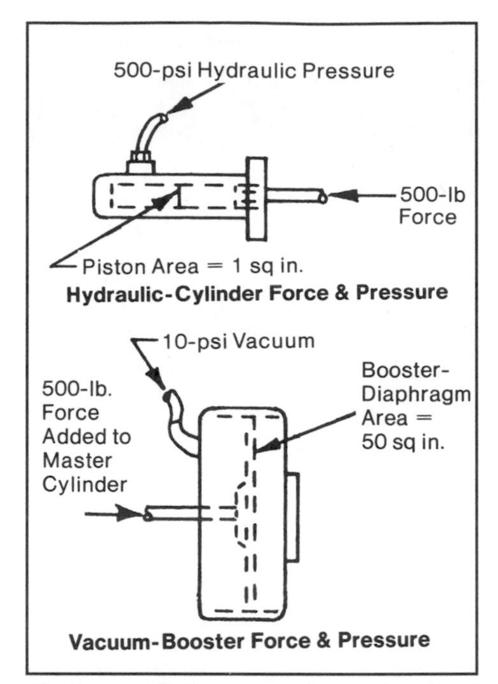

Master cylinder operates at about 1000-psi maximum hydraulic pressure. Because vacuum diaphragm operates at about 10-psi maximum pressure, booster diaphragm is 5 to 10 times diameter of master-cylinder piston to give required area.

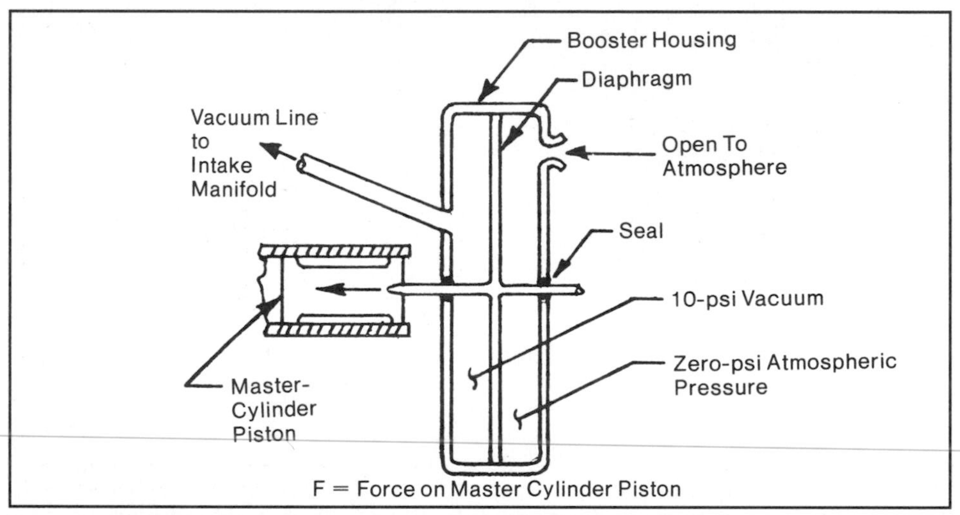

Pressure on diaphragm is difference between vacuum and atmospheric pressures. Because vacuum is below atmospheric pressure, it is given a minus value. In this example -10 psi is in vacuum side of chamber and 0 psi—atmospheric pressure—is in other side. Vacuum of −10 psi gives same force as 10-psi pressure above atmospheric, but resulting force is in opposite direction.

VACUUM-BOOSTER TYPES

Most power-brake systems use a booster operated by vacuum from the intake manifold of a naturally aspirated engine. Because manifold vacuum is highest—pressure is lowest—when the driver lifts his foot off the throttle, vacuum in the booster is maximum as the brakes are applied. This, of course, is different for a supercharged engine, where manifold pressure usually is greater than ambient air pressure.

The extra force on the master-cylinder piston is provided by a large piston inside the power booster. In a vacuum booster, this piston is called a *diaphragm.* You see the outside of the cylinder—chamber—when you look at a vacuum booster. It's the large, round object that looks like a tank.

Vacuum boosters have a large diameter because vacuum exerts a very low "pressure." Actually, the outside air exerts pressure; vacuum removes it. Ambient air pressure (14.7 psi) is one one side of the diaphragm and engine manifold vacuum the other. Air pressure is reduced on the vacuum side by engine vacuum. A perfect vacuum on one side of the diaphragm allows a maximum pressure of only 14.7 psi. This is a *perfect* vacuum. In real life, it is never that high. Compared to about 1000-psi maximum pressure in a brake hydraulic system, the vacuum-boost pressure is about 10 psi. The result is the large-diameter cylinder. Remember, total force equals pressure multiplied by the area of the piston—diaphragm in this case—whether the pressure is hydraulic or a vacuum.

The pedal linkage is connected to a mechanical linkage that pushes on the master cylinder. The driver can push the master cylinder the same as in a normal brake system. However, the pedal effort is high without assist from the booster. The booster merely adds force to the master-cylinder pushrod when the driver pushes on the pedal. The total force on the master-cylinder piston is the force from the pedal plus the booster force.

Let's take a closer look at how a vacuum boost works. To restate, a booster diaphragm supplies force by having a vacuum on one side of the diaphragm and atmospheric pressure on the other. A vacuum exists whenever pressure in a gas falls below atmospheric pressure. Most pressure gages read zero in the atmosphere, so special vacuum gages are used to measure pressure below *one atmosphere.* If most of the air is pumped out of a cylinder, atmospheric pressure tries to push inward on the outside of the cylinder. There is a lower pressure—vacuum—inside the cylinder. The main parts of a simple vacuum booster are the *power chamber* and the *control valve.* The power chamber is the diaphragm and its housing. They supply the force. The control valve lets air or vacuum into the power chamber.

Air-Suspended Boosters—All early boosters operated under no load with air on both sides of the diaphragm at one atmosphere pressure. When the brakes were applied, a valve between the booster and intake manifold was opened. Air was pumped from one side of the diaphragm. This type of booster is called an *air-suspended* booster. That means there is air on both sides of the diaphragm when it has no force on it.

Vacuum-Suspended Booster—The more modern type of vacuum booster is called *vacuum suspended.* It has a vacuum on both sides of the diaphragm when there is no force on it. Atmospheric air is let into the power chamber on one side of the diaphragm when the brakes are applied. It is easier to let air into a vacuum than it is to pump air out—particularly when it must be done quickly. Because *vacuum is an absence of air,* air naturally flows into the vacuum when the valve is opened. Also, the vacuum-suspended booster is more fail-safe than an air-suspended booster. That's another reason it is popular today.

Tandem Booster—Most vacuum boosters have one diaphragm in the power chamber. If a large force is required from the booster, cylinder diameter must be large. To make a more compact booster, some use two diaphragms. This type of booster is called a *tandem booster.* The diaphragms are mounted in front of each other in the power chamber, as are the pistons in a tandem master cylinder. They each provide part of the booster force.

Brake-Booster Operation—There are three basic vacuum-booster operating conditions:

- Brakes released.
- Brake being applied.
- Brakes holding at constant force.

The control valve is operating differently for each of these conditions. Control-valve operation is shown in the drawing, page 80.

The power chamber of a vacuum-suspended booster is divided into a *front chamber* and a *rear chamber* by the diaphragm. There is always vacuum applied to the front chamber. The vacuum line contains a check

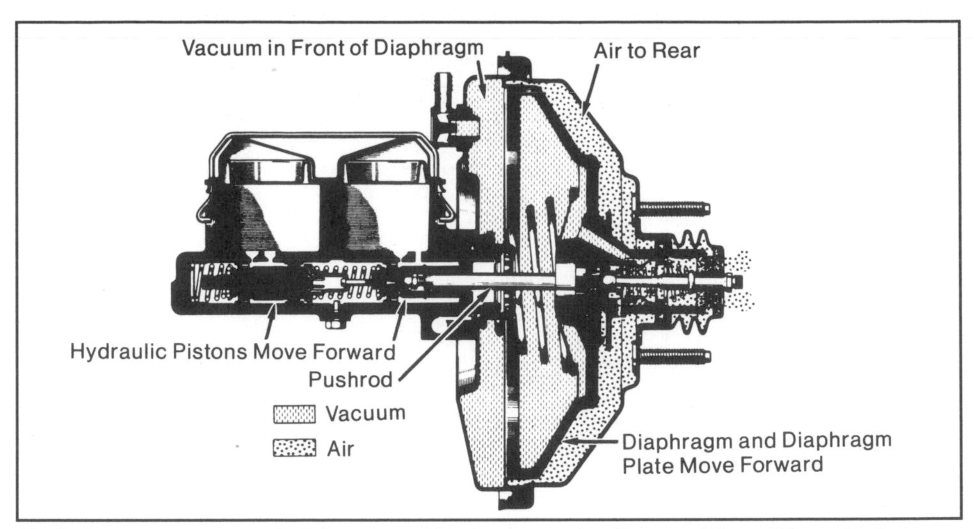

Typical vacuum-suspended booster: Diaphragm acts like a piston. When brakes are applied, air is admitted to chamber behind diaphragm, applying force to master-cylinder pushrod. Drawing courtesy Bendix Corporation.

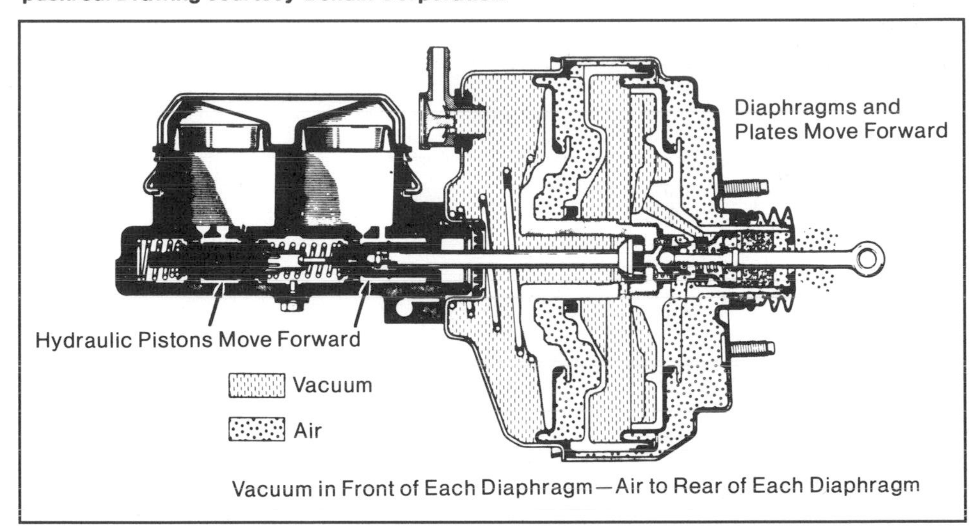

Bendix tandem-diaphragm Master-Vac booster in applied position: Vacuum acts on each of two diaphragms, doubling booster force. Unit is used where high force is required and space is limited. Drawing courtesy Bendix Corporation.

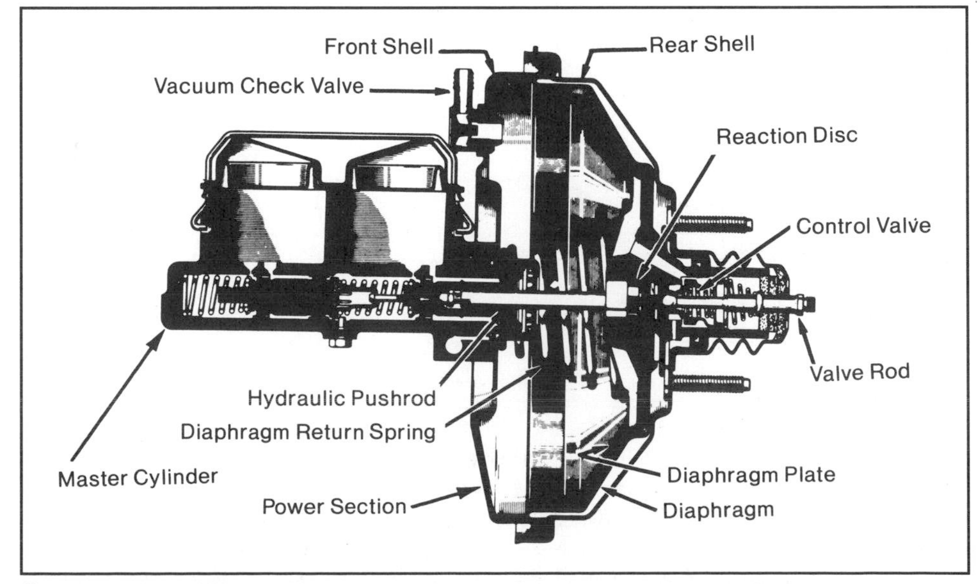

Cross-section of Master-Vac booster with working parts indicated. Control valve admits air into chamber behind diaphragm. Valve rod is moved when driver pushes on brake pedal. Drawing courtesy Bendix Corporation.

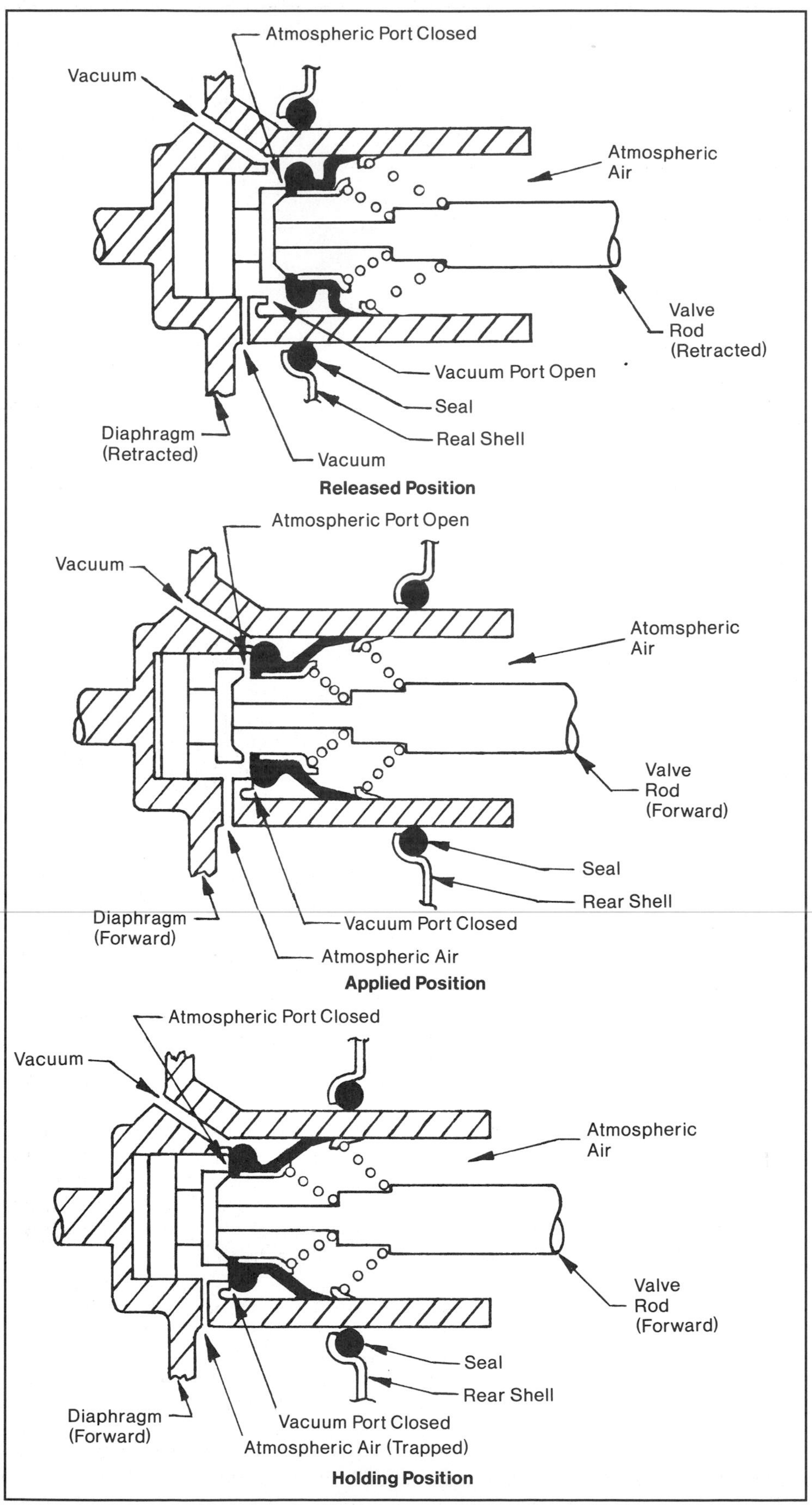

Vacuum-booster control-valve operation shown in simplified form: *Rear shell* encloses rear chamber of booster and mounts booster-and-master-cylinder assembly to fire wall. Parts move relative to each other. Valve rod moves when driver steps on pedal. Diaphragm moves when acted on by atmospheric air pressure and vacuum.

valve to prevent vacuum loss in booster when *manifold pressure* rises above *booster pressure*—or manifold vacuum drops below booster vacuum. A check valve opens to allow air to flow from booster, but closes when air tries to flow into the booster.

With brakes released, control valve opens a vacuum port that connects front and rear chambers. Pressure on both sides of the diaphragm is equalized with the brakes released. Both chambers contain a vacuum as long as the engine is running.

With brakes applied, control valve moves forward. This shuts the vacuum port between front and rear chambers and opens the atmospheric port. Air flows into rear chamber through the atmospheric port, raising pressure against the diaphragm. The diaphragm is pushed forward by the air pressure on its rear side. The diaphragm then pushes on the master-cylinder pushrod and adds to the force being supplied by the driver.

If the brake pedal is applied without moving, forces on the control valve shut both the atmospheric port and vacuum port. This maintains a constant force on the master-cylinder pushrod. If the brake pedal is pushed harder, atmospheric port is opened again and additional force is applied by the booster. *With brakes released,* vacuum port is opened: Booster force and assist disappear.

Booster Spring—Boosters often are fitted with springs to give the driver the correct amount of feel. It is important that pedal effort not be too low. This makes the brakes too sensitive. For the correct amount of feel, pedal effort should increase with pedal travel. As the spring is compressed, pedal effort increases.

HYDRAULIC BOOSTER

Some brake boosters use hydraulics rather than vacuum. Fluid is pressurized by the pump that also supplies the power-steering system.

A hydraulic brake booster is used on cars that have a poor source of vacuum from the intake manifold. Not only does a supercharged engine eliminate the manifold as a source of vacuum, many newer cars are equipped with emission controls and vacuum-operated accessories that reduce manifold vacuum. On such cars, hydraulic boosters are more practical, particularly if the car is also

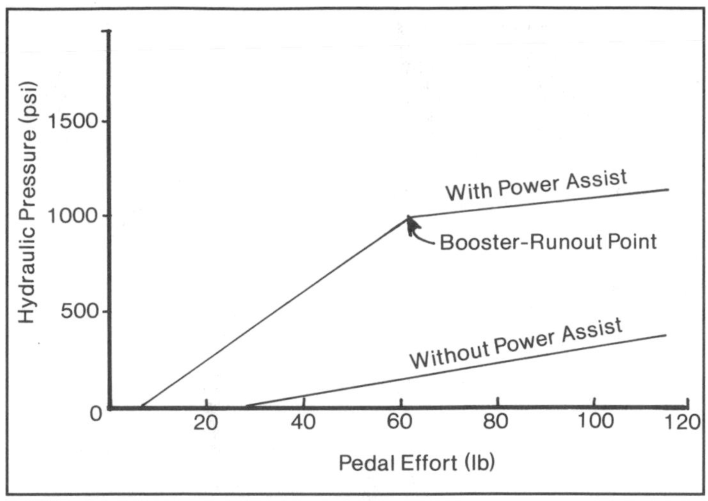

Curves show effect of power assist on pedal effort. Runout point is where power booster supplies maximum possible force. If pedal effort is higher than 50 lb, brake-fluid pressure increases the same as it would without power assist—curve with power assist is parallel to non-power-assist curve past runout point. Runout point varies according to each car. Boost force is limited by available vacuum or power-steering-pump pressure.

Bendix Hydro-Boost hydraulic brake booster mounts between master cylinder and brake pedal. High-pressure hydraulic fluid allows unit to be much smaller than vacuum booster. Hydraulic pressure is supplied by power-steering pump. Drawing courtesy Bendix Corporation.

equipped with power steering. And, there's the advantage of fitting the smaller hydraulic booster into a crowded engine compartment.

Like the vacuum booster, hydraulic booster is located between the brake-pedal linkage and master cylinder. The boost piston has hydraulic pressure behind it to add force to the master-cylinder pushrod. In the event of booster failure, the driver can push on the master-cylinder piston just like a normal brake system. However, pedal effort is much higher due to the low pedal ratio.

Basic hydraulic-booster parts are:

- Boost piston.
- Sliding *spool* valve, which regulates braking pressure.
- Lever system to operate spool valve.
- Reserve-pressure system.

The boost piston operates much like any hydraulic piston. Because power-steering hydraulic pressure is much higher than the "pressure" available from vacuum, the piston is much smaller than a vacuum-booster diaphragm. The boost piston is located between the rod from the brake pedal and the master-cylinder pushrod. When the brakes are applied, the lever on the input rod slides the valve to a position that allows high-pressure hydraulic fluid to enter the chamber behind the piston. This applies force to the master-cylinder pushrod.

Reserve pressure is stored in a cylinder called an *accumulator.* There is a very heavy spring inside the accumulator that applies a force to a piston. Hydraulic fluid contained between the piston and the end of the cylinder is pressurized by the spring. If the engine is shut off, pressure from the power-steering pump ceases. The pressurized fluid stored in the accumulator is enough to apply the brakes several times with the engine shut off. This safety feature allows the car to be stopped if the engine stalls or the car is rolling with the engine off.

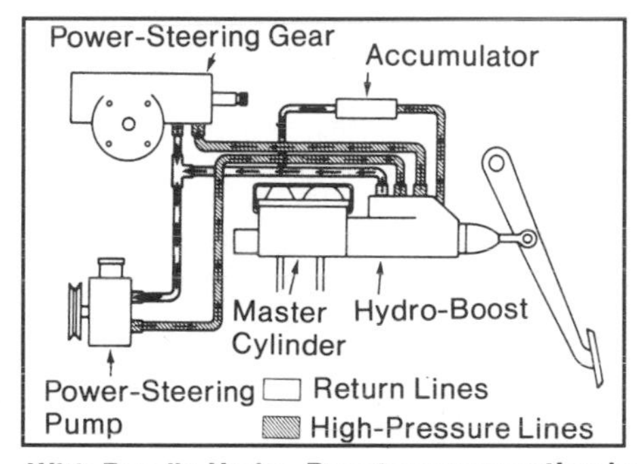

With Bendix Hydro-Boost power section in released position, most fluid from pump is routed through power section to power-steering gear. Some fluid returns to pump reservoir through return port. Spring accumulator stores sufficient fluid under pressure to provide two or three power-assisted brake applications in case fluid flow from power-steering pump ceases. Accumulator is charged with fluid when steering wheel is turned or brakes are applied. Drawing courtesy Bendix Corporation.

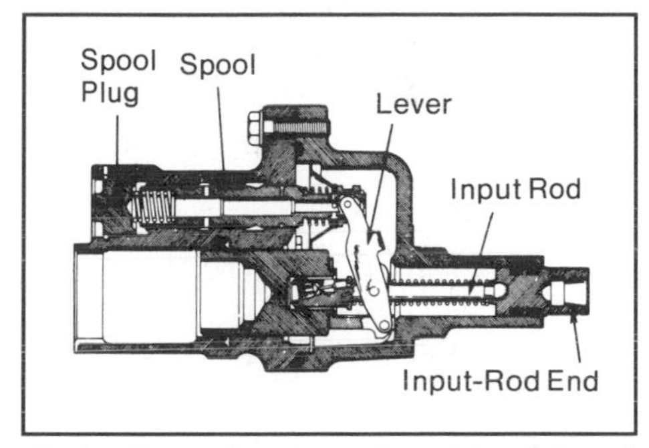

Hydro-Boost in released position (above) and applied position (below): With brake pedal depressed, input rod and piston move forward slightly. Lever assembly moves sleeve forward, closing off four holes leading to open center of spool. Forward movement of spool allows additional hydraulic fluid to enter cavity behind booster piston to pressurize area. This moves piston and output rod forward against master-cylinder pistons to apply brakes. Drawing courtesy Bendix Corporation.

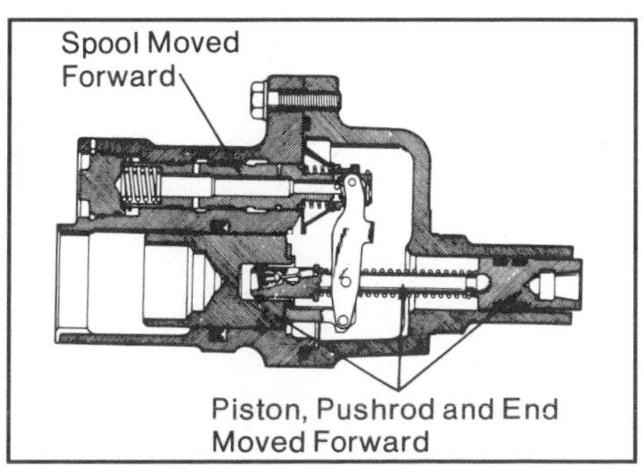

ADJUSTABLE PUSHRODS

Most boosters are designed with adjustable master-cylinder pushrods. This rod should just touch the master-cylinder piston with the brakes released. A long rod will leave the master-cylinder compensating port covered and cause brake drag. A short rod will add slop or excess pedal travel to the system. See your car's shop manual for instructions on adjusting the master-cylinder pushrod.

Other Types of Brakes — 8

Outlawed by most racing organizations, movable air brakes, such as one at rear of this 1955 300SLR Mercedes, are extremely effective at over 100 mph. Frontal area and drag coefficient are increased dramatically when air brake is activated. Photo courtesy Mercedes-Benz.

F_F = Force on Front Tires
F_R = Force on Rear Tires
W_C = Car Weight
F_D = Aerodynamic Drag
R_D = Rolling Drag
F_T = Forward Thrust
F_A = Aerodynamic Downforce

Forces on moving car: Vertical forces on the tires, F_F and F_R, are determined by car's weight and aerodynamic forces. Aerodynamic-drag force, F_D, acts on car near center of body cross section, viewed from the front. Downforce, if any, can act anywhere on body. With this car, downforce is mostly on rear wing.

In this chapter, I briefly describe brakes other than the traditional drum or disc brakes. Also, there are several different ways to operate brakes. And, there are other ways of stopping a vehicle other than friction at the tires. Although briefly discussed, I don't go into details on other brake systems. The subject is too extensive for this book.

AERODYNAMIC DRAG

Drag acts to slow a moving car. Drag forces act toward the rear of the car when it is moving forward. Drag consists of *rolling drag* and *aerodynamic drag.* Rolling drag is caused by resistance of the tires to rolling, friction in the wheel bearings, and all other drive-train friction forces.

Rolling drag or *resistance,* exists at all speeds, but increases with speed. You can experience rolling drag by pushing a car on a level road. You must keep pushing or rolling drag will stop the car. The force you use to maintain speed equals rolling drag.

Aerodynamic drag is caused by air flowing around the car body. The force on the car is zero when air speed is zero. It increases with the *square* of the air speed. This means that aerodynamic drag quadruples if you double car speed. If a car has 100 lb of aerodynamic drag at 30 mph, it will have 400 lb at 60 mph. If speed is increased to 90 mph, aerodynamic drag is increased to 900 lb—9 times the aerodynamic drag at 30 mph. At racing speeds, aerodynamic force on a car can approach the car's weight. On the very powerful land-speed-record cars run at the Bonneville Salt Flats, aerodynamic drag gets so high the tires can spin at high speed trying to overcome it!

Aerodynamic drag is determined by maximum cross-section size and the shape of the car. The cross-section size is called *frontal area*. Frontal area is approximately the height of the car times its width. Frontal area is nearly the same for most road cars, as it is for most race cars in a given class. Shape varies a great deal from one car to another. Reducing drag by improving shape is called *streamlining*. A streamlined shape has low aerodynamic drag for its size.

The shape of the car compared to a standard flat shape is described by a number called the *drag coefficient*. A high drag coefficient means a high-drag shape. A low drag coefficient means a streamlined shape. The worst car shapes have a drag coefficient of nearly 1.0. The most streamlined cars have a drag coefficient of about 0.2. Aerodynamic drag on a car varies directly with its drag coefficient. For example, a car with a drag coefficient of 0.6 has twice the aerodynamic drag at a given speed as the same-size car with a drag coefficient of 0.3.

When the driver lifts his foot off the throttle at high speed, drag on the car will rapidly slow it. Deceleration equals the total drag force divided by car weight. Above 60 mph, most drag is aerodynamic. If a 2000-lb car has a drag force of 500 lb, it will decelerate at 0.25 g when not under power. Because aerodynamic drag drops as speed drops, deceleration drops as car speed drops. Aerodynamic drag is very low at low speeds.

Except for lights atop body, custom Camaro pace car has improved streamlining on an already excellent body shape. Note smooth-flowing lines with a minimum of projections or corners.

Old sports cars are unstreamlined. Body shapes are rough, angular and have lots of parts sticking into airflow. Streamlining of cars refers to shape, not size of body. An unstreamlined car has a higher drag force on it than a streamlined car of the same size.

Following is the formula for the acceleration of an object in g's. Remember that deceleration is negative acceleration.

$$\text{Acceleration} = \frac{F}{W}$$

F = Unbalanced force on object in pounds

W = Weight of object in pounds

Remember that if the unbalanced force acts in a direction to slow the object, that force has a negative value. A negative force makes the acceleration of the object negative.

If an object happens to be a speeding car and the driver lifts off the throttle, the unbalanced force is drag force.

HOW TO CALCULATE AERODYNAMIC-DRAG FORCE

The formula for aerodynamic-drag force is simple:

Aerodynamic drag
$= 0.00256 A_F C_D S^2$ in pounds
A_F = Frontal area of car in square feet
C_D = Drag coefficient of body shape
S = car speed in miles per hour

The frontal area is the cross-sectional area of a car as viewed from the front. If you draw a front-view outline of the car on a sheet of graph paper and count the squares inside outline, that will approximate the car's frontal area. If each square on the paper is scaled at 1 inch on a side, the frontal area will come out in square inches. If you divide the number of square inches by 144, you get square feet.

Frontal area has nothing to do with the shape of the front of a car. Only the maximum cross-sectional size of the car is important. Shape affects the drag coefficient, not the frontal area. Typical drag coefficients for production cars vary from a low of 0.30 to a high of 0.60. The drag coefficients for many cars are given in magazine articles and road tests. Find such an article for your car to determine its drag coefficient.

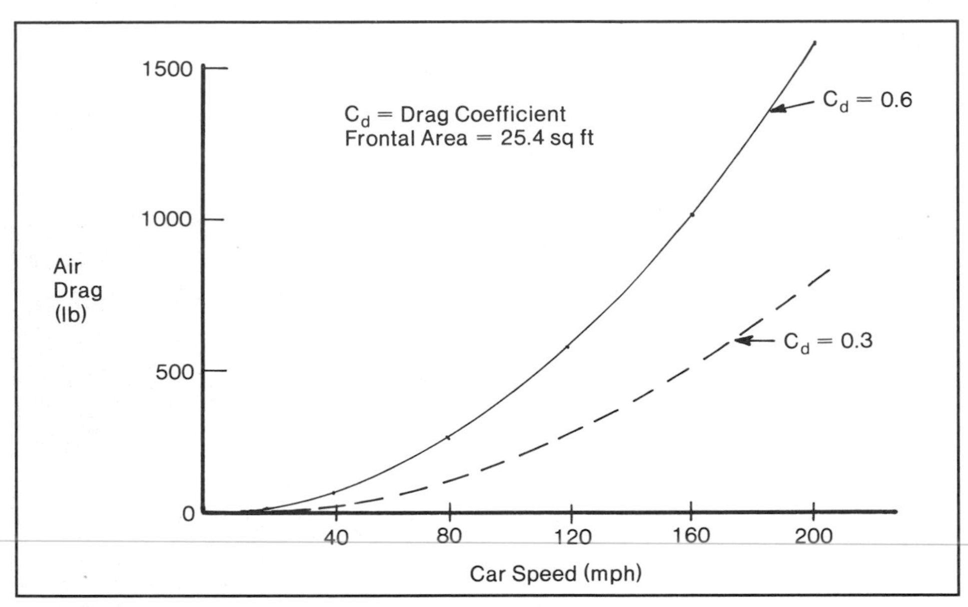

Graph shows how aerodynamic drag varies with car speed and drag coefficient. Solid line is for unstreamlined car with a 0.6 drag coefficient. Dotted curve is for streamlined car with the same frontal area, but half the drag, or 0.3 drag coefficient. Drag figures are typical for a large sedan.

The formula for deceleration can be written another way:

$$\text{Deceleration of a car} = \frac{F_D}{W_C}$$

F_D = Drag force on car in pounds
W_C = Weight of the car in pounds

In this formula, deceleration is in g's. If you know the drag force on a car, you can calculate its deceleration. Approaching it another way, if you could measure deceleration, you could calculate drag force. This is how vehicle drag can be determined without using a wind tunnel.

AERODYNAMIC BRAKING

If a car is streamlined and suddenly increases its drag coefficient for braking while at high speed, it can slow quicker. This is how aerodynamic braking works. Added drag is created by changing the shape or size of the car. The drag-coefficient change has to be large to get a significant effect from aerodynamic braking. Also, the speed must be high. It also helps if the car is light because deceleration depends on weight. Finally, aerodynamic braking is worthwhile only on a race car that brakes at very high speed.

Aerodynamic braking is efficient because there is no heat to dissipate in the brake system. Instead, aerodynamic drag heats the surrounding air as the car passes through it. The wheel brakes can be applied along with aerodynamic braking for even higher deceleration. Deceleration with both brake systems operating is greater than either system can achieve singularly!

If aerodynamic braking is so efficient, why is it used infrequently or only on a few specialized race cars?

- The braking force is useful only at very high speeds—near 200 mph.
- Aerodynamic braking will not bring a car to a dead stop. The braking force approaches zero as car speed nears zero.
- Aerodynamic brakes add weight, complexity and cost.
- Because car frontal area should increase for aerodynamic braking to be effective, the car will occupy more space on the road with the brakes applied. This may make it difficult to drive near other cars.
- In most racing classes, movable aerodynamic devices are illegal.

Aerodynamic brakes have been tried in road racing, such as on the Mercedes sports cars of the mid-'50s and early winged Can-Am Chaparral racers.

Mercedes used a flap-mounted flat on the tail of their 300SLR. This flap was raised hydraulically by operating a control valve in the cockpit, nearly doubling the frontal area of the car. It also made the airflow more turbulent. The combination of the larger frontal and air turbulence more than doubled aerodynamic drag. Because this car exceeded 180 mph on the 3-mile straightaway at LeMans, and had to slow for a 30-mph hairpin turn, the aerodynamic brake took a lot of load off the wheel brakes. Competing drivers complained about a visibility problem when the Mercedes aerodynamic brake was raised. This, however, was probably due to the superiority of the Mercedes rather than a "visibility" hazard.

The Chaparral ran a large wing over the rear wheels, which could be varied in its angle of attack. The wing ran at low angles when the car was accelerating, to give some downforce and to minimize drag created by the wing. When braking, the driver could tip the wing to maximize downforce and drag. The downforce increase on the rear tires allowed more rear-wheel braking. Because this car was highly successful, the racing rules were changed, banning movable aerodynamic devices.

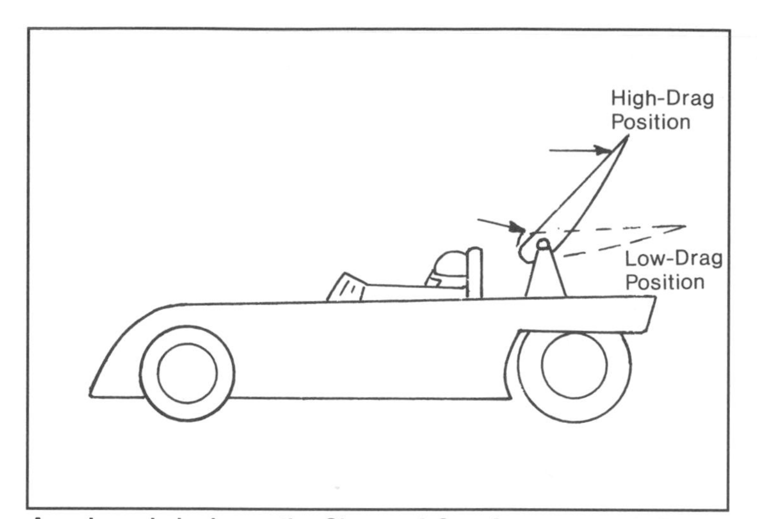

Aerodynamic brake on the Chaparral Can-Am car was similar to this. Wing in low-drag position supplied downforce to rear tires for improved high-speed cornering. When flipped to high-drag position, added aerodynamic drag increased deceleration. Movable wings are illegal in road racing today.

Aerodynamic braking has been used for years in drag racing. Racing chutes stop funny cars much quicker than ordinary brakes. Cars are traveling near 250 mph at finish line. After each run, parachute is repacked. This would not work for cars where brakes are applied more than once during a race.

Aerodynamic braking has found a home in drag racing. All the really fast drag cars, such as top fuelers, funny cars and pro stockers, use a parachute to assist stopping. Because a parachute relies on aerodynamic drag for effectiveness, it is true aerodynamic braking. Obviously, a parachute won't work in road racing or on the street because it has to be repacked after each application. In a drag race, this is no problem because the car makes one stop per run. The chute is repacked by the pit crew prior to the next run.

Some drag cars reach or exceed 250 mph and are very light. If large, the chute applies much more stopping force than the wheel brakes, thus is extremely effective in reducing speed to under 100 mph in a short distance. In fact, the only limit to how fast a car could be slowed with a chute is the inertia force that the driver must withstand when the chute opens.

The critical part of setting up a drag-racing chute is where and how the chute mounts to the frame. If mounted off-center, when opened the chute will cause the car to swerve. Ideally, the mount should be near the CG height, and centered. The maximum force on a parachute mount is many tons, so mount strength is critical. Parachute mounts have ripped off cars, leaving the driver only the wheel brakes to stop with. The mount should be strong enough to lift several race cars from it. Those who sell drag-race parachutes, such as Simpson Safety Equipment Co., can provide instructions on mounting and setting up the system.

Compressed air moves brake shoes by means of an actuating chamber mounted on axle housing. Actuating chamber rotates tube (arrow) through pushrod and lever to move brake shoes against drum. Note lack of backing plate on these truck brakes.

Pneumatic truck brakes are actuated by compressed air supplied by engine-driven air compressor. Air reservoir has emergency supply of air so brakes won't fail if compressor belts break.

PNEUMATIC BRAKES

Pneumatic brakes, commonly called *air* brakes, refer to the actuating system. Pneumatic brakes use compressed air to operate the brakes instead of hydraulics. This type of brake system is used primarily for large commercial vehicles. A truck usually uses pneumatic brakes because there are advantages over a hydraulic brake system.

Pneumatic brakes are actuated by compressed air supplied by an engine-driven air compressor. The compressor supplies high-pressure air to a storage tank, which is big enough to maintain pressure when the brakes are applied.

Compressed air is fed to the brake-actuating system by valves controlled by the brake pedal. If the driver pushes harder on the brake pedal, more air pressure is applied to the brakes. The driver is not supplying the force or air movement, so his strength doesn't matter.

Because trucks are very heavy under full load, power-assisted hydraulic brakes would be necessary. A major advantage of pneumatic brakes is they have built-in power assist. Another advantage is pneumatic brakes are not affected by small leaks in the system. Air is supplied continuously by the compressor, so unless there is a huge hole in the system, there is never a loss of pressure. Instead, the air compressor must work harder to maintain pressure in the storage tank.

The pneumatic system does not use hydraulic fluid, so it is not affected by

Aircraft multi-disc brake: Brake, which is buried in wheel, is good for one stop after landing. Brake must cool before another landing can be made.

HPBooks' very "slippery" Monza at Bonneville Salt Flats. To illustrate effect aerodynamic drag has on speed, Tom Monroe set C/Production record in car at 217.849 mph. Power was by fuel-injected 370-cubic inch small-block Chevy. A/Production record is 219.334 mph, set in a '67 Camaro powered by a 510-cubic inch big-block-Chevy—37% more displacement and 0.68% faster! This is a good-news/bad-news situation when it comes to braking. If two cars with different drag coefficients are traveling at the same speed and stop at the same deceleration rate, car with the lower C_d will require the higher braking force. Photo by Bill Fisher.

the reduced boiling point of old, contaminated fluid. Thus, the system can go much longer between servicing. This is important on long-haul trucks.

The disadvantages of a pneumatic brake system are size, complexity and cost. Also, it takes engine power to run the air compressor. System pressure is much lower than that of a hydraulic system. Thus, the diameters of all the cylinders—master and wheel—are much larger. Trucks have lots of room, so there's no problem fitting large cylinders. But size has a definite disadvantage: It takes time to build up pressure at the brakes.

There is a small delay between the time the driver hits the pedal and the brakes actually apply. Engineers can design a system to reduce, but not eliminate, this time lag.

Compressor cost makes a pneumatic brake system more expensive than a hydraulic system. However, because many commercial vehicles require compressed air to run other systems on the vehicle, the cost of the compressor is justified, considering its double-duty function.

There are other complexities in a pneumatic brake system. The compressor must have a pressure-relief valve to prevent overpressurizing the system. Also, there's a pressure-controlled unloader that takes the load off the compressor once storage-tank pressure reaches about 100 psi. There are also one-way valves to maintain storage-tank pressure if the compressor fails.

Air compressors introduce both water and oil into a brake system. These liquids are removed to prevent the pneumatic system from being damaged. Not only can the parts corrode due to moisture, but they could freeze and cause brake failure. The fluid-removal system is similar to fluid traps in a shop compressed-air system.

Pneumatic brakes are actuated by a diaphragm inside a housing, rather than a piston in a cylinder. These units are the *brake chambers.* You can see them mounted on the rear axles of large trucks and trailers. They are too big to fit inside wheels, so they operate the brakes through a shaft and mechanical linkage. A return spring inside the chamber returns the diaphragm to its retracted position when the brakes are released.

On some large off-road commercial vehicles such as earth movers, graders and road rollers, a combination hydraulic/pneumatic system is used. Referred to as an *air-over-hydraulic system,* the brakes are operated by a hydraulic system, but the hydraulic system is applied by a pneumatic system between the brake pedal and the hydraulic system. This is a type of power assist that uses air pressure rather than vacuum.

MULTIDISC BRAKES

A multidisc brake works like a multidisc clutch. There are a number of *driven* and *stationary* discs in a stack. Friction material is between the discs. The driven discs rotate with the wheel. When the brakes are applied, the stack is clamped together. The stationary discs, because they can't rotate, slow the rotating discs by friction.

Multidisc brakes are small and light compared to other types of brakes that absorb the same energy. However, because the rubbing surfaces are all "buried" inside the unit, multidisc brakes cool slowly. Thus, a multidisc brake is good for only one hard stop. Considerable cooling time is needed before the next stop is made.

The compactness of a multidisc brake makes it desirable for aircraft use. An airplane has small wheels to save space, making a tight fit for any brake. A plane only stops hard once as it lands, and has a great deal of cooling time before another landing and stop. And, the extremely high temperatures generated in aircraft brakes dictate the use of sintered-metallic or graphite-composite friction materials.

Aircraft-type multidisc brakes might work well on dragsters or land-speed-record cars, but to my knowledge they have yet to be used on such vehicles.

Dragster brake is operated by hand lever. Note pushrod between lever and master cylinder (arrow). Hand-lever movement can be more than a brake pedal, so higher pedal ratio is possible with lever-operated brakes.

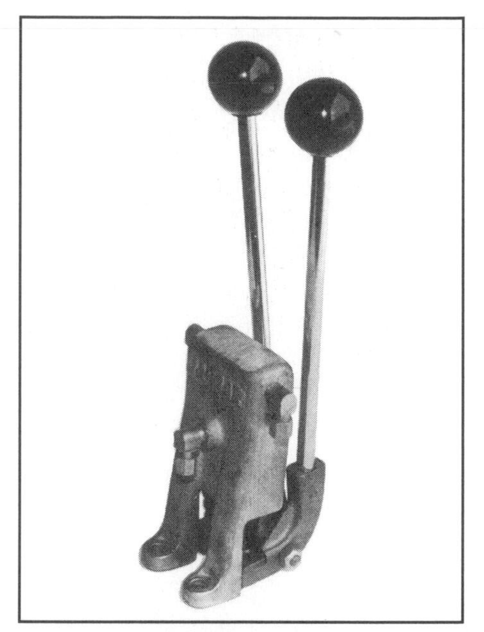

Ja-Mar cutting-brake assembly is used on sand buggies for steering control in soft sand. One operates one master cylinder when pulled; other lever operates opposite cylinder. Each master cylinder is designed to operate a single rear brake.

HAND-OPERATED BRAKES

Some specialized cars use a brake system operated by a hand lever rather than a brake pedal. A dragster is one of these. The dragster hand-lever brake operates identically to a foot-pedal brake, but with two differences:

- Maximum force with a person's hand is less than can be applied with a foot.
- Hand-lever stroke can be longer than a foot pedal, allowing a higher ratio.

Like a foot-brake pedal, the hand lever operates a master cylinder through a pushrod. Because dragsters have rear-wheel brakes only, a single master cylinder is used. Most of the stopping is done by the parachute. The wheel brakes are used at low speeds, particularly for holding the car on the starting line.

Another type of hand-operated brake is found on dune buggies. Known as a *cutting brake* or *steering brake,* it has one or two levers with two separate master cylinders to operate the right or left rear wheel.

Cutting brakes are used for steering the buggy in soft sand or tight turns. The front tires often do not have sufficient bite for steering control in these situations. Because rear tires have more bite, they can turn the buggy quickly when one wheel is braked. A conventional brake pedal is used for braking the buggy during normal operation.

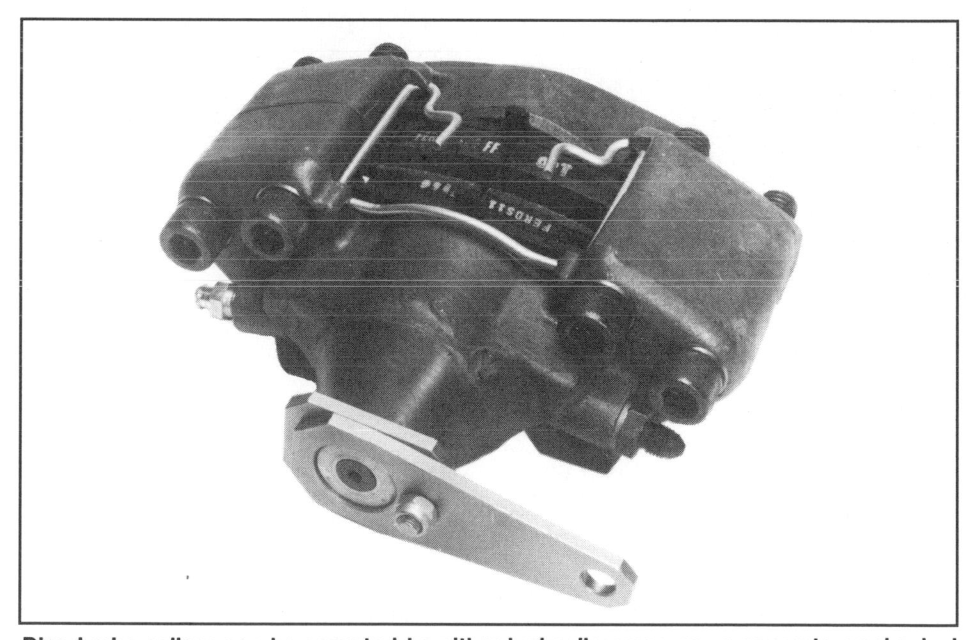

Disc-brake caliper can be operated by either hydraulic pressure or separate mechanical linkage. Used on competition rally cars at the rear wheels and operated by a mechanical linkage by a parking-brake lever, driver can pull lever to slide out rear wheels for entering a turn. This allows ultra-tight turns to be negotiated quickly in dirt or on slick surfaces. Courtesy AP Racing.

Hand-operated rear-wheel brakes also are found on competition rally cars. Unlike a cutting brake, rear-wheel rally-car brakes are operated simultaneously with one hand lever/master cylinder. Rally cars run on slick roads and often have front-wheel drive. It is handy for the driver to brake only the rear wheels to throw the car into a deliberate skid. This helps to negotiate tight slippery turns where steering control might be marginal. Separate from the hydraulic system, special mechanical linkages are often used for the rear brakes. This is similar to parking-brake linkage, but is designed for severe use.

High-Performance Brakes 9

Design of modern race-car-brake system is not arrived at through magic. Instead, carefully planned steps are required. Success starts with a written statement of what brake system must do. This is your *design criteria.* Don't lose sight of it when you're deep into the design.

If you are designing a special car for either street or racing use, you'll also have to design the brake system. It's difficult to take a complete brake system from one car and use it on a different one and have it work properly. Car weight, configuration and performance all affect brake-system design.

By brake-system design I mean putting together a group of components that work properly. I'll not attempt to discuss how to design a brake rotor or proportioning valve. Instead, I will show how to make the necessary calculations, plan the entire system, and select the components to do the job. The correct components already exist, regardless of the car. The problem is selecting the right ones for your car.

This chapter covers complete brake-system design. If you have an existing car and are trying to improve the brakes, read Chapter 12. However, if you are designing the car rather than modifying it, you have more choices. There are fewer ways you can easily change an existing car. This is why the design should be carefully planned before buying or building parts.

Brake-system design requires several steps. There are many ways to approach the design process, but here are my suggestions:

- List design criteria.
- Calculate forces on tires.
- Calculate brake torques at maximum deceleration.
- Choose brake type and mounting location.
- Determine hydraulic pressure required.
- Choose pedal-and-linkage design.
- Calculate master-cylinder diameter required.
- Design rest of brake system.

There may be some reason to rearrange these steps. For instance, you may already have the pedals and wish to retain them. That'll work fine unless you have to compromise brake-system performance to do it. But, if existing components won't give good brake performance, don't use them. The brakes are too critical to compromise because of a few extra dollars.

DESIGN CRITERIA

The first step in any design process is listing the reasons for the design and the objectives. This written document is your design *criteria.* For a brake system, several sheets of paper should be enough. You are writing the criteria so you can refer to it later. Keep your design-criteria list handy and read it often during the design process. Don't lose track of your objectives.

A brake-design criteria should answer the following questions:

- Under what conditions will the brake system have to operate?
- How long do you expect the system to last?
- Is car to be used for racing, street or both?
- What performance specifications should the brakes meet?
- If you have to compromise on

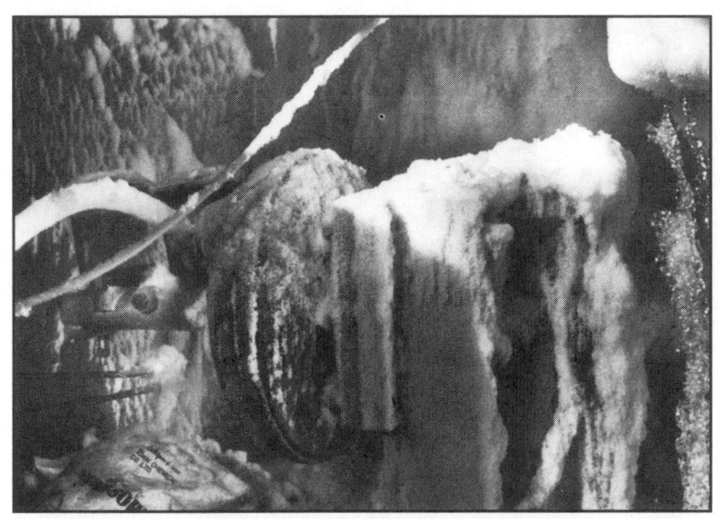

Road cars must have brake components that will withstand extreme cold. Corvette master cylinder and booster are being tested at 58F below zero (-50C). Road-car brake manufacturers must test components under extremely severe conditions. If designing a special car, you can save money by limiting conditions car has to operate under. Photo courtesy Girlock Ltd.

Easiest car to design is one with a special purpose—win sports-car races, in this case. If car also had to transport wife and kids to church on Sunday, and still win races, race-car designer would really have a tough job. Be realistic, but limit functions car must perform—design job will be easier.

specifications, what is most important? What is least important?

- What are maximum loads imposed on system? What are normal operating loads?

Operating Conditions—Think twice about this one: It is much harder to design a brake system to perform under every possible operating condition than it is to limit its use. For instance, do you want the brake system to function in the rain? Will it get used in sand dunes like an off-road vehicle? Is it used strictly for drag racing, where there's a long cool-down period after each stop? Will the car be used on icy, salted roads?

Be specific and practical about the conditions your vehicle will likely encounter. The more operating conditions you can eliminate or minimize, the easier and cheaper the job will be. The most difficult job an engineer has to do is design a car that will be driven on the street, in races, off-road and in weather conditions ranging from the burning deserts to sub-zero cold. With luck, your special car will have specialized operating conditions for the brakes. This will eliminate some extremes.

Design Life—Every car component or system has a certain life assumed when the engineer designs it. Most road-car parts are expected to last at least 50,000 miles, but often last twice that.

If every car were designed to last 40 years, the cars would be too expensive and much heavier. If something is designed for a long life, it costs more. So restrict long-life items to those that are safety-related. Because brakes are safety-related, the structure of the brake system should be extremely durable. You choose the miles, time or number or races. If you are using passenger-car brake parts and will be operating them at less than their original design loads, you can expect them to last a long time.

Some brake-system parts wear out quicker than the basic system. This includes the linings, seals and, sometimes, the drums and rotors. Parts that wear should be designed to do so in a safe manner—so they don't cause a loss of braking. Conventional brake systems are so designed. Therefore, if you use conventional design practice, the brakes will be safe when parts wear out.

What life do you expect wear-prone brake parts to have? If the linings you are using last 10,000 miles on a 2500-lb sedan, you can expect them to last at least that long on a 1500-lb sports car driven the same way. Use comparisons to predict useful life. However, if you drive your special car harder than the sedan that weighs the same, don't expect the linings to last the same. More brake-lining area or a harder lining would be needed to compensate.

Always consider the application: If stock-car designer failed to consider that car might compete in road races with right turns, driver would be in trouble. It is easier to design a car for left turns only. Likewise, it's easier to design brakes for a short track with high-banked turns than one with long straights and flat turns.

You cannot make an exact prediction of brake life. However, write down what you are expecting and keep the desired life in mind while designing the brake system. You must have a design goal.

Racing vs. Street Use—This decision must be made early in the design process. Make usage part of your criteria. The brake system that works

Production-sedan designers have the most difficult design job. I wonder if the VW brake-design engineers visualized Rabbits racing at Riverside on a hot summer day? Brakes get a severe workout compared to their everyday highway service.

Maximum loads on car may occur during an off-road excursion or some other emergency situation. Normal operating loads don't include unexpected trips in the rough stuff, but such conditions must be allowed for in a design, including driver error.

SAMPLE PERFORMANCE SPECIFICATIONS

	Car A	Car B
Type of car	Sports Car	Race car
Usage	Street & slalom	Road racing
Percent racing use	5% or less	100%
Horsepower	200	150
Car weight (lb)—average racing	2000	1100
Weight distribution (front/rear)	50/50	40/60
Maximum car weight (lb)	2400 (street)	1150 (max fuel)
Minimum car weight (lb)	2000 (racing)	1050 (min fuel)
Maximum speed (mph)	125	150
Average race speed (mph)	60	100
Maximum race duration (minutes)	5	60
Acceleration 0-60 mph (seconds)	8	6
Maximum tire grip—street	0.9	—
Maximum tire grip—racing	1.2	1.4
Maximum deceleration—street (g's)	0.9	—
Maximum deceleration—racing (g's)	1.2	2.0 (with wings)
Downforce at top speed (lb)	0	500

Performance specifications for two imaginary cars: Car A is a sports car designed for both street use and slaloms; car B is an all-out race car for a small formula-car class. Note how road-car's weight can vary. Brake system must be able to accommodate all variations, but remain balanced.

best for racing is not the best for the street and vice versa. Either system will work OK for a little while in either application, but you will have problems sooner or later.

Remember that while racing, the brakes are usually warmed up before a hard stop; they are not on the street. Street-driving conditions require protection from dirt and corrosion; racing conditions may not. Some forms of racing are much worse than street driving for both contamination and high temperatures.

The hardest thing to design is a car that is used for both street driving and racing. If you attempt this, you must make design compromises that will hurt the car's performance in both situations. Therefore, try to limit the car's operating conditions, for highest performance and lowest cost.

Performance Specs—List specifications you want to meet. Because each type of vehicle has different specs, I can't give you a list for your car. However, the above table gives a sample listing for two types of cars. Note that car weight and speed are included. Obviously, it takes more brake to stop a fast, heavy car than it does to stop a light, slow one.

Include car weight and CG location in the performance specifications. You must know these before you can design the brake system. When starting a car design, you may not know the exact weight or CG location. To start the brake-system design, estimate the weight and CG location for a car such as yours. Then you can begin your brake-system design.

After car and brake system are designed, go back and do all the calculations again using the exact car weight and CG location. If you estimated correctly the first time, few brake-design changes should be required. However, it usually takes two or three rounds of design effort before an entire car design is completed.

BRAKE-SYSTEM LOADS

As already discussed, the two loads to be concerned with are *maximum loads* and *normal operating loads.* Maximum loads on most brake components occur during an emergency. It could be a panic situation where a crash may be about to occur. The driver panics and slams his foot on the brake pedal as hard as he can, usually locking the wheels. This may result in loss of control or worse. It is the maximum-loading *condition* for the brake pedal and linkage.

There may be different maximum-load conditions for parts of the brake system other than the pedal and linkage. A brake-reservoir bracket experiences maximum load when the

Formula car ran head-on into barrier with brakes locked. Forward-projecting master cylinders pushed bulkhead rearward and broke off brake pedal due to resistance of driver's foot. Because of severe leg injuries, '85 Indy Car rules prohibited forward-mounted master cylinders. Later rules prohibit pedals mounted forward of front-wheel center line.

MAXIMUM LOADING FOR BRAKE-SYSTEM COMPONENTS

Component	Maximum-Load Condition	Maximum Load on Component
Brake pedal & linkage	Panic stop—wheels locked	600 lb forward & 200 lb side load on pedal pad.
Hydraulic system	Panic stop—wheels locked	Calculated hydraulic pressure with 300 lb on pedal.
Brake caliper	Panic stop—wheels locked	Calculated hydraulic pressure with 300 lb on pedal.
Caliper bracket	Maximum deceleration	Maximum calculated brake torque.
Brake rotor	Maximum deceleration	Maximum calculated brake torque.
Fluid-reservoir bracket	Maximum bump & cornering	5-g downward load & 2.5-g side load on reservoir.

Exact values and loading conditions will vary from one car to another. Make sure you select the worst loading condition for each component—that is, the condition most likely to cause a failure. Notice that the worst condition is not the same for each component in the brake system. Hydraulic-system components are designed to withstand 300 lb on the pedal, rather than the 600-lb load the pedal is designed for. The pedal load includes a factor for sudden load application, which does not apply to the hydraulic system.

car hits a bump. Assume that the driver runs off the road and hits a hard bump. The bracket should be designed for similar maximum-load condition as the suspension and frame. This maximum-load condition is different than the maximum-load condition for the brake pedal. See the sample list of maximum loads in the above table. Write down the maximum-load conditions for all system parts and calculate the loads.

For all parts of the system receiving loads from the brake fluid, the normal operating load is during maximum-deceleration. Slamming your foot on the pedal and locking the wheels does *not* give maximum-deceleration. In a maximum-deceleration stop, the wheels are not locked, but the brakes develop maximum torque. The driver pushes less on the pedal than he is capable of. This is a normal stop in a race or a *controlled* fast stop on the road.

All brake-system parts must be made to *operate properly* at normal operating loads. And, they must *not break* while under maximum load.

Maximum Loads—Each car and car designer will have different requirements for what loads to use for designing the brake system. I use the following when designing a race car:

75-lb pedal force at maximum deceleration—normal operating condition;
600-lb forward and 200-lb sideways maximum force on pedal—panic-stop conditions.

Factor of Safety—The maximum loads discussed in the previous section include the effects of suddenly applied forces. However, they *do not* include a factor of safety. As previously discussed, the safety factor accounts for the *difference* between the maximum load on a part and its breaking strength. Never design a part to break just as it reaches its maximum load. Not even an airplane design is that marginal.

The safety factor accounts for unknowns that cannot be calculated. One is needed because parts sometimes have hidden flaws, they corrode after long use, and loads are often difficult to calculate accurately.

I recommend using a factor of safety of 3 on critical components. This means that the breaking strength of the part should be at least three times the maximum load on the part. You can test a part to see if it is strong enough, if you can afford to destroy one part. In a breaking test, apply three times the maximum load the part was designed for. If it doesn't break, you have a good design. But, destroy the part after the test and use a brand new one in the car. The test may have caused damage. Remember, you tested it to *three times* the highest load that it should ever experience.

Loads—Brake-system components also are loaded inertially when the car is accelerating, decelerating, cornering or hitting bumps. Inertia loads are expressed in g's. One g is an inertia load equal to the weight of a part, assembly or car; 2 g is a load equal to two times the weight, and so on. I use the following loads to design brackets and structures that hold parts together:

- Maximum bump load = 5 g except for off-road cars or dirt-track racers. Use 10 g or more for race cars that normally hit large bumps.
- Maximum cornering load = 2.5 g on a car without wings or at speeds below 100 mph. Cars with wings above this speed will have higher loads.
- Maximum deceleration = 2.5 g on a car without wings or at speeds below 100 mph. Cars with wings above this speed will have higher loads.
- Maximum acceleration = 2.0 g on all cars except dragsters. Use 5.0 g or higher for dragsters.

On small parts such as brackets holding other parts to the frame, vibration must be considered. A small, lightweight part is subjected to higher vibration loads than a large, heavy part. The amount of vibration depends on engine design, suspension stiffness, engine mounts, road

Maximum loads can occur in two directions at once. Here, stock-car driver cuts a close apex. Resulting bump load from going over curb on right front wheel occurs at same time as maximum cornering load.

Toughest type of vehicle to maintain brake balance under all loading conditions is nose-heavy, front-wheel-drive pickup. Load-sensing proportioning valve between body and rear axle of Dodge Rampage adjusts brakes so brake balance is maintained under all load conditions.

surface, and many other factors. If in doubt, design small parts to withstand twice the maximum loads just given. If anything falls off the car, increase its design load when doing the repair. The only way to measure vibration loads is with expensive sophisticated equipment, which most people can't afford.

CALCULATING TIRE FORCES

To start a brake-system design, first determine forces acting on the tires during a stop at maximum deceleration. This includes the effects of both weight transfer and aerodynamic forces on the car. To do the job right, you need to know some specific facts about the car:

- Weight of car including driver and other loads.
- Wheelbase length.
- CG height and fore-and-aft location.
- Maximum tire grip.
- Aerodynamic forces on body.

A car has to be almost completely designed before this information can be determined accurately. Therefore, all the above information may not be available at the time the brake system is first designed. Consequently, you must make assumptions—educated guesses. Start by making the brake system adjustable if you don't start with accurate information. Later, during car testing, you can make adjustments to the brake system.

Weight—The weight of a car greatly affects the brake-system design. The car will have to stop in its heaviest form, traveling at maximum speed. If the car is for racing, the calculation is easy. Most race cars have nearly a constant weight except for fuel weight. So, design race-car brakes assuming full fuel load.

On a road car, weight can vary considerably. The hardest vehicle to design brakes for is a truck. Trucks weigh much more fully loaded than when empty. In addition, their centers of gravity move greatly when loaded. The design of a brake system for such a vehicle should be based on its maximum weight. However, brake balance must be considered at the no-load/full-load extremes. Unless the truck has an adjustable brake balance or a special anti-skid device, balance will probably be incorrect when empty. Most trucks lock the rear wheels under hard braking if reasonably well balanced when loaded.

CG Location—CG location can be estimated or calculated with reasonable accuracy. Most sedans are slightly nose-heavy when empty and closer to 50/50 weight distribution when fully loaded. This puts the CG at the midpoint of the wheelbase at full load. The majority of road sports cars also have 50/50 weight distribution. Most race cars are tail-heavy, with drag racers being the most tail heavy and stock cars the least. You can get a good feel for weights and CG locations from automotive-magazine road tests.

CG height is harder to guess at. Most sports-car CGs are slightly above wheel center. A race-car CG is lower than on a road car. If you know where all the heavy parts of the car are located, you can calculate CG height with reasonable accuracy. But, if you don't know where they are, it is probably too soon to design the brake system. Establish the car design first, then come back to the brake system.

As illustrated, simple chart can be used to calculate CG location. Find the weight of each major component of the car and its planned CG location measured from the ground and distance from the front-axle center line.

Multiply the weight times a distance—as long as it's from the same point—and write the number in the chart. If total car weight seems too low, you left out something. Miscellaneous small items are hard to estimate. However, try to estimate their weight when designing the brakes. Also, don't forget liquids and people. The sum of the individual weights is obviously total car weight. The sum of the weight times distance column divided by total weight gives CG location.

Tire Grip—Tire grip is important. It determines maximum deceleration of the car. On a car without aerodynamic downforce, maximum deceleration of the car *equals* maximum tire friction coefficient. On a car with aerodynamic downforce, deceleration is higher.

Tire grip varies considerably. Road condition is as important as the tires themselves. Racing tires on dry pavement can have a grip of over 1.0.

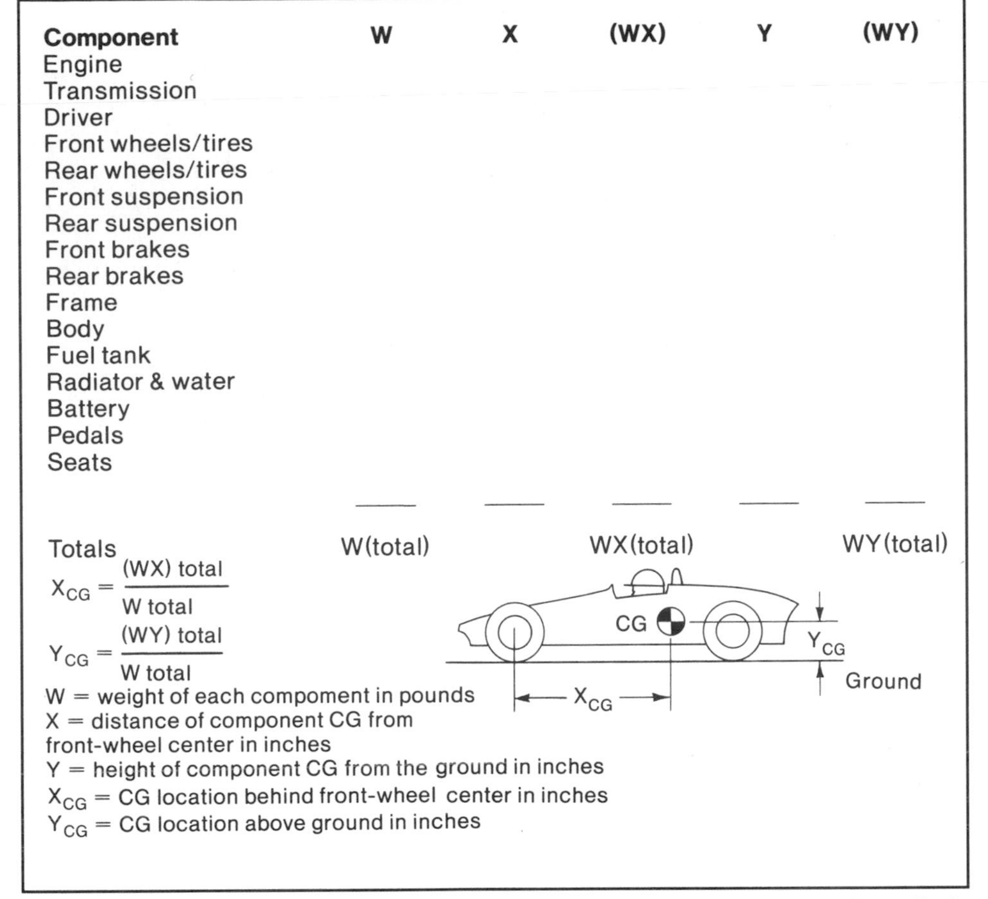

Component	W	X	(WX)	Y	(WY)
Engine					
Transmission					
Driver					
Front wheels/tires					
Rear wheels/tires					
Front suspension					
Rear suspension					
Front brakes					
Rear brakes					
Frame					
Body					
Fuel tank					
Radiator & water					
Battery					
Pedals					
Seats					
Totals	W(total)		WX(total)		WY(total)

$X_{CG} = \frac{\text{(WX) total}}{\text{W total}}$

$Y_{CG} = \frac{\text{(WY) total}}{\text{W total}}$

W = weight of each compoment in pounds
X = distance of component CG from front-wheel center in inches
Y = height of component CG from the ground in inches
X_{CG} = CG location behind front-wheel center in inches
Y_{CG} = CG location above ground in inches

CG location can be determined by filling out chart and performing calculations. Each component in a car has its own weight and CG location. List weight and CG location for each component as shown. If component is forward of front-wheel center, use a negative number for X—in this case W times X will be negative. When adding totals for WX, make sure you add negative numbers correctly or WX total will be too large.

Road tires seldom exceed 0.8. You'll have to choose the tires before you can estimate maximum grip. If you guess too high, weight transfer will be too high.

Aerodynamic Forces—Most race cars decelerate quickly at high speed because of aerodynamic downforce from wings and the shape of the body. Downforce varies with car design and speed. On cars that have huge wings or ground effects, downforce may exceed the weight of the car! Consequently, you should know something about downforce before you can design a brake system. First, the supplier of your wing(s) should be able to give you downforce information. However, on a ground-effects car, testing in a wind tunnel is usually the only way to get this information. You can't estimate anything by just looking at the car.

At highway speed, aerodynamic forces are small. So ignore aerodynamic forces on a road car unless you have accurate information. Again, be ready to change the brake system during testing if things don't work as you estimated.

In addition to the amount of aerodynamic force, you must know *where* this force acts on the body. The point where the center of all aerodynamic forces acts is called the *center of pressure.* On a wing, the center of pressure is roughly a fourth of the way back from the front edge of the wing. On a car body, it could be anywhere. If you don't have any information, assume the center of pressure is at the center of the body—not the wheelbase. Designing a ground-effects car without testing to find the center of pressure may cause big trouble. Many handling problems on ground-effects cars have been caused by a center of pressure being in an unexpected place.

Add Up the Numbers—Now, with all the estimated weights or loads, calculate tire forces. To make life simple, assume that friction force on a tire equals vertical force times its grip. For a tire with 1000 lb of vertical force on it and a grip of 0.8, friction force is 800 lb. If the grip were raised to 1.1, that tire could deliver 1100 lb of friction force. The friction force is what stops the car—it should be as high as possible.

See drawing, page 94, that shows the forces acting on a car during braking. For this calculation, I assume the right and left tires at both ends of the car are loaded equally. This assumption is more accurate than other data usually used for this calculation. The formulas with the drawing are used to calculate tire forces. If you are assuming some aerodynamic lift, put a negative number into the formula for downforce.

When designing a car to run on various surfaces or different types of tires, change the grip value accordingly. Then perform a calculation for each grip. Grip might be as low as 0.3 for a slick surface to perhaps 1.4 for a hot racing tire on a dry track. Resulting answers will be the range of tire forces you must deal with.

CALCULATING BRAKE TORQUE

Brake torque is the friction force on the tire, multiplied by the rolling radius of the tire. For design of the brake system, *always use maximum* brake torque. Maximum torque occurs when:

- Vertical force on tire is maximum.
- Tire grip is maximum.
- Largest-radius tires are used.
- Car is stopped at maximum possible deceleration.

If a car has a nut on the center of each wheel, you could measure brake torque with a large torque wrench. While pushing on the brake pedal with an effort needed to give maximum deceleration, turn the wheel with the torque wrench and measure torque *as the wheel turns.* Even though this test is done rarely, this example illustrates what brake torque means. A brake system must be able to deliver the needed brake torque. Otherwise, the driver won't be able to stop the car at maximum deceleration with an acceptable pedal effort. He may not have the strength to achieve maximum deceleration!

Calculate maximum brake torque using the formulas given. Use the values of grip, car weight and aerodynamic force that give maximum friction force on the tire. Calculate

Modern cars with wings and ground effects generate much more downforce than ordinary cars. Indy Car cornering and braking loads are much higher than for a modified road car. You must know speed and downforce to correctly calculate loads generated by tires.

Design of ground-effects car should be left to experts. Wind-tunnel testing is the only sure way to determine aerodynamic loads on car. Note ground-effects exit tunnels at rear of car. Photo by Tom Monroe.

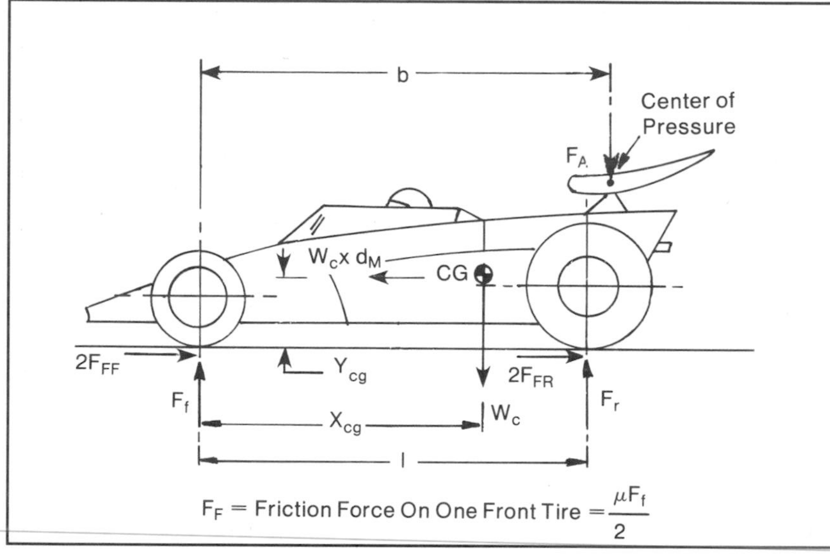

l = wheelbase of car (in.)
W_c = total weight of the car (lb)
F_f = vertical force on both front tires (lb)
F_r = vertical force on both rear tires (lb)
F_A = aerodynamic downforce (lb)
μ = grip (coeffcient of friction) of tires
d_M = maximum deceleration (g's)
b = distance from front axle to center of pressure (in.)
X_{cg} = distance from front axle to CG (in.)
Y_{cg} - height of CG above ground (in.)

$$F_f = W_c\left[1 - \frac{X_{cg}}{l} + \frac{\mu Y_{cg}}{l}\right] + F_A\left[1 - \frac{b}{l} + \frac{\mu Y_{cg}}{l}\right]$$

$$F_r = W_c + F_A - F_f \qquad d_M = \mu + \frac{F_A \mu}{W_c}$$

F_F = Friction Force On One Rear Tire = $\frac{\mu F_r}{2}$

If you can estimate or calculate all data shown on drawing, you can calculate forces on tires. From tire forces, friction forces are computed. Friction forces are then used to calculate brake torque during maximum deceleration. It is assumed F_A is the only aerodynamic downforce.

both front- and rear-tire forces.

Maximum brake torque is calculated by the following formulas:

Front-brake torque = $F_{FF} r_{FT}$ in inch-pounds
F_{FF} = Friction force on front tire in pounds
r_{FT} = Rolling radius of front tire in inches

Rear-brake torque = $F_{FR} r_{RT}$ in inch-pounds
F_{FR} = Friction force on rear tire in pounds
r_{RT} = Rolling radius of rear tire in inches

Rolling Radius—The *rolling radius* of a tire is the distance from the center of the wheel to the pavement *with the tire loaded.* You cannot get an accurate rolling radius by measuring an unloaded tire. The tire deflects when supporting a car.

Rolling radius depends on the load on the tire, tire size, inflation pressure and stiffness of the tire casing. So measure rolling radius with the car on the ground. Make sure car weight is the same as yours will be. Also check inflation pressure; correct it if necessary.

SELECTING & POSITIONING BRAKES

Take the following into consideration before selecting brakes:

- Disc or drum brakes.
- Brake diameter.
- Brake mounting—inboard or outboard.

Each decision must be made before selecting a brake.

There are a number of brake sources. Several manufacturers have brakes for special applications. Or, you could use standard road-car brakes. Your choice depends primarily on the intended use. For instance, is your car for street use or racing? If it's a race car, use brakes designed for racing. Brakes from a road car are usually too heavy for best racing performance. And racing brakes on a street car may be an unnecessary expense.

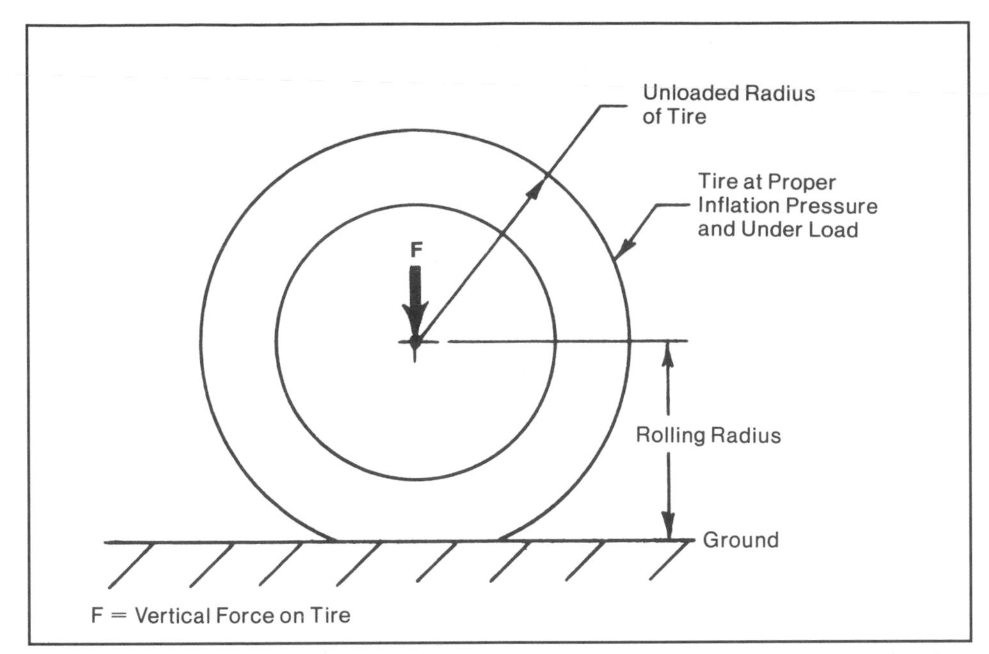

Rolling radius of tire is distance from wheel center to ground with car weight on tire. Tire must be at correct inflation pressure.

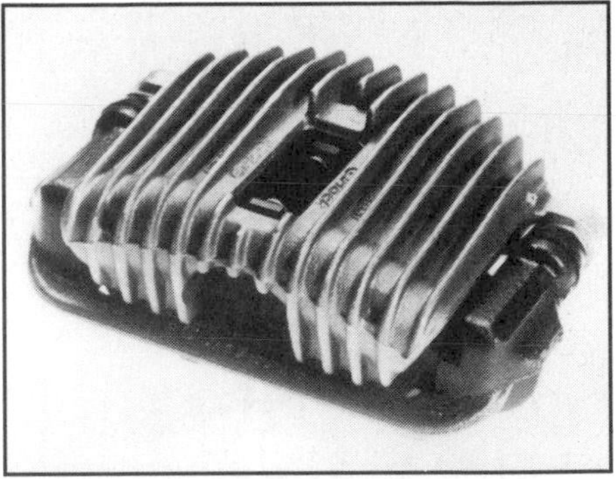

While most road-car brakes are too heavy, some are worth considering for high-performance use. Cast-iron 1984 Corvette sliding caliper was designed more for high performance than low cost. It may be a natural for road car with 4-wheel disc brakes. Photo courtesy Girlock Ltd.

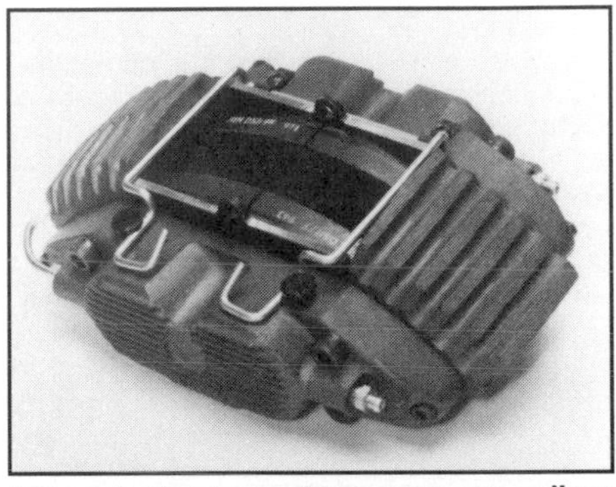

Although fixed, aluminum race-car caliper appears similar to Corvette caliper in above photo. AP CP2751 caliper was designed for IMSA road racers using 13-in.-diameter rotors. These require minimum 14-in.-diameter wheels or inboard brakes. Photo courtesy AP Racing.

Disc Brakes or Drums?—Because disc brakes are usually superior to drum brakes, most people choose discs for high-performance applications. For racing, there is no other choice if you want to win. Drum brakes have many limitations. You will need a special reason to use drum brakes for racing—such as rules *requiring* them.

On a road car, a combination of disc and drum brakes may be best. Drums may be standard on the rear axle you will use. Also, road cars must have parking brakes—drums are usually set up for parking brakes. Although disc brakes on the rear are good for a road car, they may introduce added complexity and cost. A disc brake that's not designed for a parking brake may be difficult to adapt.

Now that I've posed the disc/drum-brake question, I'll assume you are using disc brakes on all four wheels. Most high-performance cars are so designed. If you choose drum brakes, the design procedure is similar. The difference is that you'll have to consider the servo action of a drum brake. Servo action is discussed in Chapter 2.

Brake Diameter—Disc-brake diameter is the outside diameter of the rotor. Choose the largest-diameter brakes that will fit in the wheels, if the brakes are mounted in the wheels, not inboard, page 98. The only exception to this is on a very light car or one with large-diameter wheels. Most high-performance cars benefit from the largest possible rotor that will fit.

Maximum rotor diameter is about 3-in. smaller than wheel diameter. For example, a 10-in. rotor should go inside a 13-in. wheel or a 13-in. rotor in a 16-in. wheel. Caliper design determines how big a rotor can be used for a given wheel.

For pavement racing, use a large-diameter wheel with a low-profile tire rather than a smaller wheel and a high-profile tire. With a given rolling radius, the large wheel allows the use of a larger brake rotor; and the low-profile tire corners better.

Disc-Brake Selection—Rotors and calipers should be selected as a matched set. By using a caliper and rotor designed to work together, you don't have to worry about a dimensional mismatch between the two. So choose the caliper and rotor intended for the same road car or from a race-car-brake manufacturer.

Calipers and rotors must meet the following requirements:

- Rotor must be massive enough to absorb heat from at least one stop from maximum speed at maximum vehicle weight.
- Rotor and caliper must fit mounting space on car.
- Brake must deliver maximum brake torque required without excessive fluid pressure.

Race cars generally use largest possible brake diameter. On this Indy Car, caliper almost rubs against inside of wheel.

One-Stop Temperature Rise Road-Car Rotors—Rotor weight must be high enough to absorb all kinetic energy stored in the moving car. The worst condition is stopping a car traveling at top speed with the car at its maximum weight. Some road cars have brakes that probably don't meet this requirement. This is probably be-

Exception to the rule: Brake is considerably smaller than wheel allows on front of sprint car. Demands on brakes of car raced on dirt are much less than if raced on pavement.

Garage area or pits is best place to look at and photograph race-car brakes. Take advantage of years of expensive trial-and-error development to select rotor sizes and designs. Talk to the mechanics. If they have brake trouble, consider their problems during your brake design. Indy Car brake was photographed at Phoenix International Raceway—one-mile paved track that puts high demands on brakes. Photo by Tom Monroe.

cause some road-car designers think it's unreasonable to drive at the top speed of a car at maximum weight and suddenly do a hard stop. I don't think it's unreasonable, even though driving at top speed is illegal and unsafe on the road; but some people do it.

From Chapter 1, the kinetic energy of a moving car is determined as follows:

$$\text{Kinetic energy} = \frac{W_c S^2}{29.9}$$

in foot-pounds
W_c = Weight of car in pounds
S = Speed of car in miles per hour

After calculating the kinetic energy of a car, determine how much energy goes to the front brakes and how much to the rears. A good guess is to proportion kinetic energy according to the brake torques. This relationship is only approximate, but the calculation for rotor weight does not have to be exact. Use the following formula:

$$K_F = \frac{K_T T_F}{(T_F + T_R)}$$

K_F = Kinetic energy to front brakes in foot-pounds
K_T = Total kinetic energy of car in foot-pounds
T_F = Front brake torque in inch-pounds
T_R = Rear brake torque in inch-pounds

$$K_R = K_T - K_F$$

K_R = Kinetic energy to rear brakes in foot-pounds

The kinetic energy calculated is for a *pair* of brakes—two front and two rear. Divide the answers by 2 to get kinetic energy absorbed by each brake.

Now, compute the temperature rise of the rotor after one stop using formula on page 11. Use weight of single rotor or drum for W_B. This is the average temperature of the entire rotor, not the surface temperature.

If the average rotor temperature is near 1000F (538C) use a heavier rotor. This means a thicker rotor if you have chosen the largest rotor diameter possible. The type of brake linings used will determine how hot the rotor can be and still have the brakes work.

Race-Car Rotors—If you are designing a race car, and the brakes are used many times during a race, a calculation for one stop will not mean much. Before each stop, brake temperature will be much higher than 100F (38C). For a road racer, assume the rotors are about 500F (260C) when the brakes are applied. Also, each brake application is not from top speed to full stop as assumed in the previous calculations.

Racers have experimented with different-size and -weight rotors for years. Although some work, others can overheat. After years of trial-and-error, a brake design has been developed for each racing class that works for most cars under normal conditions. Go to the track and look at the brakes on cars similar to yours. If the rotor you selected is greatly different in size and weight, you are probably wrong. If in doubt, go larger. Use this method to select rotors on a race car, unless your car is unique.

One word of caution: If other cars in your class are using water-cooled brakes—page 147—they either have undersized rotors or improper air ducting. Water cooling is a crutch and should not be used unless larger brakes cannot be fitted. In other words, if the competition is using water cooling, go to a larger or thicker rotor than they use.

While at the track, you can make interesting observations about brakes. If you are at a flat, paved oval track where hard braking is used going into turns one and three, time various cars early in a long race and late in the same race. If lap times increase on some cars and not on others, chances are the cars running slower near the end are "running out of brakes." See what the various cars are doing differently.

At a road course with long straights followed by tight turns, such as turn five at Elkhart Lake, the importance of good brakes is evident. Walk along the straight to where the cars start braking and see how much deeper cars with good brakes go. Good brakes can mean as much as a 200-ft advantage going into a slow corner off a fast straight. Imagine how much horsepower and handling it takes to make up 200 feet!

Race-car-brake manufacturers can be a big help when selecting rotors for a race car. They are aware of all applications for their products and what works best. They also are made aware of it when their brakes don't work. Manufacturers work with race-car designers to develop the best rotor design for each racing class.

Contact companies such as Hurst, JFZ, Tilton, Wilwood or their dealers before you buy rotors for your special race car.

Swept Area—Another aid in selecting the proper rotor size is the *swept area* of the brake. The chart on page 28 shows swept area for various cars. Consider using a swept-area-to-weight ratio for a car of nearly the same power and speed as yours. If you are designing a race car, use a swept-area-to-weight ratio for race cars, not for road cars. In both cases, if long lining life and fade resistance are more important than brake weight, use a higher swept area.

Higher swept area means you must use a larger-diameter rotor. This may also mean using larger-diameter wheels, if the wheels limit rotor diameter. If you simply cannot get enough swept area because of the wheels, you can resort to using inboard brakes, page 98. An inboard brake is mounted *inboard* on the frame or transaxle, transferring braking force through a driveshaft to the wheel.

Vented or Solid Rotors—When looking at cars like yours, note whether vented or solid rotors are used. Usually, rotor weight and cooling requirements determine if vented rotors are needed. Vented rotors are used where heavier rotors are needed. A very thick solid rotor does not cool adequately. Because of this, it is unusual to see a solid rotor thicker than 1/2 in.

If your car needs vented rotors, refer to the rotor section in Chapter 3. Various types of vented rotors are available. Your brake supplier will be able to compare the merits of one type versus another. I suggest using vented rotors with curved vents if you can afford them. The cooling advantages are worth the expense.

CALIPER SELECTION

Caliper and rotor selection go together because they often come from the same manufacturer. He will supply a caliper that works with a certain-diameter and -thickness rotor. Make sure the caliper fits in the desired area. Some calipers are more compact than others. Pay attention to the details. You must not machine metal off a caliper to provide clearance to a wheel or suspension member.

Also, use a caliper that will mount on your suspension or chassis components. Some calipers are easily adapted to existing components. Consult a brake manufacturer. He may offer a ready-made caliper mounting bracket.

If you are also designing a new suspension, be sure to consider the caliper and provide space large enough for it. A proper caliper is every bit as important as the suspension.

Don't buy a caliper that must use a flimsy mounting bracket. Make sure the bracket is very stiff and strong. Extra "beef" on this part helps the braking. A mounting bracket that twists can cause poor brake performance and uneven pad wear.

The brake torque delivered by a caliper and rotor must be as great or greater than what is required for maximum deceleration. The maximum possible brake torque is limited by the maximum operating pressure the caliper can withstand. If this maximum pressure is exceeded, the caliper can flex excessively, the seals can leak and, in severe cases, the caliper can fail. Never design a brake system that exceeds the manufacturer's recommended maximum-pressure rating.

Caliper & Rotor Placement—When fitting brakes to a new car, it's important to position the caliper in the best position around the rotor, and the rotor and caliper in the best lateral position.

First, let's consider caliper position around the rotor. Assume you are looking at the left-front wheel from the outside. Theoretically, the caliper could go at any position from 1 o'clock to 12 o'clock. There are many things to consider, such as existing mounting bosses, bleeding, spindle deflection, air-duct location, and suspension and steering clearances.

If you have an existing car, mounting bosses on the spindles—or uprights—may dictate caliper location. Bleeding is best at 3 and 9 o'clock. Usually, calipers positioned at 12 and 6 o'clock must be removed, rotated and have a spacer inserted to bleed. Spindle and axle-shaft deflection during cornering can cause caliper-piston *knockback* with calipers at the 12- and 6-o'clock positions. This should not be a problem with full-floating hubs or sliding calipers.

Because air ducts enter from the front of a brake, 3 o'clock is usually best for the caliper so the ducts can enter the center of the rotor at 9 o'clock. Although suspension clearances differ from car to car, there are usually ball joints and uprights at 12 and 6 o'clock. Steering arms are either at 3 or 9 o'clock. Usually, when everything is considered, 3 o'clock is best and 9 o'clock is second best, but any position will work. It all depends on what is best for your car.

For lateral positioning, the things to consider are existing mounting bosses, and wheel and suspension clearances. With an existing car, you may be able to use existing bosses and hole locations, but you will probably have to fabricate an adapter plate. It must be rigid. I like 1/2-in. steel plate.

Caliper did not fit wheel correctly, so metal was removed from outboard edge. Don't do this to get a caliper to fit. Caliper flexing, cracking or other problems may result if excess metal is removed. Instead, select calipers and wheels that clear without modification.

Position each caliper so it has at least 0.080-in. clearance to the wheel and 1/4 in. to moving suspension parts. Be sure to check the front suspension with the steering at full lock, right- and left-turn. If no caliper position can be found that clears at full lock, consider steering stops. Except for dirt oval-track cars, few race cars need as much steering angle as a road car.

Rotor Offset—Once lateral caliper position is found, the rotor position is fixed. Determine rotor-hat offset next. Hat offset is the distance from the hub mounting surface to the rotor mounting surface. Never order hats until the caliper and rotor are positioned. Rotor hats with almost any offset can be supplied off the shelf or machined to specification by your brake dealer or manufacturer.

Rotor-Hat Holes—For some reason, many hats come with large holes around their circumference. These holes help air flow to the outboard surface of solid rotors and ventilated rotors not using sealed air ducts. With sealed ducts, these holes allow cooling

Inboard rear brakes are used on this Formula Ford. Brake cooling is excellent, rotor diameter is not limited by wheel size, and handling on rough surfaces is improved. Inboard brakes are not used on ground-effects cars because they reduce tunnel width and resulting downforce.

Sprint-car rear axle: Single inboard brake on solid axle is possible because of exposed axle shaft and locked rear end. Not only is weight reduced by one-half, when one wheel hits a bump, brake moves smaller distance than wheel, reducing brake's *effective* unsprung weight. Improved handling results. Photo by Tom Monroe.

Caliper mounts to quick-change side-bell and rotor-to-axle adaptor clamps to axle shaft: Actual sprint-car installation is at right. Note bearing housing—*cage*—with control-arm brackets at outer end axle. Photo at left courtesy AP Racing; photo at right courtesy Tilton.

air to escape from the center of the rotor before it flows through the rotor. Consequently, these holes should be plugged. When ordering new hats, specify with or without holes, depending on your rotor and cooling system. If existing hats have holes and you wish to plug them, aluminum tape with high-temperature adhesive works well.

Inboard Brakes—For one reason or another, you might want to consider mounting the brakes inboard. Not only does this allow larger rotors, cooling is better because the brakes are not shrouded by the wheels. Wheel-weight reduction also improves handling on rough surfaces and brake torque does not go into the suspension.

Why aren't many brakes mounted inboard? To begin with, they complicate a car design. For example, mounting provisions must be made for the rotor and caliper on the chassis. Then an axle shaft and joints, such as those for an independently sprung, driven wheel, must be used. Finally, the spindle must rotate in the upright. All this increases complexity, cost and weight. To top it off, most front-wheel-drive cars don't have room for inboard brakes. Race cars using ground effects need the narrowest body possible between the side ducts; inboard brakes make the body wider and reduce the aerodynamic downforce on the car. Use inboard brakes if it makes sense, but keep in

TOTAL PISTON AREA FOR EACH TYPE OF CALIPER

Type of Caliper	A_T = Total Piston Area (sq in.)
Fixed mount—one piston on each side of rotor	$2A_p$
Fixed mount—two pistons on each side of rotor	$4A_p$
Fixed mount—three pistons on each side of rotor	$6A_p$
Floating mount—one piston on only one side of rotor	$2A_p$
Floating mount—two pistons on only one side of rotor	$4A_p$
Floating mount—three pistons on only one side of rotor	$6A_p$
A_p = area of one piston (sq in.)	

Chart gives total piston area for each type of caliper. Floating—sliding—caliper has one piston doing job of two in a fixed caliper. If you don't know the type of caliper, remember that floating-mount calipers have piston(s) on only one side. Area A_p of each piston is in square inches. Determine A_p for your particular caliper.

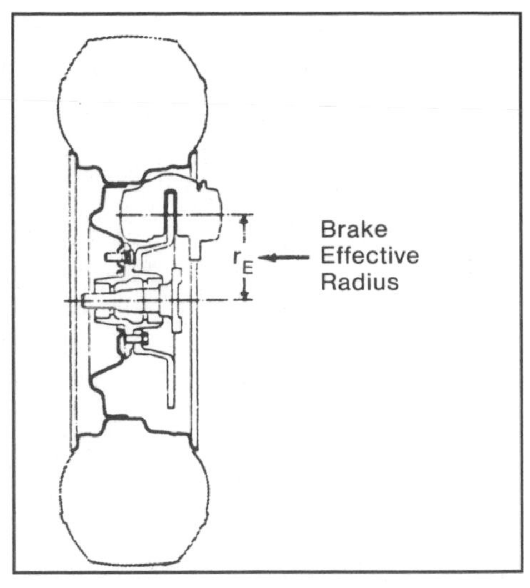

Effective radius of disc brake is measured from center of pad to center of rotor. Make sure you have correct rotor/caliper combination to determine effective radius.

mind the disadvantages of the concept.

DETERMINE HYDRAULIC-SYSTEM OPERATING PRESSURE

Once required brake torque is known and you've made a caliper selection, it's possible to determine fluid pressure. Some brake catalogs give the relationship between hydraulic pressure and brake torque for a given caliper and rotor. However, remember that the torque-versus-pressure relationship depends on brake-lining friction coefficient.

To find the fluid pressure required for your car, you need the following information:

- Brake-lining friction coefficient.
- Total piston area for each caliper.
- Rotor effective radius.
- Brake torque at maximum deceleration.

If you have the friction coefficient for your linings, use it. Otherwise, assume it's 0.3. Most linings are this approximate value. Start by calculating caliper-piston area:

Area of piston $= 0.785\, D_P^2$ in square inches

D_P = Diameter of piston in inches

Total piston area is the area of one piston multiplied by the number of pistons for a fixed-mount caliper. If you have a floating caliper, multiply the area of one piston by *twice* the total number of pistons to get total *effective* piston area. This is illustrated above.

If you've selected a caliper with non-circular pistons, such as those from Alston Industries, you must use an *effective piston area* in the formulas. Effective piston area for each caliper is given in the caliper manufacturer's catalog. If you use a formula that requires using piston diameter, you must compute *effective piston diameter*. This is the diameter of a circular piston that has an area equal to the piston area of the non-circular caliper. All formulas requiring piston diameter will work using this method.

Brake *effective radius* is the distance from the center of the rotor to the center of the brake pad. Once you know the effective radius, use the following formula to compute fluid pressure at maximum brake torque.

$$\text{Maximum hydraulic pressure} = \frac{T_B}{\mu_L A_T r_E}$$

in pounds per square inch

T_B = Brake torque in inch-pounds

μ_L = Coefficient-of-friction of brake linings

A_T = Total area of caliper pistons in square inches

r_E = Effective radius of brake in inches

Once maximum pressure is calculated, make sure the caliper can operate at this pressure. Usually, the maximum operating-pressure rating of the caliper is specified by the brake manufacturer. It varies from about 1000 to 1500 psi, depending on the caliper. The caliper cannot safely operate continuously above maximum-rated pressure because seal or caliper-bridge fatigue failure may occur. However, an occasional locked-wheel panic stop is OK.

If calculated fluid pressure is too high for the caliper, the solution is simple. You can do one of the following:

- Use a caliper with more piston area—bigger pistons or more of them.
- Use more than one caliper on each rotor.
- Use a larger rotor diameter, if there's space.
- Use a caliper that has a higher operating pressure.

If you use more than one caliper per rotor, each caliper supplies its rated torque. Thus, two calipers on each rotor doubles the torque that brake can deliver. For a given torque, fluid pressure for a two-caliper disc brake will be half that for a single caliper; required fluid displacement will be double.

Operating Pressure—In general, you should use an operating pressure that's well below maximum-rated pressure for a caliper. Leaks and excessive deflection can result from excess fluid pressure. If maximum operating pressure is, say, half of maximum-rated pressure for the caliper, the extra margin of safety in the brake system is sufficient.

PEDAL & LINKAGE DESIGN

After selecting and positioning the brakes and calculating maximum hydraulic-system operating pressure, the next step is to design the pedals and linkage. Various types of pedals are discussed in Chapter 6. As stated earlier, use a hanging pedal if possible. This allows the master cylinders to be mounted high in the car for easy servicing. Also, brake balance on balance-bar setups is easier to adjust.

The disadvantage of hanging pedals is the extra height required. On a low-profile race car, there may not be sufficient room for hanging pedals. Consequently, you'll have to use floor-mounted pedals.

Two calipers on rotor doubles brake torque for a given hydraulic pressure—an easy and effective means of lowering pedal effort or fluid pressure. Smaller single-piston caliper is stiffer than a dual-piston caliper, too. Photo by Tom Monroe.

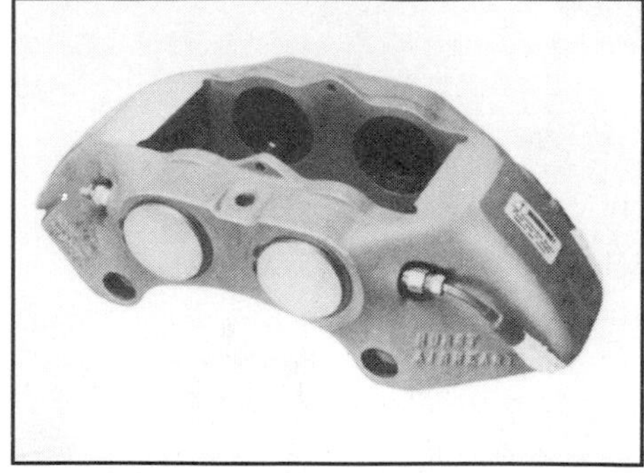

Hurst/Airheart 200X2 caliper is largest one offered by this manufacturer. Brake torque is rated at 23,000 in-lb brake torque at 1200-psi hydraulic pressure. Double-piston caliper is popular for use on racing stock cars. Photo courtesy Hurst Performance.

Ideal race-car brake-pedal setup; balance bar with remote adjuster. Remote-adjuster cable is routed to left of balance bar under clutch pedal. Floor-mounted pedals are necessary because of low-profile body height.

Decide at this point what type of brake-balance method to use. There are several choices available for a race car:

- Balance bar.
- Adjustable proportioning valve.
- Combination of the two.

Road cars usually don't use balance bars because they are difficult to incorporate into existing pedals. Also, brake balance for road use is not as predictable as for racing. This is because road cars run on many types of surfaces. And, adjusting a balance bar while driving on the road is simply not practical. On a race track, conditions are stable or reasonably predictable. Also, the driver constantly uses the brakes hard and can sense the need to adjust balance.

Balance bars are used on race cars for two main reasons:

- Provides stiffer pedal feel than a proportioning valve.
- Separate master cylinders are easy to change to make large brake-balance changes. An adjustable proportioning valve has limited adjustment.

If you plan on using a balance bar, consider using an adjustable proportioning valve, too. A balance bar can be adjusted to accomplish the desired front-to-rear brake balance, but the proportioning valve does it automatically over a limited range of tire-grip values. The disadvantage of using a proportioning valve with a balance bar is the added weight, cost, complexity and pedal travel.

Unless you like climbing into a race-car cockpit upside down on your back, make sure the balance bar can be adjusted from a convenient position. This may affect the mounting position you choose for the pedals. But, pedal and balance-bar maintenance are critical to keeping a brake system in top operating condition. A balance bar with a remote-adjustment cable and knob may be the solution.

Brake-pedal ratio and master-cylinder size(s) should be selected at the same time. The object is to achieve the maximum operating pressure you established with an acceptable pedal effort using the following steps:

- Select pedal ratio.
- Select pedal effort at maximum deceleration.
- Calculate forces on master-cylinder pushrod(s).
- Select master-cylinder diameter(s) that gives proper maximum operating pressure in hydraulic system.
- Check geometry of the pedal and linkage at extremes of travel.
- Design pedal bracket and master-cylinder mount.

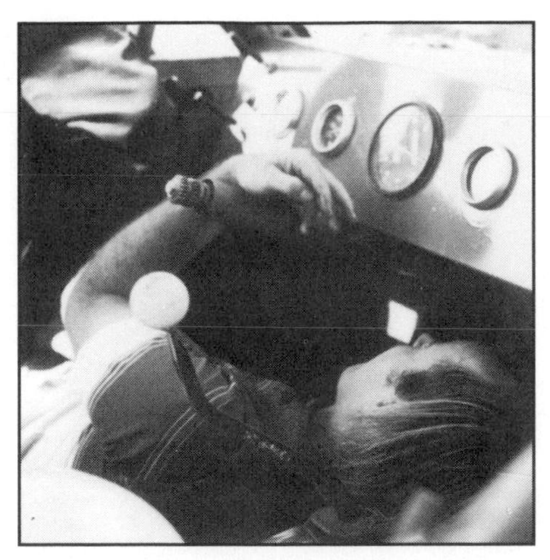

Making under-dash brake-balance adjustment forces mechanic into this uncomfortable position. Hanging brake pedal with balance bar and remote adjuster is more practical.

• Change anything necessary and start over.

Select Pedal Ratio—There is nothing magic about the pedal ratio. Many values are possible. However, if you don't know what to use, try a pedal ratio of about 5.0.

If you buy a set of pedals, use the pedal ratio built into them. Unless the resulting master-cylinder size(s) or linkage movement is incorrect, you should have a good design. Don't design a new pedal unless you have to. It is time consuming and difficult to do correctly.

Some points to consider when selecting a pedal ratio:

• With a given pedal ratio, is the master cylinder(s) mounted in a convenient place?

• Can you get master cylinder(s) with enough stroke so the pedal hits the floor before the master cylinder(s) bottoms?

• Does the pedal position feel comfortable?

Selecting Pedal Effort at Maximum Deceleration—Most people can stop a car that requires 100-lb pedal effort. However, this is high for a road car. And, it may not allow enough safety margin for brake fade for street or racing. I recommend that you use a 75-lb pedal effort for maximum deceleration. If you use less pedal effort, the brake pedal will feel less rigid and will have more travel.

Master-Cylinder Pushrod Force—If your car uses a tandem master cylinder or a single master cylinder, the force on the pushrod is found by multiplying the pedal ratio by pedal effort.

For example, if pedal ratio is 10.0 and pedal effort is 75 lb, force on the master-cylinder pushrod is 750 lb.

If you are using a balance bar, the force just calculated above is the pedal force that acts on the balance bar. It is divided between the two master cylinders according to the formulas on page 75.

I recommend that you assume the balance bar divides pedal force equally between the two master cylinders. This allows the maximum amount of balance-bar adjustment in either direction. It also makes the design easier. If you do this, divide equally the resulting pushrod force found by multiplying pedal effort by pedal ratio. This gives the force on each master-cylinder pushrod. If you do not center the balance bar, calculate the force to each pushrod.

Now that you have force on each pushrod, you also have the force applied to each master-cylinder piston. The forces are the same.

MASTER-CYLINDER SIZE VS. PISTON AREA

Nominal Bore Size	Diameter In.	Area Sq In.
5/8 in.	0.6250	0.3068
11/16 in.	0.6875	0.3712
19mm	0.7480	0.4394
3/4 in.	0.7500	0.4418
20mm	0.7874	0.4869
13/16 in.	0.8125	0.5185
21mm	0.8268	0.5369
22mm	0.8661	0.5892
22.2mm	0.8740	0.5999
7/8 in.	0.8750	0.6013
23mm	0.9055	0.6440
29/32 in.	0.9063	0.6451
15/16 in.	0.9375	0.6903
24mm	0.9449	0.7012
25.4mm	1.0000	0.7854
1 in.	1.0000	0.7854
1-1/32 in.	1.0313	0.8353
26.6mm	1.0472	0.8613
1-1/16 in.	1.0625	0.8866
1-1/8 in.	1.1250	0.9940
28.6mm	1.1260	0.9958
1-1/4 in.	1.2500	1.2272
31.8mm	1.2520	1.2311
1-5/16 in.	1.3125	1.3530
1-11/32 in.	1.3438	1.4183
1-1/2 in.	1.5000	1.7671
1-3/4 in.	1.7500	2.4053

Listed are diameters of popular mass-produced master cylinders and their areas. Note that some metric sizes are close to standard inch sizes. Be careful to get correct sizes when buying replacement parts. Many cylinders have diameters marked on them.

Master-Cylinder Size—Now that you know the desired maximum operating pressure and force on the master-cylinder piston, it's simple to calculate the required area for the master-cylinder piston:

$$\text{Master-cylinder-piston area} = \frac{F_{MC}}{P}$$

in square inches
F_{MC} = Force on master-cylinder piston in pounds
P = Hydraulic pressure in pounds per square inch

A list of piston areas for commonly available master-cylinder diameters is given in the accompanying table. Most cars use 3/4—1-in.-diameter master cylinders. If your calculated area falls between two standard sizes, choose the master-cylinder diameter closest to the required area. You can make small adjustments later with the balance bar.

If the required master-cylinder area is smaller than what's available, go to a larger numerical pedal ratio or

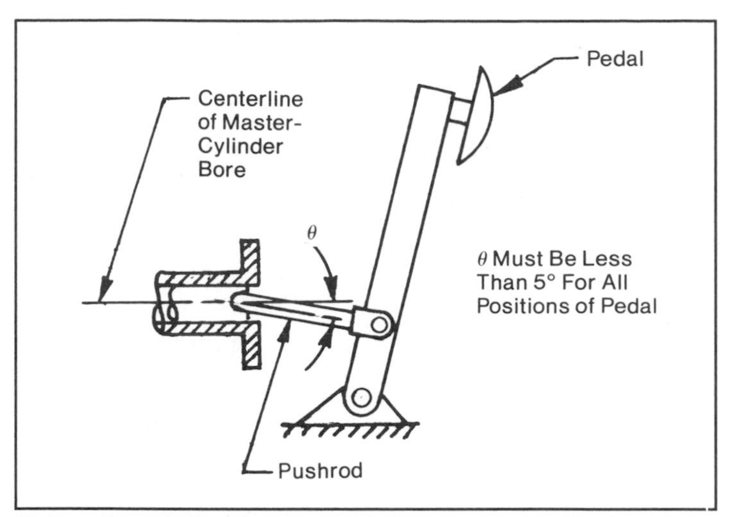

Make a drawing to determine angle between pushrod and master-cylinder centerline. Angle must not be excessive, or binding and rapid cylinder wear will result. Limit angle to 5° maximum. If angle is too large, increase length of pushrod or shorten total pedal travel. Angle should be nearly zero with pedal in position where maximum deceleration occurs. On most cars with hanging pedals, this occurs with pedal at about 2 in. from floor.

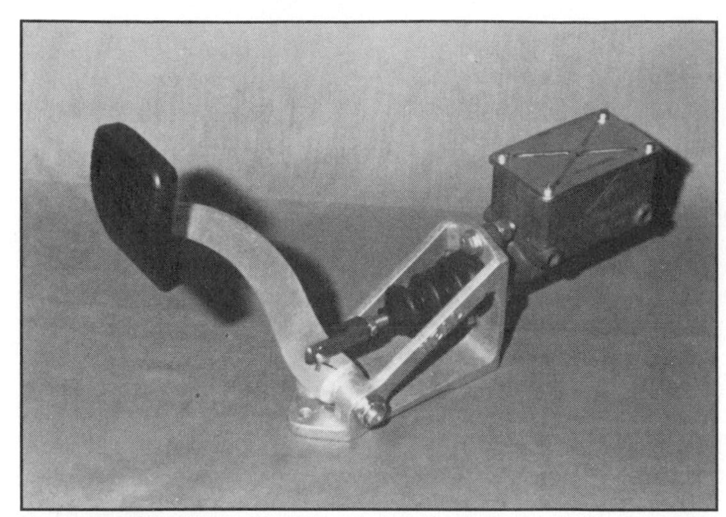

Floor-mounted Neal pedal bracket ties master cylinder to pedal pivot in one assembly. Only forces on chassis are those delivered directly by driver's foot. Larger forces on master cylinder and pedal pivot are taken by bracket.

power brakes. However, don't use power brakes on a race car unless absolutely necessary. There are easier ways of avoiding this, unless your race car is extremely heavy.

If you select a 5/8-in.-diameter master cylinder, be aware that you cannot go to a smaller size later—there are none available. It is better to use a 3/4-in. or larger cylinder so you can make a change, if needed. For example, if new tires give more grip than what you allowed for, you'll need smaller master cylinders to keep the same pedal effort for maximum deceleration. So, leave yourself some room to make changes. To increase the master-cylinder diameter and maintain pedal effort, use a larger pedal ratio.

Pedal & Linkage Geometry—A brake pedal must be designed to fit the driver. As the driver pushes the pedal, the ball of his foot should stay in contact with the pedal pad. Measure your foot or the driver's—if the sizes are significantly different—and check pedal positions.

Also, look for a car with pedals in the "perfect" position for the driver's feet. Measure the relationship between the floor, pedals and seat. Copy the design of a good car if you find one that's suitable. There is no perfect pedal design, because each person's body has slightly different dimensions. So, design your car to suit the driver.

If you have doubts about pedal position, make a full-size model of the pedals and seat. Position them as they will be in the car and see how they fit. Now's the time to change things that are not comfortable. Check the pedal in the extreme positions of travel. You should have no more than 6 in. of total travel at the pedal pad. Try it to be sure.

After you make a drawing of the pedal and its mounts, check the linkage at the extreme positions. There should be no binding. Pay particular attention to the angle the pushrod makes with the master cylinder. If the pushrod goes through extreme angles to the bore of the master cylinder, it can bind. This causes severe problems, including breakage. A longer pushrod may be required to reduce the misalignment angle. A *maximum* of 5° misalignment between pushrod and master cylinder is recommended by brake manufacturers. See the accompanying drawing for an illustration of this problem.

Pedal-Bracket & Master-Cylinder Mount—On many cars the pedal bracket and master-cylinder mount are the same. This makes a strong, rigid system with minimum of overall weight. Race-car pedal brackets manufactured by companies such as Tilton, Neal or Winters use this concept. A pedal bracket is difficult to design, so buying one of these sturdy, lightweight brackets makes sense. These parts are designed for racing applications and are easy to use.

I discuss pedal-bracket design in Chapter 6. However, remember this important point: The bracket must not deflect an excessive amount when the pedal is pushed hard. A pedal bracket mounted in the center of a thin sheet-metal fire wall is *not* satisfactory. This, or a flexible pedal bracket, is worse than air in the hydraulic system. With a flexible pedal bracket, you cannot pump up the pedal or bleed the trapped air to cure the problem!

If you stress-analyze the pedal bracket, remember to use maximum load—not normal operating load. The maximum load occurs during a panic stop when the driver pushes with all of his might. If the bracket is strong enough for this, it will work fine during maximum deceleration.

Sample Design Problem—Following is a sample race-car brake-design problem. The steps are in the same order presented.

First select the disc-brake system that gives the required brake torque at a 500-psi operating pressure at the front brakes and 400-psi at the rear. These calculations are explained in the first part of this chapter.

A proven pedal system has an 8.0-to-1 pedal ratio. I wish to design for a pedal effort of 75 lb at maximum deceleration. The pedal force acting on the balance bar is calculated:

Force on balance bar = $P_E R_P$ in pounds
P_E = Pedal effort in pounds
R_P = Pedal ratio
Force on balance bar = (75 lb)(8.0)
Force on balance bar = 600 lb.

Accident waiting to happen: Steel brake line must be secured to frame with a bracket where the flexible hose attaches (arrow). With this setup, steel line is subjected to bending and fatigue as the suspension moves up and down. Eventually, the line will fracture, causing brake failure.

Brake-line mounting bracket is simple plate, located near suspension pivots (arrow). This minimizes hose bending and pulling as wheel moves. Line and hose attaches to bracket using an aircraft *bulkhead fitting.* Jam nut clamps bulkhead fitting to bracket.

Assuming the balance bar divides pedal force equally between the front and rear master cylinders, the force on each pushrod is 300 lb.

Although the master cylinders have equal forces, different size cylinders are required for front and rear brakes. This is due to the difference in the operating pressure of the front and rear systems.

From the formula on page 101:

$$\text{Piston area} = \frac{F_{MC}}{P}$$

F_{MC} = Force on master-cylinder piston in pounds
P = Hydraulic pressure in pounds per square inches

For the front brakes, master-cylinder piston area is:

$$\text{Piston area} = \frac{300 \text{ lb}}{500 \text{ psi}}$$
$$= 0.6 \text{ sq in.}$$

For the rear-brake master cylinder:

$$\text{Piston area} = \frac{300 \text{ lb}}{400 \text{ psi}}$$
$$= 0.75 \text{ sq in.}$$

The closest standard-size master cylinder is selected from the table on page 101:

Front master-cylinder diameter = 7/8 in. (area = 0.601 sq in.)
Rear master-cylinder diameter = 1 in. (area = 0.785 sq in.).

Note that the front master cylinder is almost perfect and the rear cylinder is a bit too large. I'll use the balance bar in testing to fine-tune the system.

Also note that the smaller master cylinder is considerably larger than the minimum size. This is good in case I guessed wrong at the start of the design. I can easily change hydraulic-system pressure by using smaller cylinders. When a problem occurs, usually a smaller master cylinder is required to reduce pedal effort.

After finishing the calculations, I draw a side view of the pedals to check linkage geometry. Because I selected a standard pedal design from a manufacturer, this step is probably not required. Pedal sets should be designed to work without binding.

Custom-fabricated reservoirs are secured with hose clamps and rubber cushions in between. Reservoirs will slip if clamps loosens.

Reduced diameters at centers positively locate fluid reservoirs in mounting clamps. Clamps would have to loosen considerably before reservoirs could slip.

COMPLETE BRAKE SYSTEM

There are now other steps required to complete a brake-system design. The major items remaining are:

- Route brake lines on chassis.
- Attach flex hoses to chassis.
- Locate and mount remote master-cylinder fluid reservoir(s) (if used).
- Locate proportioning valve (if used).
- Design mounts for calipers, rotors and other brake hardware.
- Design cooling-duct system, if required.
- Confirm that bleeders are accessible and mounted high.

After the car is completed, the brake system must be tested before serious use, Chapter 10.

Testing

10

Testing is required to get maximum performance from any car, but particularly race cars. This includes brake testing. George Bignotti and Tom Sneva discuss effects of changes to car between practice runs while tire engineer measures tire temperatures. Photo by Tom Monroe.

Brake dynamometer, designed to run continuously, is driven by a 289-cu in. Ford V8 through a gearbox. Power delivered to brake is equivalent to a sedan under a hard stop. Datsun drum brake is being tested.

As you now know, a great deal of the brake-system design process is based on assumptions. For this reason, a brake-test program is absolutely necessary. Often, when a new race car is designed and built, testing is neglected in favor of getting the car to its first race. This is false economy. Testing at a race takes a lot longer than at a planned private test session. I've seen many people use up a whole racing season to get a car working properly when they could've sorted out the car in one full day of testing.

Brake testing falls into two categories: laboratory testing and track testing. Laboratory testing involves running a single brake on a testing machine called a *brake dynamometer.* See photo on the back cover. This type of testing is similar to testing an engine on an engine dynamometer. Because a brake is out in the open, it can be studied and observed while it operates. Brake-dynamometer testing is usually done by the big brake manufacturers, but if you are so inclined you can make such a device for yourself.

Most brake-testing time will be spent on the track. You take the car to a private test track and test the brakes. Measuring brake performance and pushing them to the limit should be a part of the test. Many tests that can be done very easily during track testing would otherwise be difficult or impossible during a race. Even a practice session during a race weekend is not appropriate for testing. Other cars on the track prevent you from doing what is needed for a good test.

BRAKE-DYNAMOMETER TESTING

Brake manufacturers use brake dynamometers for laboratory testing. A single brake is mounted on and driven by the dynamometer. Instruments measure and record such things as brake temperature, rpm, brake torque, hydraulic pressure and stresses. The dynamometer is powered by an engine or electric motor. Usually, a large high-speed flywheel is used.

Dynamometer testing involves revving up the flywheel until there is a predetermined amount of energy stored in it. The driving motor is disconnected and the brake is applied to slow the flywheel to a predetermined rpm or to a dead stop. The amount of stored energy in the flywheel is known. It is equivalent to a fraction of the kinetic energy stored in a moving car, or about one-third of the total energy for testing one front brake and one-sixth for one rear brake.

Dyno testing tests a brake's energy-absorbing capacity for one or a series of complete or partial stops. The flywheel can be stopped slowly or rapidly, depending on how much hydraulic pressure is applied to the brake. Sophisticated dynamometers are computer controlled and can be programmed to duplicate brake applications, speeds and decelerations encountered in city traffic, a mountain descent, or laps around a race track.

Another type of dynamometer applies constant power to the brake. This steady-power input can be at a particular rpm. As the brake heats up, its torque changes. Torque can be measured with instruments. Friction change with temperature at a constant rpm can be determined with this setup.

Temperature Measuring—The secret of proper brake-dynamometer testing is the instrumentation. A good setup will have numerous temperature-measuring devices mounted in critical locations on the brake. Generally, *thermocouples* are used for this.

A thermocouple is a joint of two wires with dissimilar metal. When the joint, or *thermocouple junction,* is heated, it generates a small electrical voltage. This tiny voltage can be measured and converted to indicate a specific temperature with a *pyrometer.* If the thermocouple junction is in contact with the item to be measured, a remote temperature measurement is possible.

HOW MUCH HORSEPOWER GOES INTO THE BRAKES?

Horsepower is the rate at which energy is delivered. The heat energy received by the brakes during a stop can be expressed as horsepower. For example, if a brake is driven by an engine on a brake dynamometer, power into the brake equals the power output of the engine. In a car, the kineic energy of the moving car supplies the power. However, that kinetic energy originally came from the engine.

One horsepower equals 550 ft-lb of energy delivered each second. The total power into all four brakes during a stop is given by the following formula:

$$\text{Horsepower} = 0.00268 W_C d_M S$$

W_C = weight of car in pounds
d_M = maximum decleration in g's
S = speed of the car in miles per hour.

Because the speed of a car changes during a stop, power to the brakes also changes. Deceleration may change during the stop due to changes in downforce, and this also changes power to the brakes. Obviously, the highest power is for a fast, heavy car with lots of downforce and big racing tires.

Below are sample calculations that may explain why brakes are designed the way they are for various types of cars.

Type of Car	W_C (lb)	S (mph)	d_M (g's)	Power(HP)
Large sedan, street tires	4000	60	0.8	515
Small sedan, street tires	2000	60	0.8	257
Racing stock car, racing tires	4000	200	1.2	2573
Fast race car with high downforce	2000	200	2.5	2680
Small race car with low downforce	1000	120	1.2	386

Thermocouples can be used on any stationary brake part. To measure moving parts, such as a rotor, a sliding contact or more-sophisticated instruments are needed. Often, rotor temperature is measured with a thermocouple brought into contact with the rotor immediately after the brake stops.

A hand-held pyrometer connected to a thermocouple probe is a useful device for measuring brake and tire temperatures at a race track. When racers use the term pyrometer, they usually mean *hand-held pyrometer* or *thermocouple.*

There are many interesting comparisons to make when dynamometer testing. Here are some possible tests:

- Comparison of brake friction materials for fade resistance, wear and torque.
- Determining the most critical part of braking system when excessive temperatures are reached.
- Finding the maximum temperature a brake can withstand without failure.
- Comparison of brake-fluid performance at high temperatures.
- Testing different types of calipers, brake drums or other hardware.
- Comparison of different rotor designs.
- Testing effect of caliper-piston insulators.
- Finding maximum torque the brake is capable of achieving without structural failure.

COMPONENT TESTING

Manufacturers do a lot of brake-system-part testing. They test master cylinders and wheel cylinders on test rigs that automatically apply the brakes over and over. This tests durability of the parts.

By cycling a brake cylinder rapidly, testing takes much less time than track testing. In addition, components subjected to internal pressure are tested for bursting strength at pressures much higher than normal. It must be proven that each item has the strength to withstand the maximum possible pressure and have a significant factor of safety.

If you wish to pressure-test components, it is not difficult to set up a test. You must get a hydraulic pump or build a lever to operate a

Complete line of temperature-measuring instruments is offered by Omega Engineering, Inc. Digital meters shown are designed to be used with thermocouples. By combining a meter with a thermocouple mounted in a special probe, you can custom-build your own pyrometer. Such instrumentation is useful for testing on a brake dyno or race car. Photo courtesy Omega Engineering.

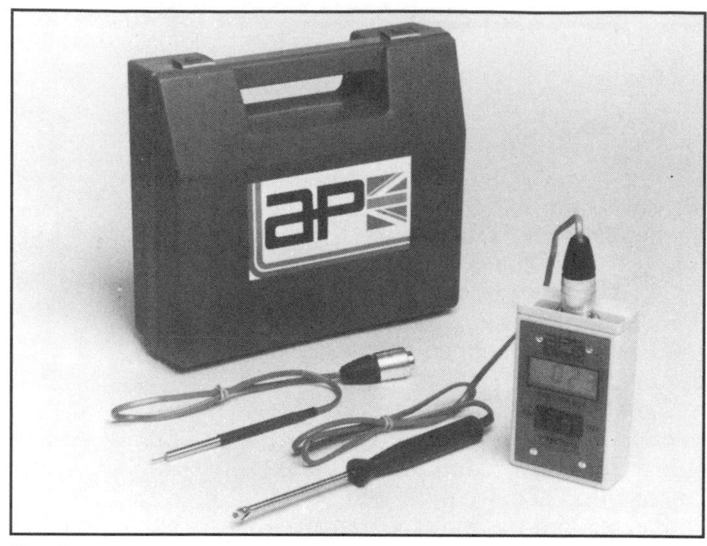

Handy pyrometer kit is manufactured by AP Racing and sold by Tilton Engineering. It comes with two probes; one for soft materials such as tires, and one for hard materials such as brake rotors. Instrument can be used to measure brake temperatures on open-wheel cars. Photo courtesy AP Racing.

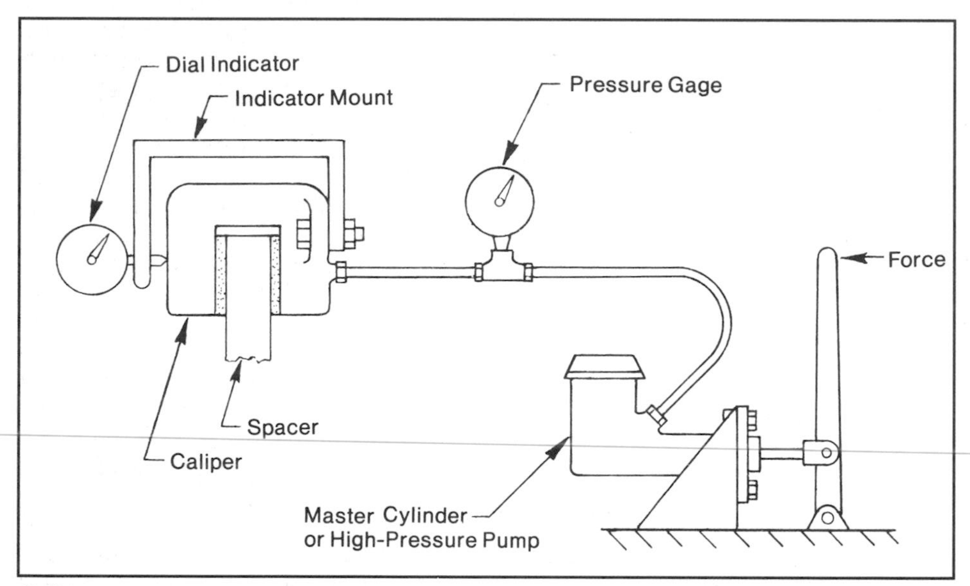

Caliper pressure-test setup: Pressure gage should have minimum 1500-psi range; 2000 psi is preferable. If maximum pressure is exceeded, gage will be damaged. Spacer gives something for pads to push against. Note how dial indicator is mounted: Bracket is attached to one side of caliper and gage is set square against opposite side so it will read total caliper deflection. Test can also be performed on the car if a pressure gage is installed in brake line. Remove gage before driving car to avoid excessive pedal travel.

brake master cylinder. The lever can be operated by hanging weights on a pulley system. You can then apply an exact load to the master-cylinder pushrod and maintain it so you can make measurements.

System Deflections—It is interesting to test the deflection of hydraulic-system components while the system is pressurized. Of particular interest is how much a caliper deflects outwardly. You can measure this deflection with a dial indicator mounted to the caliper. You must also measure either hydraulic pressure in the system or force on the master-cylinder piston. System pressure can then be calculated. A hydraulic-pressure gage with a large-enough scale will do it.

An engine oil-pressure gage won't have enough capacity. Relatively inexpensive gages reading to 2000 psi are available at industrial-supply stores. Marsh Instrument Company's J4878 pressure gage is a 2000-psi gage that uses minimum displacement.

Put a tee in the hydraulic line and install the gage. Make sure air cannot be trapped in the gage or other high spot in the system. Pressurize the system and record caliper deflection at incremental pressures. Draw a graph of pressure versus deflection. It should be approximately a straight line. If it's not, something in the system has slack, which is being taken up, or something is failing. Check to see if the deflection returns to zero when pressure is removed. If some deflection remains without pressure, something has bent or crushed.

You can measure brake-hose and line deflection as well. These deflections are small, but you should be able to measure them with a dial indicator while using extremely high system pressure. Take a number of measurements just to be sure.

THE TEST TRACK

The only way to get the most from a brake system is to test it on the car. You don't have to test on a race track. Unlike chassis testing, brake testing can be done in a straight line. Therefore, you may have more options.

Each area of the country has different facilities and laws. List possible testing locations so you don't forget any. Pick the one most suitable for your purposes. Don't ignore the legal limitations, such as noise ordinances. People next to your "test track" may not appreciate the sound of screeching

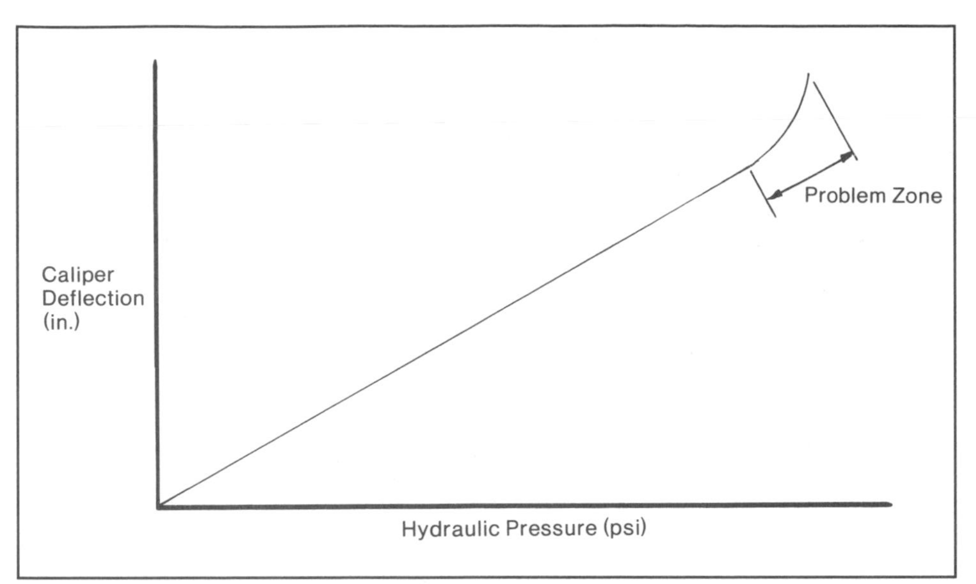

Use data obtained from caliper-deflection test to plot pressure-versus-deflection curve. Curve should be a straight line. If deflection increases at a faster rate at high pressure, something in the system is failing. Don't risk trouble by using brakes in the *problem zone*.

When testing a race car's brakes, do it on the same type of track it will run on. This track is best for testing an oval-track dirt car. It's not much good for testing a road-racing car.

Stock car has pressure gage mounted on brake line next to each master cylinder. By recording pressures, an exact numerical record of brake balance is available. Don't leave pressure gages in hydraulic system because they increase pedal travel. Pressure gages are for testing only.

WARNING

High-pressure testing is used only to find the limits of a design and determine its safety factor. If you are using a hydraulic pump capable of 3000 psi or more, you may be able to test items to failure because most brake systems operate below 1500 psi. There is some danger associated with such testing, so be careful. Air trapped in the system can escape with *explosive force*. If parts break, they can go flying at high velocity. Isolate people from the items being tested, and make sure to bleed all air from the system. *Always* relieve pressure before looking at the parts.

If you pressurize a part to three times its maximum rated capacity and it does not fail or distort, it is well designed. However, all parts tested to excessive pressures should not be used. They should be destroyed or marked permanently for easy identification. There may be damage hidden from view that makes the part unsafe to use.

If you are making a test rig for pressures up to 3000 psi, go to an aircraft-surplus shop for parts. Aircraft hydraulic systems operate at 3000 psi. Consequently, their high-pressure hoses and fittings can take this pressure. Check with the shop to be sure you get parts rated for 3000-psi pressure. Hand pumps that can pressurize a system to 3000 psi are available. You'll also need pressure gages for 3000-psi capacity.

Before using surplus lines and fittings, wash them throughly with alcohol to remove dirt, moisture and aircraft hydraulic fluid. Aircraft brake fluid is totally different from automotive brake fluid. Always start with clean components.

tires, even if you think your testing is strictly legal.

Some possible locations for brake testing are:

- Road-racing track rented for testing.
- Drag strip rented for testing.
- Large parking lot.
- An airport, particularly one with light air traffic or unused runways.
- Straight road away from residences, intersections and with little traffic.

Race Tracks—Obviously, the best place to test a race car is a race track like the one you will compete on. Even for testing a road car, a race track has the advantages of safety and privacy. Some race tracks are available during the week for a nominal fee. You should be prepared to bring whatever safety equipment the track requires, even if you consider brake testing much safer than going flat-out.

Chances are that track management and insurance companies won't make any distinction between one type of driving and another. Usually, they'll require the driver to wear a helmet and use other standard racing safety equipment. You and all those involved may also have to sign a waiver that relieves the track of all legal responsibility if someone is injured.

If you are testing a road-racing car, you should test on a road course. The same holds true for other types of cars. Oval-track cars should be tested on an oval track—dirt cars on dirt and pavement cars on a paved track—dragsters on a drag strip, and on down the line. This will allow you to run the car around or up and down the track many times in succession and test the brakes for temperature and wear under simulated racing conditions. This will not be possible at other types of test tracks. If you can't test on a race track, you'll have to use actual racing for "real-world" testing.

Although a drag strip is ideal for testing a drag-car brake system, one is also useful for testing other types of cars. You can do one stop from moderately high speed at a drag strip. Or, if the track allows, you can run several stops in succession by running back and forth. Usually, there is

Drag strip can be used to test brakes at medium to high speeds. Dragster reaches top speed in quarter mile, so its brakes get a realistic workout. Typical road car will only reach about two-thirds of its maximum speed. Regardless, drag strips are usually safe and handy for brake testing.

Small airports are potential brake-test sites. However, you must first get permission from the airport manager.

plenty of room to bring the car to a safe stop.

The advantages of a drag strip are that it will be clean, smooth and level. These are important considerations. The problem with a drag strip is its length. You usually can't get a car near its maximum speed in a quarter mile, unless you are running a drag racer. Most road cars reach about two-thirds maximum speed in a quarter mile. This represents about one-half of the maximum kinetic energy a car can deliver to the brakes. Thus, testing at a drag strip leaves some doubt about the ultimate capability of a brake system.

If you can't rent the drag strip, enter the car in a drag race. You may be allowed to run *time only,* where you don't engage in actual competition. If you apply the brakes hard at the end of the strip, you can get some idea of how the brakes work at medium-high speeds.

Drag racing will limit testing because carrying an observer during each run may not be allowed. Also, you won't be able to get as many runs as in a private test session.

Parking Lots—Parking lots are sometimes available on weekends. The problem is finding a lot big enough for testing brakes. Most are too small. Also, you'll run into problems with dirt, bumps and nearby people. And, it may be difficult to get permission to use a really large lot because they are usually owned by large corporations. The larger the company, the more difficult it will be to talk to the individual who makes the decisions.

If you find a parking lot to run on, be aware of the legal and safety requirements. Always run a muffled exhaust, unless you are really out in the "sticks." Test during hours when people will not be disturbed. If you get permission to use the lot, try to get it in writing. Invite the person in charge to attend your test session. Not only is this good PR, but if the police come, he can give them a good story. Often, a well-meaning bystander will call the police, because he's not aware you've obtained permission.

Airports—A large testing surface can be found at an airport. The runways are always long and usually smooth, level and relatively clean. The best place to try is a small airport, which may be used occasionally by private planes. They may have a runway not in use, or possibly an access road that can be closed temporarily for testing. It helps if the airport manager likes race cars. Don't be afraid to ask. You might be pleasantly surprised.

Airports have the advantage of minimal noise problems and a lot of room. You may have to sign a waiver to release the airport management from responsibility. Also, they may want proof of insurance. If you are professional in your approach to testing, it will help a great deal.

Public Roads—If you have a road car, you can try a public road as the last resort. It may be legal to stop a car in a straight line if you do it safely. If you do not speed, make excessive noise or appear to be driving recklessly, you may be able to convince a police officer to allow you to test. It helps to be far from houses and other people.

Race Testing—Brake testing while running a race can be done as a last resort. If you do this, pick an event that offers maximum practice time and minimum competition. For example, a local drag event is much better than a championship event. For road racers, try a solo-I or time trial rather than a race. In solo-I events, the cars run one at a time. The practice sessions are often quite long. Another option is to enter the car in a race-driving school. But first, get permission from the sponsoring organization.

PLANNING BRAKE TESTS

Plan your tests *before* you get to the track. Testing time is valuable. If you have to do any work on the car other than what you've planned, you'll probably run out of time. Don't waste time trying to think up what to do next at the track. *Plan ahead.*

Start by writing down your test plan. List what you'll be testing for. Include a step-by-step plan of how you'll do the testing. While writing the test plan, think of all the test instruments, tools and parts needed to do the job. List these separately. A sample test plan is shown on page 109. It will give you an idea of what's needed. Determine exactly what's needed in your plan.

Data Sheets—In addition to the test

SAMPLE LIST OF BRAKE-TESTING EQUIPMENT

- Tool box for general work on car.
- Safety equipment for driver, fire extinguishers.
- Spare parts for engine and chassis.
- Spare tires and wheels.
- Brake linings, brake fluid, special brake-maintenance tools.
- Pyrometer, temperature-indicating labels, temperature-indicating paint.
- Stopwatch.
- Toppling-block setup.
- Yarn for tufts, tape, scissors.
- Video camera, still camera, portable TV and VCR.
- Fuel, oil, water, air tank.
- Jack, jack stands, creeper.
- Materials for modifying brake ducts, racer's tape, rivets and rivet gun.
- Notebook, paper, pen.

SAMPLE TEST PLAN

1. Inspect brakes and record starting data.
2. Overall brake-performance test.
   a. Record comments on performance.
   b. Inspect and adjust brakes as required.
3. Balance adjustment tests.
   a. Record deceleration before setting balance.
   b. Test brakes and adjust balance.
   c. Continue until maximum deceleration is attained.
4. Cooling-duct tests.
   a. Tuft tests to confirm airflow in duct.
   b. Measure brake temperature after one stop.
   c. Run car for measured time and measure cooling effect.
   d. Modify ducts and test cooling again.
5. Alternate lining tests.
6. Alternate fluid tests.
7. Long-distance testing.
   a. Check temperature.
   b. Check wear rate after test.
   c. Record driver comments and results of brake inspection.
8. Bed-in spare linings for next race.
9. Set brake balance with wings adjusted to maximum downforce settings.
10. Wet down track, install rain tires, and balance brakes for wet.

HPBooks® BRAKE TESTING DATA SHEET Car ________ Date ______

1. Brake System Specification.
   A. Balance-bar setting ________________
   B. Proportioning-valve setting ________________
   C. Lining material: Brand ________ (Front) ________ (Rear)
   D. Brake fluid: ________ Brand ________ Type
   E. Rotor diameter & thickness: ________ (Front) ________ (Rear)
   F. Caliper brand & type: ________ (Front) ________ (Rear)
2. Brake performance.
   A. Deceleration readings.
   B. Comments on each test.
   C. Tires used, size and brand ________ (Front) ________ (Rear)
   D. Road condition ________ (Weather) ________
   E. Comments ________________

3. Temperature testing.

	(Front)	(Rear)
A. Maximum caliper temperature	________	________
B. Maximum rotor-edge temperature	________	________
C. Maximum hub temperature	________	________
D. Maximum lining temperature	________	________
E. ______ temperature	________	________

   F. Comments ________________

4. Wear testing.

	(Front)	(Rear)
A. Lining thickness—beginning	________	________
B. Lining thickness—ending	________	________
C. Wear (beginning/ending)	________	________

   D. Test duration ______ miles, ______ laps, ______ stops @ ______ mph
   E. Test track ______ Conditions ______
   F. Comments ________________

plan, make several copies of the blank data sheet on this page for recording test results. Pertinent information about the exact brake-system setup should be included: type of linings, fluid, balance-bar setting, and similar information. These will be valuable facts to determine what setting caused what to happen.

Recorded comments on brake performance versus temperatures are vital. If you are recording deceleration rates, make a form for that, too. These test-data forms can also be useful at a race to record car settings and compare them to actual performance on the track.

Lists—You can never bring enough tools or parts to a test session. If you leave a part or tool at home, you can be sure it's the one you'll need. Murphy's Law always applies.

Make a list of tools, supplies and instruments. Add to these lists anything needed for car maintenance or repair. Bring spare parts for things likely to break. If you have room, also bring everything that's not likely to break.

Organize parts and supplies into sturdy boxes and label them. This will make taking everything much easier. If possible, assign one pit crew member to keep track of things, and put everything back immediately after use. A well-organized team is a must for successful testing and racing.

Put your test plan, test-data sheets, and parts and supplies lists into a notebook consisting of a three-ring binder or similar record-keeping system. Bring a clipboard to write on. Keep the notebook with you during all tests and races; put all technical information in it. The notebook will become your "bible" after a year or so. Because test data is valuable, make copies of everything when you get

Pyrometer is used to measure brake temperature on open-wheel race car. This is difficult on full-fendered cars. When checking temperature, quickly hold pyrometer probe firmly against rotor until meter needle stops rising. Peak temperature during braking cannot be measured with hand-held pyrometer because some cooling will occur. Peak temperature can only be measured with an onboard pyrometer, or temperature-indicating label or paints.

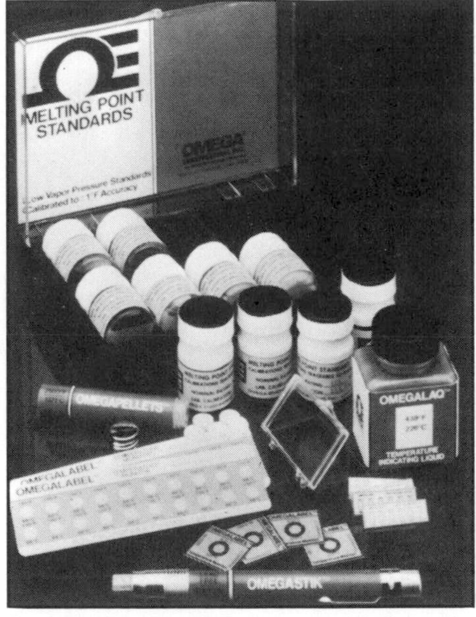

Omega Engineering Inc. specializes in temperature-measuring devices. Some of their temperature-indicating paint and labels are shown here. Photo courtesy Omega Engineering.

home. Store the copies somewhere other than in the notebook. Notebooks get lost, particularly at the races.

Finally, bring as many *qualified* helpers as possible, and all the equipment you think you'll need. Include the required safety equipment. Even if you are testing in a straight line, bring a crash helmet and driving suit. You can never be too careful.

BRAKE-TESTING INSTRUMENTS

Certain instruments are needed to get maximum benefit from a test session. Of primary importance is brake temperature. It can be measured with either a hand-held pyrometer, remote pyrometer or with *temperature indicators* placed on the brake.

Pyrometer—The pyrometer is a special thermometer. It has a probe that can be placed against an object. Temperature is then read from a meter. A pyrometer is commonly used to measure race-car tire temperatures. It can also be used to measure brake temperature up to the limit of the meter scale. The pyrometer-probe tip must be held in firm contact with the object being checked so heat can *soak* into the probe. When the meter stabilizes, the reading is taken and recorded.

An excellent dual-purpose pyrometer is AP's CP-2650-8 sold by Tilton Engineering. This instrument has a 0—800C (32—1472F) range. It is useful for checking both tire and brake temperatures. The liquid-crystal display allows it to be used with ease in the sunlight. There are two probes—a sharp one for tires and a flat one for hard surfaces such as brake components. I highly recommend such an instrument for the serious racer. It can also be used for engine tuning, such as checking which exhaust pipe is running hotter or cooler than the others.

Remember when using a hand-held pyrometer that the object being checked will cool rapidly. Therefore, it must be measured quickly after maximum temperature is reached. Unfortunately, it is impossible to measure the peak temperature of an iron brake rotor with a hand-held pyrometer. But you can come close if you work quickly.

Temperature Indicators—You may find that it's easier to use *temperature indicators* rather than a pyrometer. There are several types of indicators. Some are paint, changing color when they reach a certain temperature. Different color paints change color at different temperatures, so by using various colors, the peak temperature can be found.

One line of temperature-indicating paint is available from Weevil Ltd. in Europe. Also, Tempil has a complete line of temperature indicators, including Tempilaq temperature-indicating paint. Another is sold by Tilton Engineering. Usually, these paints are applied to the periphery of the brake rotor. They can be used on the caliper body as well.

Another temperature indicator is a stick-on label. These are sold by Tempil (Tempilabel) and Omega Engineering Products (Omega Label), respectively.

These adhesive-backed labels have white spots that turn black at incremental temperatures, as indicated next to each spot. If some spots on the label turn black and others do not, a peak temperature was reached between the values of the highest-temperature black spot and the following white spot. If all spots turn black, the highest temperature on the label was exceeded. In such a case, you need the next higher-temperature label available. If no spot turns black, the temperature of the lowest spot was not reached. In this case, go to a lower-temperature label.

Have a wide range of temperature indicators for your first test. You probably have no idea what the temperature ranges will be. Once you get some experience, you'll know which temperature indicators to use.

If you want to be able to get accurate peak and non-peak temperature readings without stopping the car, you'll need a *remote pyrometer,* a *multi-*

Excellent temperature indicators are manufactured by Tempil. Tempil's Tempilaq paint melts when heated to rated temperature. Tempilabel spots turn black when heated to rated temperature. Photo by Tom Monroe.

As with other temperature-indicating paints, AP Racing's indicators can be used to check rotor or brake-drum temperature. Temperature indicators will tell you the highest temperature reached, if you use the right range of paints. Courtesy AP Racing.

position switch and thermocouples to put in contact with the test objects. A full line of remote temperature-measuring equipment, including thermocouples and leads, is available from Omega Engineering Products.

Their 12-volt model 650K digital-readout pyrometers are relatively compact and are easy to read. All digital-readout pyrometers require external power. Omega's series-7000 dial-readout pyrometers are self-powered, compact and less expensive than digitals, but not as easy to read accurately.

Nationwide Electronic Systems' Slimline II digital-readout pyrometer IDT-7726 is compact, 12-volt, d-c powered, and is easy to read. It requires a 100-0193-01 d-c isolator module. Be sure to specify type-K thermocouples when ordering. Also specify whether you want a pyrometer that displays temperature in Fahrenheit or Centigrade.

Thermocouple Installation—Thermocouple attachment is important. The thermocouple junction must be in contact with the test object. Otherwise, readings will be low. High-temperature epoxy works well for attaching thermocouples to calipers. Thermocouples can be pinched under the head of a screw. Or, silver soldering works well on some metals. Whatever you use, be sure the attachment doesn't insulate the thermocouple from the test object.

Rotor temperature is difficult to measure with a thermocouple. The thermocouples must be silver-soldered or crimped into a small flat probe, insulated and lightly spring-loaded against the rotor friction surface. Brake manufacturers who measure rotor temperature with *rubbing* thermocouples make their own; they are not for sale.

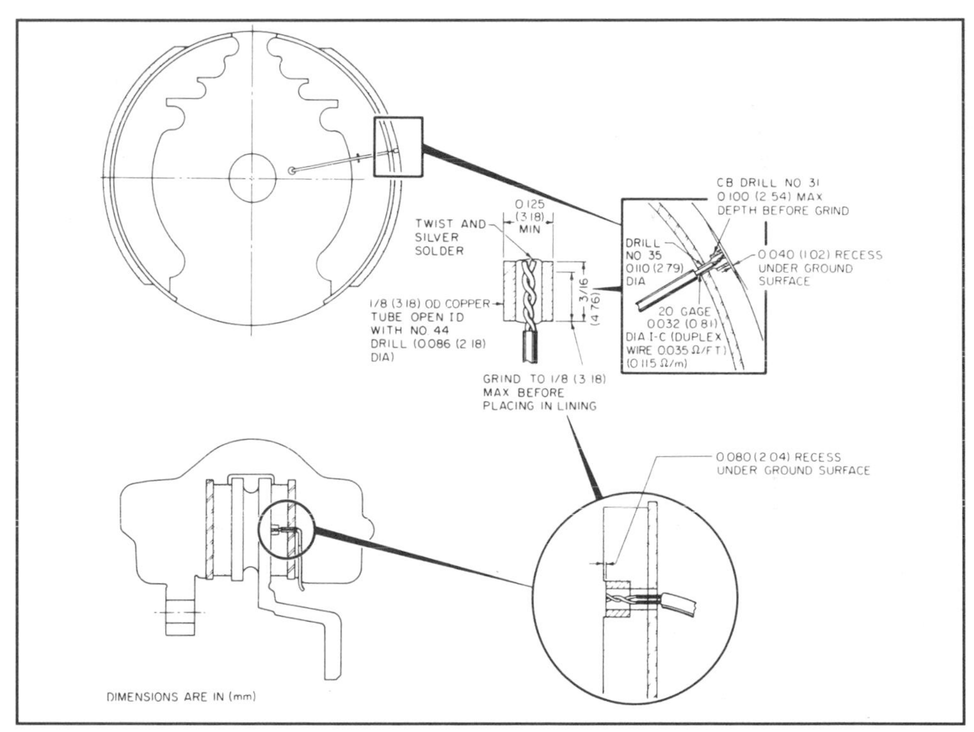

Drawings show how to install thermocouple plugs in disc and drum-brake linings so brake temperatures can be monitored while running. For this purpose, you can make your own *type-K* thermocouples by twisting chromel and alumel wire together and silver soldering the twisted end in a copper tube. You can get this wire from companies such as Omega Engineering. Reprinted with permission © 1984 Society of Automotive Engineers, Inc.

A practical alternative used by most American brake and lining manufacturers is to install the thermocouple in the lining material close to the friction surface. Temperatures measured in this way are slightly lower than actual rotor temperature. However, the difference is consistent and the installation is much easier and more durable than the rotor-rubbing thermocouples.

The thermocouple end is stripped back approximately 0.500 in., twisted, inserted into a 3/16-in.-long,

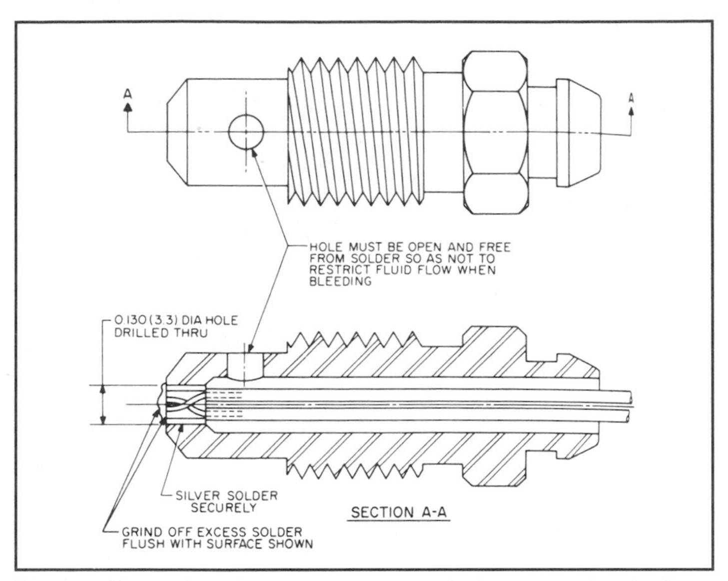

Drawing shows bleed-screw thermocouple for measuring brake-fluid temperature. The screw is first drilled through, then twisted thermocouple is silver-soldered in place. Reprinted with permission © 1984 Society of Automotive Engineers, Inc.

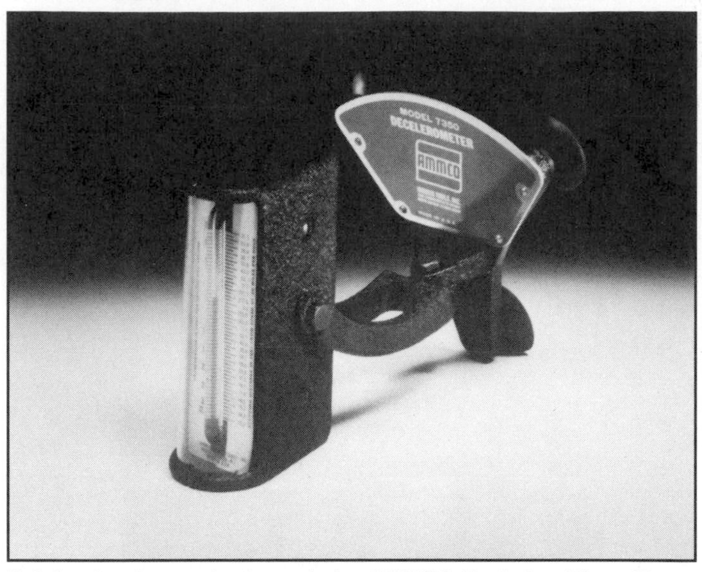

U-tube decelerometer is made by AMMCO Tools Inc. It's calibrated in feet per second per second (ft/sec/sec) rather than g's (1 g = 32.2 ft/sec/sec). Instrument reads down scale for deceleration and up scale for acceleration. Deceleration scale is used for brake testing; acceleration scale can be used for engine tuning and shift-point determination. Instrument can be mounted to windshield with suction cups or bolted to instrument panel. Photo courtesy AMMCO Tools Inc.

1/8-in.-OD X 0.086-in.-ID copper tube and silver-soldered in place to form a plug. The plug is installed in the friction material. Typical disc- and drum-lining installations are shown in the drawing, page 111.

If the plug hole is drilled through the lining and shoe with a number-31 (0.100-in.) drill, the plug can be reset easily when the lining wears down to the plug surface. If the thermocouple plug is not reset, readings will fluctuate between rotor and lining temperature.

The drawing at top of page shows how to install a thermocouple on a bleed screw for measuring brake-fluid temperature in a caliper or wheel cylinder.

With a multi-position switch, many thermocouples can be connected to the pyrometer so they can be read individually. When running thermocouple leads from the brakes to the chassis, band them to the flexible hydraulic line with plenty of slack so they don't break when the suspension moves.

Pressure Gage—Brake-fluid pressure can be measured with a simple pressure gage. Marsh Instrument Company's J4878 gage has low displacement and is used widely. Another pressure gage is also available from Hurst Performance.

A pressure gage can be used in the shop or at the track. If used at the track, mount it where the driver or observer can see it easily. Fill the gage with fluid before mounting it. And always put a valve at the gage so it can be shut off when not in use and so the gage won't lose its bleed when removed. The line to the gage can be bled by loosening the valve jam nut while applying slight pressure to the brake pedal. When fluid begins to leak, retighten the nut to complete the bleeding.

Hydraulic pressure relates pedal effort to actual force on the brakes. When used with a decelerometer, the pressure gage can be used to determine brake-system effectiveness, or to plot pedal effort versus vehicle deceleration.

Camera—Photos are a good source of test data. Use a camera that provides instant prints. Photograph the car just as the driver slams on the brakes at a prescribed braking point. You may be able to detect whether the front or the rear wheels lock up first. It's sometimes tough to tell which locks up first by observation. To help, paint a large X on the wheels and tires with water-soluble white paint.

Even better than using a still camera to detect brake lockup is a video camera. It can be played back immediately after making a shot. A video camera can also be used to record instruments in the car during a stop. To do this, the camera must be mounted securely in a bracket. A camera is an accurate data recorder—much more so than a person using his eyes.

Decelerometer—A decelerometer is valuable for testing the stopping capability of a car. One can be as simple as *toppling blocks* or as complex as a recording *g-meter*. See accompanying illustration and photos. Buy the best decelerometer you can afford. Use the Tapley meter in a car in which a passenger can ride along and observe. Perhaps a camera can replace the passenger if you care to rig one up. The Tapley meter indicates actual deceleration rates as the car is stopping.

AMMCO Tools' model 7350 U-tube decelerometer is easy to install and adjust. It works well. This instrument is used widely by the American automobile and brake industries, and many state vehicle-inspection stations in the U.S.

The simplest decelerometer is the toppling blocks. Set up a tray with rectangular blocks as shown. The blocks are lined up along a ledge on the tray. When the car stops, the taller blocks topple over at lower deceleration rates than the shorter ones. Deceleration rate is between the tallest block that toppled and the shortest one that didn't.

Toppling blocks can be knocked over by bumps or vibrations, so make

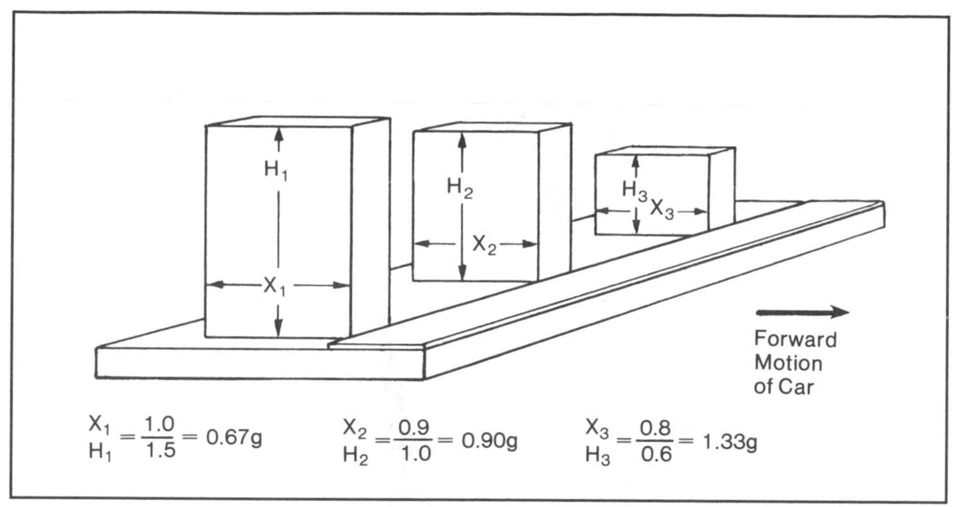

Dimensions of toppling blocks determine maximum deceleration they can take before toppling. Blocks dimensions don't have to be precise as long as they are square. Measure height and width, of each block and calculate its toppling deceleration as shown. Each block is positioned against a thin ledge on a flat tray to allow it to topple when the car decelerates. Tray can be angled to compensate for vehicle-dive angle.

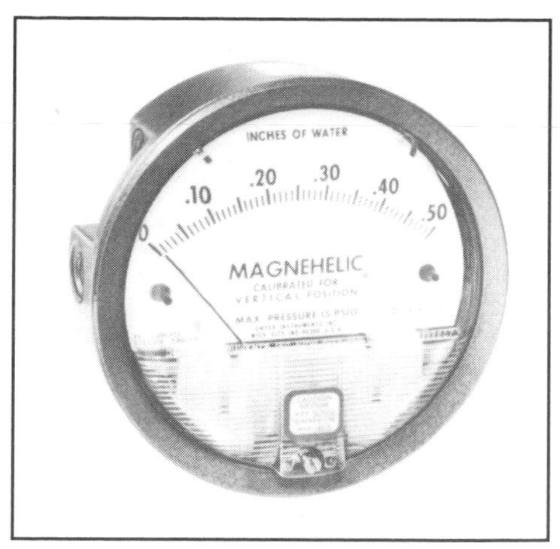

Dwyer's 2010 magnehelic gage is easier to install and use than manometer for measuring air pressure.

at least three runs to verify the results. If you don't get consistent results and the brakes definitely work the same each time, try another test method. However, be aware that variations in results can be caused by brake-temperature changes or other changes.

Tire-Test Instruments—Don't forget that tire grip affects braking. Thus, to test brakes, also test the tires. You may wish to try other tires during the test and change brake balance to suit the different tires. Bring along the extra tires and instruments needed to test them. A tire pyrometer and an air-pressure gage are the usual minimum instruments for tire testing.

Air-Pressure Gages—One last instrument to consider is a gage for measuring low air pressures in brake-cooling air ducts and at the exterior surfaces of the car. These pressures are not measured in psi, but in inches of water (in.H_2O). For comparison, the pressure 27 in. below water surface is 27 in.H_2O, or 1 psi.

Gages for measuring low pressures are very useful for developing brake ducts, engine-induction ducts, water- and oil-cooler air ducts, driver-cooling ducts and even car body-surface aerodynamics. U-tube *manometers* are used often for measuring low air pressures, but a dial gage is easier to read and install in a car.

Dwyer Instruments' 2010 Magnehelic gage reads up to 10 in. H_2O. This small 5-in. round gage can be mounted temporarily on an instrument panel or in the center of the steering wheel of a cramped formula car. Dwyer also carries all types of manometers.

BRAKE-PERFORMANCE TESTS

There are a few basic-performance tests that can be made without much trouble or expensive instruments. All brake systems should deliver minimum performance regardless of the car type.

To run a basic brake-performance test, pick a distance between stops, such as 0.4 mile, or a time between stops, such as one minute. Accelerate from rest at wide-open throttle to your best speed—to 55 mph for a road car or 80 mph for a small race car—and maintain that speed until it is time for the next stop. Distance or time between stops can be adjusted to increase or decrease test severity. Apply the brakes firmly so the car stops rapidly without locking the wheels. After the car stops, record some comments on your data sheet:

- Did the front or rear wheels lock?
- Did the car swerve or pull to the side?
- Was there a change in deceleration as the car slowed?
- Were there noises from the brakes?
- Was the pedal firm?
- Was the stop smooth or was there any vibration?

Immediately repeat the test and record your comments. Do this until you notice a change in brake performance or until about ten stops have been completed. It's important to be consistent with each stop so your test is repeatable for later comparisons. After you finish testing, *carefully* touch the wheel where it bolts on. Record the following comments:

- Are front and rear wheels about the same temperature?
- Can you smell the brakes? Both sets?
- Is there black dust on the outside of the wheels?
- Is grease leaking from the front hubs?

The comments you've recorded are for comparison with other cars or with other tests. There are no numbers, because you are not using instruments. However, you can record your comments about specific parts of the brake-system performance. The effect of brake-system changes should be recorded as part of learning how to maintain and adjust a brake system.

The overall performance test should be followed by some work on the brakes. Usually, there'll be one aspect of performance that you wish to improve. See the trouble-shooting section following Chapter 12 for suggestions on how to correct specific problems.

TRACK TESTING

A race car or high-performance road car should be track tested. However, the time and effort that goes into testing at a track is not worth it for the ordinary transportation car. Performance tests just described done on a country road should suffice. In this section, I discuss track testing aimed at getting the highest possible performance from a brake system.

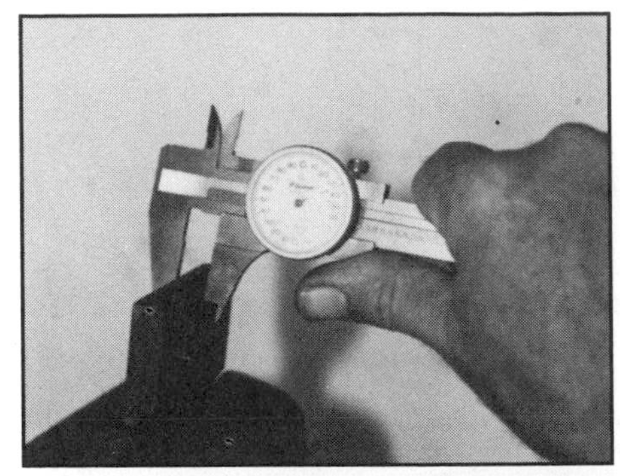

Measure lining thickness with dial caliper or micrometer before and after each test session to determine wear rate. Brake wear can be related to number of stops or laps; particularly important for planning strategy for long-distance racing. Be sure to measure pads in same place each time.

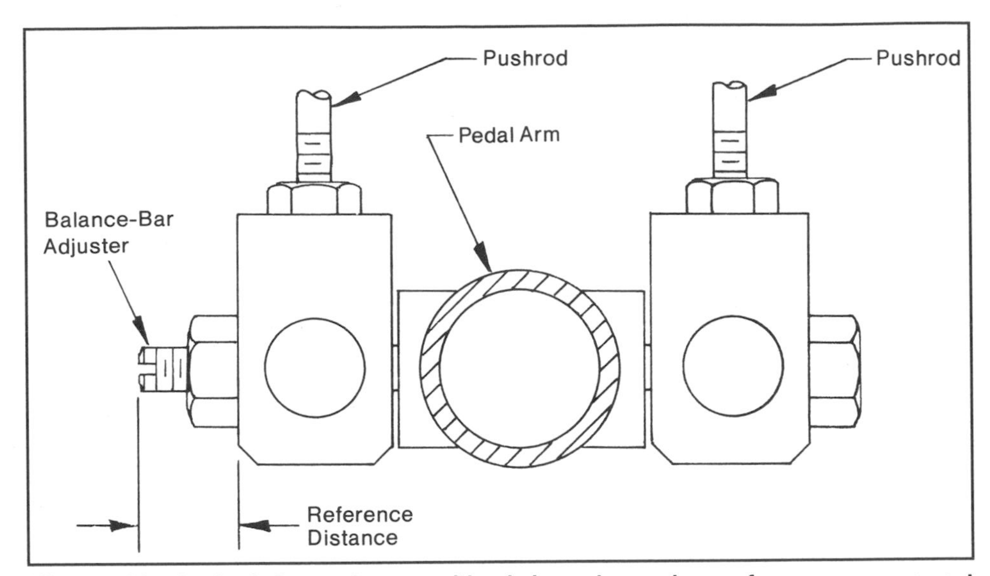

When making brake-balance changes with a balance bar, make a reference measurement so you know where you started. With this measurement, you can quickly return to your starting balance, if required.

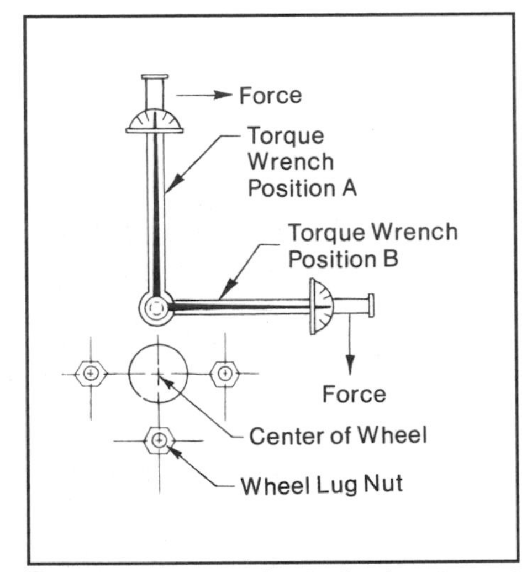

To measure torque about wheel center, put torque wrench in position B. Wrench in position A will read low because distance from wrench handle to wheel center is greater than length of wrench. Don't exceed torque capacity of lug nuts or bolts.

Locked wheels is indicated by the smoke. Excessive wheel locking can flat-spot tires, making brake-balance testing expensive. It takes a sensitive driver and careful observation to keep from ruining tires.

Recommended track tests are as follows:

- Brake-balance adjustment.
- Deceleration.
- Effectiveness.
- Temperature measurements.
- Brake-lining comparisons.

Lining Thickness—Before you start testing, check brake lining thickness. Remove the linings and measure their thicknesses with a micrometer. It's OK to use a vernier caliper. Record each lining thickness on your data sheets. Also indicate where on the lining each measurement was taken.

Disc-brake pads wear on a taper, so measure them in the middle and at each corner. Measure drum-brake linings at each corner and at the inside and outside edges of the center. Measure the combined thickness of the lining and shoe plate. Subtract shoe-plate thickness to get lining thickness.

The purpose of these measurements is to check for lining wear. By recording the total number of hard stops during testing, you can calculate the amount of wear per stop. You can then calculate about how long your brake linings will last.

Brake Balance—If the linings are not bedded in, do it before testing. This will also warm up the engine and chassis. Check page 40 for details about how to bed in brakes.

If you are testing a race car, it should have an adjustable balance bar. A balance bar is usually mounted in the pedal and is adjusted from one end by loosening a jam nut and turning the bar with a screwdriver.

If you are unfamiliar with the race car, you may not know where to set the balance bar. Start setting it at the center of its adjustment and perform a simple brake-torque test. Jack up the car and put it on jack stands so the tires are off the ground. Have someone sit in the car and lightly apply the brakes. With the helper holding the brake pedal, turn a front, then a rear wheel using a torque wrench on a lug nut as shown in the accompanying drawing. Record the results. If you can't measure brake torque with a torque wrench, use your hands and estimate it.

Start with front-brake torque at about 50% greater than rear-brake torque, if the car has 50—60% of its weight on the rear wheels. If the car is extremely nose-heavy, set front-brake torque at about 100% more than rear-brake torque. But if it is extremely tail-heavy, start with front- and rear-brake torques approximately equal. This also confirms that both front- and rear-brake systems work.

If front-brake torque is less than rear, adjust the balance bar.

The tricky part of balance testing is figuring out which end locks up first. You'll need some helpers to do this. As suggested, this is the time to use a still camera or video camera. With an instant-replay video camera, you can observe at leisure what happened.

A driver will have difficulty telling which end of a car locks up without flat-spotting the tires. Instead, he has to ease off the pedal the moment a wheel starts to lock up. If he doesn't,

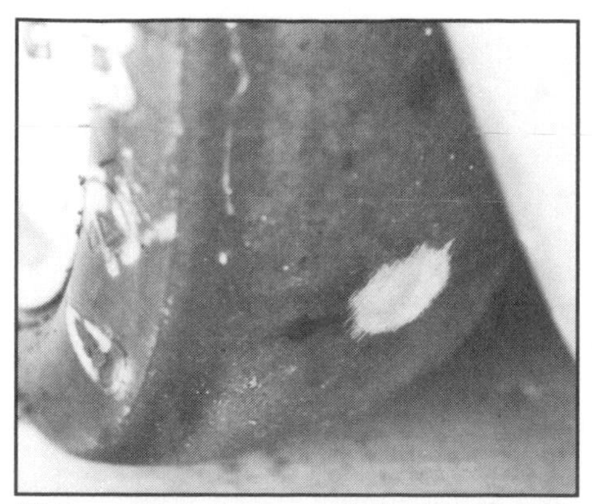

White spot on tire is cord showing through where rubber was worn off. Locking brakes risks flat-spotting tires and also going through to cord. Tire is ruined and must be replaced.

Wing at rear of sports racer improves rear-tire traction for improved handling and braking at high speed. As speed drops, so does wing downforce, increasing the chance of rear-wheel lockup when braking for tight corners.

Ground effects and rear wing provide maximum aerodynamic downforce. Wing also increases drag dramatically. NACA ducts optimize airflow to brakes with minimum drag.

he can grind right through the tires. Flat spots can ruin a tire. Be careful!

Do the first brake-balance test at moderate speed—about 40 mph. If you go too fast, the car will be hard to observe safely. If you're not using a video camera, position people on both sides of the track. Use two observers on each side if possible. Tell them to try and see if the front or the rear wheels lock up first. Don't lock up all four wheels. The test will then mean nothing, so don't brake too hard.

If the front wheels lock first, adjust the balance for more braking. Make sure you turn the adjuster in the correct direction! Record the number of turns on the adjuster in your data sheet and retest. Continue testing until you are satisfied with brake balance.

The front wheels should lock before the rears. Rear-wheel locking causes instability. If the car is turning or has any side forces on it, it will spin. Front-wheel locking causes a loss of steering control, but the car usually goes in a straight line. The final setting will depend on the type of car and the driver's preference.

If your car has wings or aerodynamic downforce from ground effects, brake balance may be different at high speeds. A rear wing loads the rear wheels as speed increases. On a car that has only a rear wing, if the brakes are evenly balanced at 40 mph, the front wheels will lock first at 100 mph. It is impossible to balance the brakes for all speeds on such a car. Compromise.

Race-car brakes are used more at high speed than at low speed. The minimum braking speed on the track is going into the slowest turn. Pick an average braking speed for the particular track and set the balance for that speed. If the driver strongly prefers one type of balance over another, the brakes may be balanced perfectly at top speed and increasingly unbalanced as speed drops. A lot of experimenting is needed to arrive at the best compromise.

Remember that brake balance is affected by the following:

- Tire-grip changes—could be caused by either track-surface changes or tire-temperature changes.
- Aerodynamic-force changes, such as a wing-angle change.
- Brake temperature—fade may occur at one end of the car before the other end.
- Change to different-size or -compound tire.
- Change to different-hardness brake lining.
- Change to the car weight distribution.
- Change in master-cylinder size.

Because it is difficult for a driver to detect small brake-balance changes, don't be concerned about small adjustments. However, if you make a

Severe changes in track conditions is why brake balance needs to be adjusted by the driver. If Geoff Lee's Formula Ford was set up for the dry, front wheels would lock in the wet. To correct, balance adjuster is changed to give more rear-wheel braking and less front. During testing, you can deliberately wet the track to find correct rain settings. Photo by Steven Schnabel.

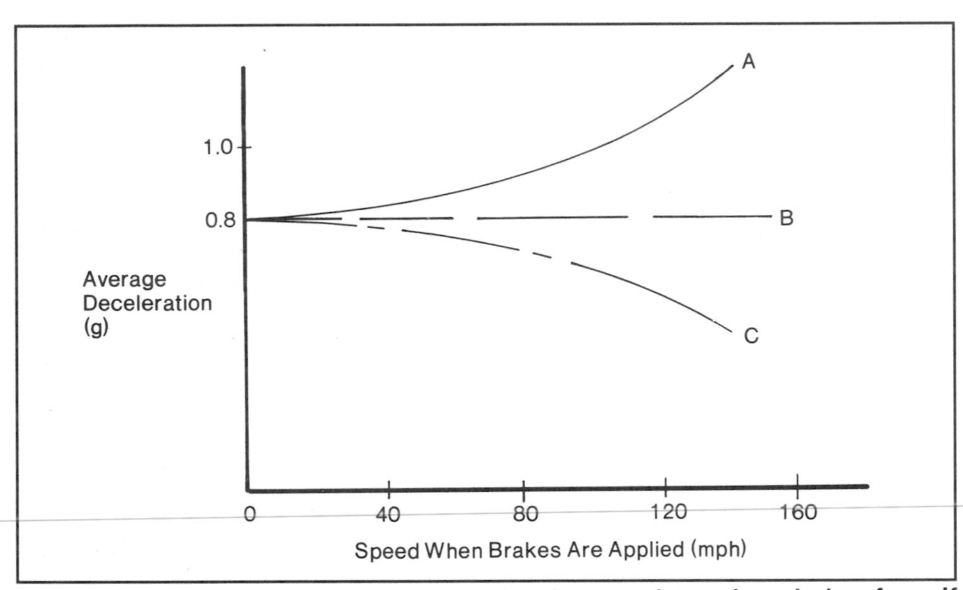

Average deceleration tests will tell you something about a car's aerodynamic downforce. If deceleration increases with higher-speed stops, stopping is improved with downforce, as shown by curve A. Curve B reflects a car with little or no downforce. Curve C is a car with problems—it has lift at high speed. Most road cars have less deceleration at higher speeds, but not as severe as curve B shows.

big adjustment, brake balance should be checked by testing. The most important factor is tire grip, caused by either changing track conditions or by the tires. Tires often lose grip with age, wear or temperature change. Test the car with the setup you will be running in the race.

If your car has a cockpit-adjustable balance bar, check it during a test session at the track. The driver should change balance to see how it feels. Also, observers should record what happens during hard braking. Run only a few laps, make only one change, and then discuss with the observers what they saw. If you make more than one change before testing, you won't be able to pinpoint what change to the brake system caused which observable change in brake performance.

If the track surface is dirty or wet, a balance test will be useless unless you plan to race on a similar surface. Instead, make a crude attempt at setting balance if the brakes were never set. Record maximum deceleration so you can judge how much to change the balance when traction improves. After a few brake-balance tests, you'll be able to relate balance settings to deceleration. Remember, *as deceleration increases, so must front-wheel braking.*

If your car has well-balanced brakes and you want to change a master cylinder or go to a different type of proportioning valve, this can be done without changing balance. However, you must measure and record pressure in both hydraulic circuits before changing parts. This can be done by installing pressure gages downstream from the front master cylinder and proportioning valve or the rear master cylinder. A convenient place is at the front- and rear-caliper bleed screws.

Modify two spare bleed screws by brazing the cross-hole shut, drilling through the center hole and brazing an adapter fitting to the end of the screw. Install one modified screw in a front caliper and one in a rear caliper. Connect a gage to each modified screw with flexible brake hose or steel tubing and bleed the gages. While someone applies the pedal to give 800-psi front pressure, read the rear pressure. You can now make the system changes. With the gages installed, adjust the balance bar or proportioning valve so front and rear pressures read the same as before the change. Be sure to keep this gage-and-bleeder-screw setup because it is an excellent tool for sorting out balance-bar, master-cylinder and valve problems later.

DECELERATION TESTS

Average Deceleration—This test is used to determine how rapidly a car stops. The test can be done with minimum effort using a stopwatch. If you know the time it takes a car to stop from a known speed, calculate deceleration:

$$\text{Average deceleration} = \frac{S}{22t} \text{ in g's}$$

S = Car speed before the stop in miles per hour
t = Time it takes to stop in seconds

To do this test you must know the speed of the car. If the car has a speedometer, have the speedometer calibrated at a speedometer-repair shop. If there is no speedometer, have a speedometer shop put the car on their speedometer-testing machine, and record the miles per hour that correspond to various tachometer readings. Make sure you record the gear ratio, driving-wheel tire size and what gear the car is in when making this test. Speedometer and tachometer calibrations are necessary because they usually read in error, causing errors in calculations.

When stopping from very high speed, deceleration changes during the stop. Therefore, average deceleration must be compared to other stopping tests done at the same speed.

Bring the car up to speed and hold it steady. Start the stopwatch the instant the brakes are applied; stop the watch the instant the car stops. Timing can be done by a passenger in the car or by an observer. Several tests should be run to minimize errors, but let the brakes cool between tests. If you are getting about the same results each time, your testing methods are good. However, it's difficult to do this test accurately, so don't be surprised if you can't get consistent results.

To reduce error, begin your stops from a higher speed. Start the stop watch as the speedometer reaches the desired speed. If a trackside observer is timing, you'll need to signal him when to start timing.

It's interesting to measure the average deceleration when braking from various speeds. If your test methods are consistent, you can plot a graph of deceleration versus car speed. This will indicate the amount of aerodynamic downforce. If average deceleration increases with stops from higher speeds, the car has increasing downforce at higher speeds. Most cars have less deceleration at high speed for several reasons. The wheels rotating at high speed take away some deceleration the instant the driver hits the brake pedal. Also, many cars have aerodynamic lift at high speed, so they have less traction.

While performing deceleration tests, also measure stopping distance for each speed. Providing you know car speed and stopping distance, deceleration can be calculated from the following formula:

$$\text{Average deceleration} = \frac{S^2}{29.9d} \text{ in g's}$$

S = Car speed before the stop in miles per hour
d = Distance it takes car to stop in feet

This test is much harder than using a stopwatch because it's difficult to hit the brakes at a certain spot. Observers can help confirm that braking really starts where the driver intends.

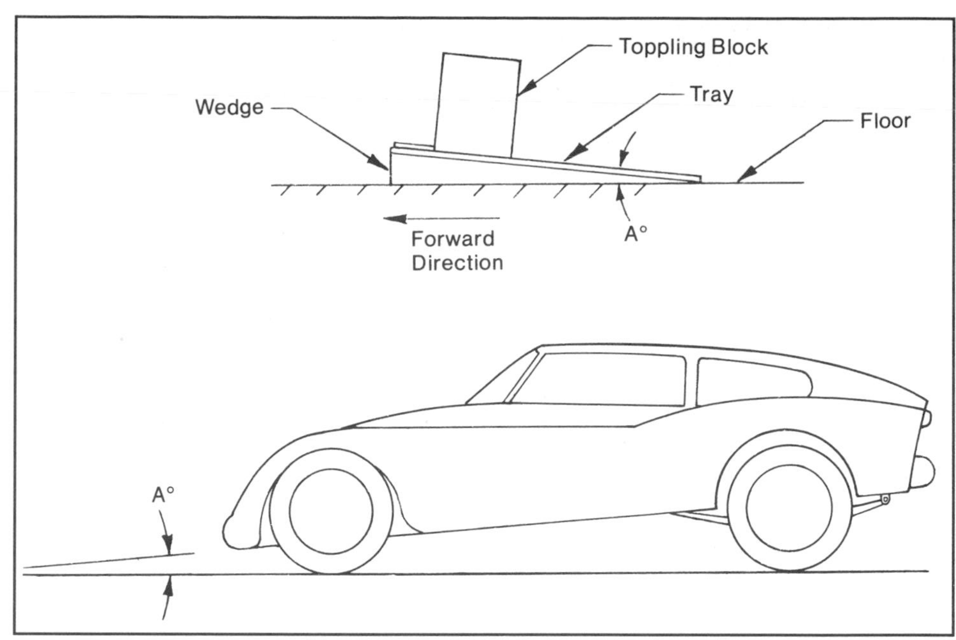

To compensate for nosedive during a toppling-block test, first photograph car from the side during hard braking. From photo, measure dive angle A of car as shown. Make a wedge at this angle and place it under toppling block tray. Use a protractor/level to check tray angle in car.

Again, accelerate the car to a constant speed and make several practice tests. If you do both time and distance measurements on each test, compare the results. If you measured the time and the distance accurately, you should get the same deceleration using the different methods. On the other hand, errors in measuring will give inconsistent results. Because speed is squared in the stopping-distance formula, the speed measurement is critical.

If you are having trouble getting accurate deceleration measurements, try this: Mark a line on the road or set up markers where braking should begin. Start the stopwatch as the car crosses the line. To apply the brakes exactly at the line, approach it with your foot off the throttle and positioned over the brake-pedal pad. This reduces driver-reaction time. It takes about 1/2 second to move your foot from the throttle to the brake pedal.

Have a passenger read the speedometer or tach. It may be necessary to reposition the instrument for easier reading. Or, if you have a camera that can be tripped electrically—necessary with a single-seater—you could use the brakelight switch to trip the camera and take a photo of the speedometer the moment the brake pedal is hit.

To get the speed right when the brakes are applied, run some tests without hitting the brakes. Back off the throttle at a mark on the road before crossing the line and observe the speed as you cross the line. Adjust initial speed so you cross the line at the desired speed.

If you are going to run a number of deceleration tests, you should invest in a decelerometer.

Maximum Deceleration Tests—You can measure maximum deceleration during a stop. You can also test deceleration at particular speeds without stopping. To do this, use a decelerometer or toppling blocks.

Decelerometer—Follow the instructions that come with the instrument. Most decelerometers require a passenger to take readings. On a single-seat racer, this is not practical. Instead, use a video camera mounted in the car to record decelerometer readings.

Toppling Blocks—The toppling-block method doesn't require a passenger. Toppling blocks—see drawing—can be built in a home workshop. Make a flat surface for the blocks to rest on. A flat tray that can be mounted firmly in your car is ideal. Don't use the car floor. Chances are it's not flat enough for accurate measurements.

Set up the blocks, gradually bring the car up to speed and hit the brakes. You will topple the blocks immediately, so you don't have to lose much speed. With the blocks, you'll find maximum or peak deceler-

Temperature-indicating paint is best way to measure rotor temperatures. Paint can be applied to outer edge of rotor and observed after test run. Experiment with different temperature ranges to get good test data. On an open-wheel car, a pyrometer can be used, but you must act fast to avoid having rotor cool off before measurement is taken.

ation during that range of speed when the brakes were applied. By running a number of tests, you can plot a graph of deceleration versus car speed. Again, this is an indication of vehicle downforce. If deceleration is higher at higher speeds, aerodynamic downforce is present and vice versa.

Each block falls over at a certain deceleration. Maximum deceleration during a stop will be indicated by the shortest block that fell over. The driver can tell immediately what deceleration he is getting by looking at the blocks.

There is some error in the toppling-block test. If the car nosedives while stopping, the blocks will tend to fall more easily than if the car stopped level. Most race cars don't nosedive enough to matter, but the error may be significant on a high-CG, softly sprung road car.

To compensate for nosedive, take a photo of the car from the side while it is stopping at maximum deceleration. Using the photo, measure the angle of tilt of the body. Compare it to a photo taken from the same angle of the car sitting level. Once the dive angle is determined, adjust the tray for the toppling blocks. Tip the tray up at the front, at the same angle the car nosedives. This will compensate for the error caused by nosediving. On a car with little nosedive, position the toppling-block tray level on the car floor.

Check Changes—When making deceleration tests, now's the time to check how deceleration is affected by changes other than to the brakes. For example, you can combine the deceleration test with aerodynamic tests to see how downforce affects braking. Combine this with brake-balance changes to achieve maximum deceleration. Remember, maximum deceleration occurs when brake balance is set so all four wheels lock simultaneously.

Although tire grip is the most important item affecting deceleration, it is also affected by a grade in the road or the wind. So test in both directions and average the results to eliminate these effects. On a windy day, a large error is introduced. Wind forces vary as the *square of the wind speed.* On a windy day, drag forces will change deceleration significantly.

EFFECTIVENESS TESTS

Brake-system effectiveness is the amount of deceleration obtained for a particular pedal effort. A brake system that is more effective stops a car with less pedal effort.

Effectiveness tests are useful for comparing brake systems and also for comparing linings. These tests must be run with a continuous-reading decelerometer, such as a U-tube decelerometer or recording g-meter and a line-pressure gage. With a recording g-meter, these tests can be run by the driver alone. With a U-tube decelerometer, an observer is required.

To run an effectiveness test, drive the car at a predetermined speed—such as 60 mph—and apply the brake pedal until a 100-psi pressure is reached. At that point, deceleration is read and recorded. Then accelerate the car back to 60 mph and repeat the test in increasing 100- or 200-psi increments until lockup is reached. Be sure to allow sufficient time between each pedal application for the brakes to cool.

Following each test, calculate pedal effort required for each pressure. Plot this data on graph paper with deceleration in g's on the vertical axis and pedal effort on the horizontal axis. For easy comparison, if you test another car or with different lining on the same car, plot the second test on the same graph. For a quick comparison and no graphs, run the tests at just one line pressure and compare decelerations.

Effect of scoop on brake temperature can be tested by enlarging scoop. If you make scoop larger and temperature drops, use this information to design an improved scoop. Use the smallest scoops to give acceptable brake temperatures because big scoops increase air drag.

BRAKE TEMPERATURE TESTS

Brake temperature is important to know. If the brakes get too hot, changes may be necessary. By measuring the temperature, you can check the success of a change. Usually, improved airflow to the brakes will be the most effective change.

Measure brake temperature with a hand-held or remote pyrometer. You can also use temperature-indicating paint or labels. The hand-held pyrometer gives the temperature at the time of the measurement. The remote pyrometer with thermocouples gives continuous temperatures. Temperature indicators give maximum temperature experienced at a specific location.

The disadvantage of a hand-held pyrometer is it takes time to stop the car, get to the brake, and make the measurement. By that time, temperatures have dropped. Therefore, the hand-held pyrometer will be useful only if you can get fast access to the spot to be measured. For example, it'll work on open-wheel race cars, but is almost impossible on full-fendered cars that don't have large openings in the wheels.

There are many ways to determine temperature. Whatever you do, plan the test first and *follow the written plan.* You must have good records to get any useful information. It doesn't do

Temperature-indicating labels are for use on disc-brake calipers. Note the word CAUTION starting at 320F (160C). Temperature indicates fluid may be boiling inside caliper. Photo courtesy AP Racing.

Temperature-indicating paint at different locations on caliper: Paint on outboard-piston housing is most important because it indicates fluid temperature where cooling air is minimal. Caliper should not exceed 350F (177C) at this point.

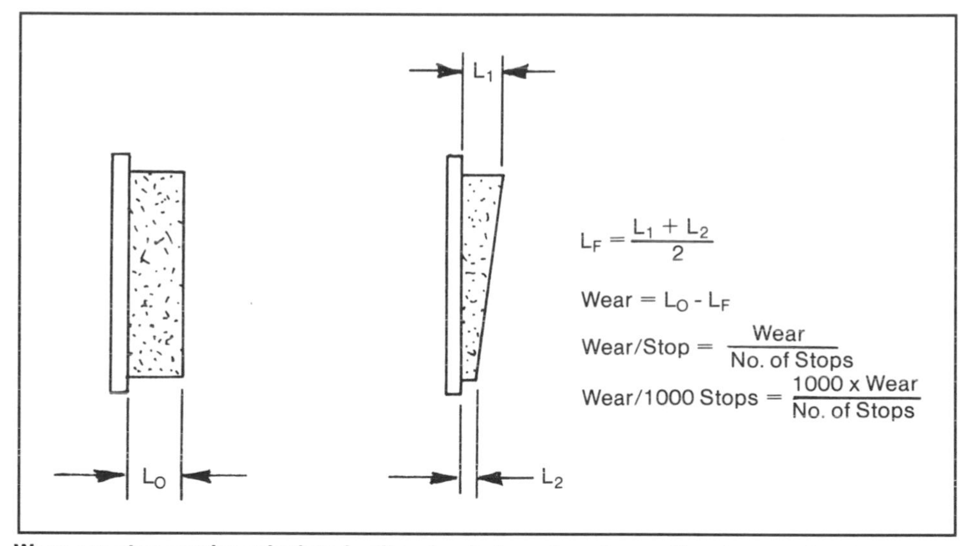

Wear per stop can be calculated using simple formulas shown. If pads wear in a taper, use average final thickness T_f to compute wear. Usually, wear per 1000 stops is more useful than wear per stop. Lining-wear calculations are very important for long-distance racing. Wear per lap or wear per hour may be most important so pad changes can be planned at fuel stops. Have crew make quick lining checks during tire changes to double-check calculations.

any good to say, "My brakes get to 1000F." You must know how many stops were made, from what speed, and how much time elapsed between each stop. Also, a car that sits still between stops is a lot different than if it was moving at 100 mph for the same time period.

If you don't have a remote pyrometer, use the next best thing—apply temperature indicators to your brakes. The paint and label indicators go on the edges of the brake rotor and caliper body, respectively. Usually, rotor temperature gets twice that of the caliper.

Accelerate to a predetermined speed and make one full stop. Check the temperature indicators and record the maximum-indicated temperature. If you have a pyrometer as well, use it. Temperature rise in one stop can be compared to a calculated number using the formula on page 11. The difference between calculated-temperature rise and the measured rise is the energy—heat—lost elsewhere, such as to the caliper, wheel, brake hats and air drag.

If you have a road-racing car, make a number of fast laps around the track at racing speed. After about 10 laps, bring the car in and measure brake temperature. After this many laps, brake temperature should stabilize. To check this, make several multilap tests, checking temperature after each set of laps. Eventually, you'll find how many laps it takes for the brakes to reach maximum temperature. They will not get any hotter unless there's a major track, car or driver change. With a remote pyrometer and thermocoupled linings, this can be done easily in one series of laps.

Pay particular attention to caliper temperature near the piston(s). If the temperature gets close to the brake-fluid dry boiling point, watch for trouble. A caliper running at over 500F (260C) will experience trouble. The seals eventually leak even if the fluid doesn't boil. So try to limit caliper temperature to 350F (177C) or less, to give longer seal life.

To evaluate differences between different components or setups, such as cooling-duct systems or rotor types, it is best to test one configuration on the left and one on the right. Then, switch sides and rerun the test.

Caution: When running tests with different setups on each side of the car, watch out for uneven braking. The driver can lose control!

No matter how careful you are, there are always test-to-test differences that are eliminated by running both configurations at once. Rerunning a test with the configurations switched and averaging results eliminates side-to-side differences.

When evaluating parts designed to run in one direction only, such as curved-fin rotors, be sure to switch parts when you switch sides.

Brake-Wear Test—Brake wear can be tested only after many stops. Earlier in this chapter I suggested measuring lining thickness before testing. If you do this, you can measure lining thickness afterward to judge wear. The amount of wear is interesting. But to be useful, you need to determine wear *per stop.*

To calculate wear per stop, you must know the number of stops or laps on the pads. You should be able to find this in your records if you recorded everything in your notebook as suggested.

After you finish testing, measure brake-lining thickness. Measure the

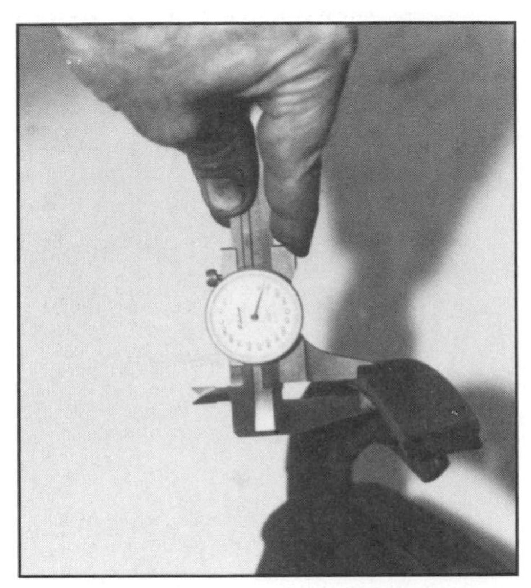

When measuring lining thickness to compute wear rate, take several measurements and average them. Because brake shoes and pads wear unevenly, wear rates are based on average wear over the lining. To compute minimum lining thickness, look for thinnest spot.

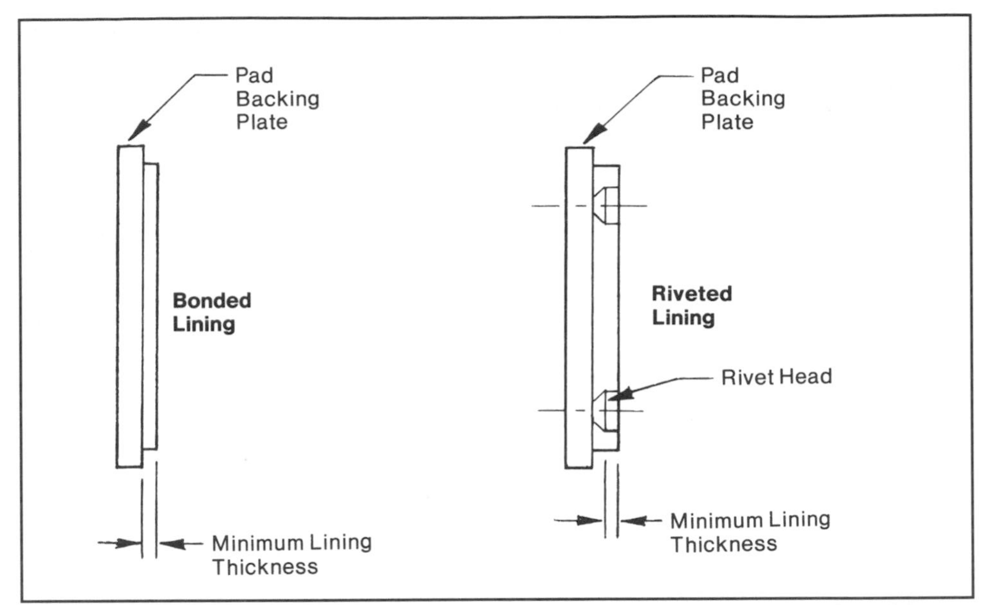

To compute number of stops available from a new lining, you must know its minimum-allowable lining thickness. This is different for riveted and bonded linings. Use depth-gage end of dial caliper to measure available wear of riveted lining. More wear will result in rotor or drum scoring.

linings in the same places as before and compare thicknesses to the original measurements. Wear is the original thickness minus the thickness after testing. Look carefully at the lining to see if it is worn on a taper. Make a diagram to record the information, as shown in the drawing.

$$\text{Lining-wear per stop} = \frac{W}{N}$$

in inches per stop

W = Total amount of lining wear in inches
N = Total number of stops

The answer from this formula will be very small. So, multiply it by 1000 to get wear per 1000 stops. This figure will be easier to use.

Once you have wear per stop, calculate the number of stops it takes to wear out the linings completely. To do this, find the minimum lining thickness allowed. Your car manufacturer or brake manufacturer should be able to supply this information. Obviously, you shouldn't allow the lining to wear so much that the metal backing plate contacts the rotor. Or, if you have riveted linings, it's the distance the rivets are recessed into the lining that you should be concerned with.

Most car manufacturers give minimum lining thickness in their shop manual. If you can't find this figure, assume you can wear the linings to within 1/16 in. of the plate or rivets.

Subtract minimum lining thickness from the thickness of a new lining to get maximum allowable wear. You can then calculate the maximum number of stops:

$$\text{Number of stops allowed} = \frac{W_M}{W_S}$$

W_M = Maximum lining wear allowed in inches
W_S = Lining wear per stop in inches per stop

Remember that brake wear increases with temperature. And, if you increase engine horsepower, the brakes will get hotter because you'll be stopping from higher speeds. Wear per stop will also increase. Aerodynamic changes also affect lining temperature and wear. Therefore, keep records of lining thickness each time you inspect the brakes. Also record the number of stops between inspections so you can keep track of changing wear rates from track to track and setup to setup.

Wear rates will change dramatically from track to track. One with long straights and tight flat turns is much harder on brakes than one with short straights and high-banked turns.

Lining-wear calculations are important in long-distance road racing. If you wear out linings and ruin a rotor, a lot of time will be lost in the pits changing it. By determining lining wear during practice sessions and calculating the number of stops or laps allowed, you can predict how many laps you can run before changing pads. Be sure to have enough sets of pads bedded-in to go the distance before the race starts.

This sort of calculation is useful only for race cars. Usually, you can determine both wear rate and the exact number of possible stops before a pad change is necessary. The brakes are always used as hard as possible, so the wear rate is predictable. On the street, the number of stops depends on the type of driving. City driving takes more stops than freeway driving. Fast driving in the mountains is a lot different than taking the family to the drive-in. Therefore, calculating wear per stop is impossible. So, for street use, keep a record of the car mileage each time the brakes are inspected. This, along with a record of lining thickness, will give you an idea of how many miles your brakes will last.

Maintenance 11

Proper maintenance is essential for brake-system safety and performance. If you have a good maintenance program, you can usually find a problem before it becomes serious. Don't wait until brake problems occur before doing something—you may encounter a life-threatening situation.

The brake system is one of the most critical parts of a car. Luckily, brakes are very reliable, even with poor maintenance. Brake trouble can usually be sensed at the pedal, so the driver knows it immediately. For example, the pedal gets lower and lower if the brakes need adjusting. And it goes soft if air gets into the system through bad seals. Worn-out linings make horrible noises as metal contacts metal. Most problems become evident before a total brake failure occurs.

One of the most dangerous problems is a line or hose failure. This potential problem can go unnoticed until the brakes are used hard. Then, a line or hose rupture can occur with the pedal suddenly dropping to the floor. If the car has a modern dual-braking system, the driver can usually stop the car. But, if the system has only two wheels braking and is in need of maintenance, it may be impossible to stop quickly. Only a good preventive-maintenance effort will prevent such problems.

Schedule—A good maintenance plan requires a schedule. If the car is raced, always do some brake maintenance before each race. Inspect the system for wear and damage. Also bleed the fluid before each event. If you fail to do this, brake performance will suffer.

On a race car, major brake-system maintenance is done each season—or more often if problems occur. Major maintenance may include replacing the seals and surfacing or replacing the rotors. Wheel-bearing lubrication is usually done at this time.

On a road car, your shop manual will give a recommended brake-maintenance schedule. Usually, it suggests inspecting the fluid level and checking the linings for wear after certain mileage has accumulated. You can modify this schedule if you find problems, but the manufacturer's recommendation is a good place to start. Here is my brake-maintenance schedule for a road car:

Every 5000 miles: Inspect hoses and lines; check fluid level.

Every 10,000 miles: Inspect linings for wear.

Every 25,000 miles or every year: Replace brake fluid.

Every 50,000 miles: Overhaul calipers; wheel cylinders, master cylinder.

This maintenance schedule does not cover everything. It just outlines the most common items and approximates brake-maintenance needs. Your car's needs depend on its design and how it's used. If the car is not driven often, you should inspect the fluid and overhaul the hydraulic system at lower mileage than recommended. Although lining wear should not be a problem, corrosion and seal leakage may be, because time becomes more of a factor.

Brake maintenance on race car should be done after each time car is raced. Mechanic prepares to disassemble and inspect brakes on this Indy Car after a practice session. He is using brake cleaning fluid to remove contamination. Cleanliness is vital when working on brakes.

BRAKE ADJUSTMENT

Brake adjusting is not needed with many cars. Disc brakes are self adjusting and modern drum brakes usually have automatic adjusters. If your car has such features, forget brake adjustment except when installing new drum-brake linings. Often, a drum brake must be adjusted manually to allow the automatic adjuster to work the first time. Automatic adjusters only maintain clearance—they don't make large adjustment changes.

On drum brakes without automatic adjusters, periodic adjustment is needed. As linings wear, clearance between the shoes and drum increases. This results in more pedal travel. Also, the parking brake requires more travel. You can usually tell when the brakes need adjusting by the increase in pedal travel. However, if you wait too long, the pedal may go to the floor if brake fade occurs. Therefore, adjust the brakes often; don't take risks.

Although duo-servo brake uses automatic adjuster, brakes must be manually adjusted after brake-lining replacement. Blade-type tool is used to turn adjuster through slot in backing plate (arrow).

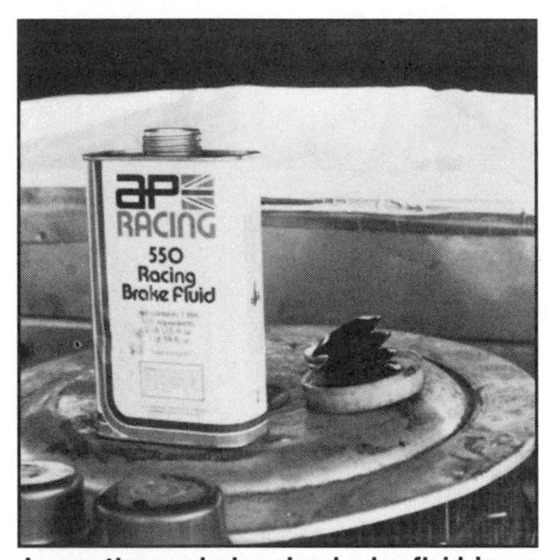

Inspecting and changing brake fluid is an important part of brake maintenance. Race car is getting a filling of AP Racing fluid. Note reservoir cap with rubber bellows. It isolates fluid from the air when cap is in place, but allows fluid to rise and fall.

Most modern road-car brakes are discs on the front and drums in the rear. With this dual-braking system, only the drum-brake system can go out of adjustment. Typically there is a brake-warning-light switch in the system that senses excess movement of the rear-brake fluid. When the brakes need adjusting, more displacement is required than the master cylinder can supply; the brake warning-light switch turns on the warning light. This switch will also detect fluid loss caused by a leak in either system.

Adjusting the rear brakes generally requires jacking up the car, supporting it, and getting underneath. There are many types of brake adjusters. One is a square bolt head projecting from the backing plate. Most domestic-car adjusters require inserting a tool into a slot in the backing plate and turning the star wheel. There may be one or more adjusters on each brake, depending on the design.

Follow the shop-manual instructions for brake adjustment. Be sure not to overtighten the brakes. A dragging brake will overheat and can cause damage, excess wear, lower mileage or fade.

Make sure the parking brake is not on when adjusting rear brakes. Otherwise, you won't be able to. The parking brake is applied to the rear brakes, which are usually drums. You will think the brake is fully adjusted if you try it with the parking brake on. Don't let this fool you.

Normally, rear drum-brake adjustment will take care of parking-brake adjustment. Except on some four-wheel, disc-brake cars, you should never need to adjust the parking brake unless it won't work *after* adjusting the rear brakes. Parking-brake adjustment is discussed on page 135.

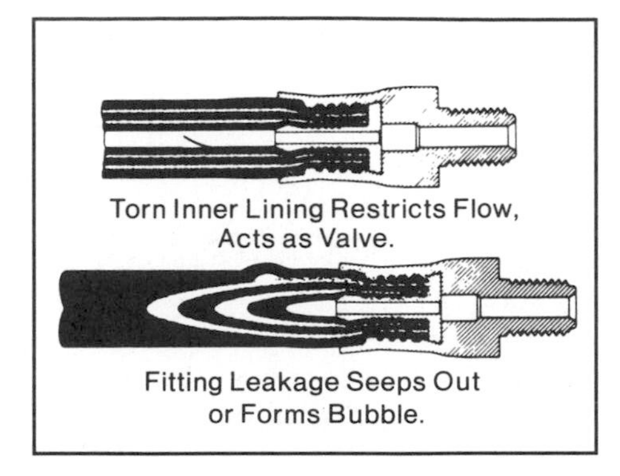

Bottom: Blistered brake hose is ready to fail. Replace such a hose immediately! Top: Hidden brake-hose failure restricts flow to or from brake. Thin layer of inner lining can lift off surface and restrict flow of fluid to brake. This may cause brake to apply later on one side, causing car to swerve. Old brake hoses should be replaced as a safety precaution. Drawing courtesy Bendix Corporation.

HYDRAULIC-SYSTEM INSPECTION

Besides brake adjustment, the most common maintenance item is checking the hydraulic system. To be safe, check the fluid level about every 5000 miles. Fluid leakage can cause total brake failure on a single brake system. It can also cause problems on a dual system. If you discover that fluid level has dropped each time you look at it, locate the problem and fix it. Fluid shouldn't leak from a brake system. Typically, a leak can be traced to a defective wheel-cylinder seal, caliper or master-cylinder. Or, a leak may occur at a fitting or hole in a line or brake hose. Look for drips and wet areas near the hydraulic system. Fluid leaks should be fixed immediately—procrastination can be dangerous!

In addition to fluid level, you should check all brake hoses and lines. This can be done while you're performing other maintenance under the car, such as adjusting the brakes or doing a lube job. Inspect the brake hoses at both ends of the car. Look for cracks, chafe marks, or other damage. A lump on a hose indicates failure. Fluid can seep through the first few layers of reinforcing cord and form a lump or blister on the hose. If a hose blisters, replace it immediately. Don't chance sudden brake failure.

Occasionally, a pinhole in a brake hose can allow air to get into the brake fluid. The hole can be the type that leaks only when the brakes are released, drawing in air, creating bubbles. When pressurized, this leak can seal and not let fluid out. If your brake hoses are old and weathered, replacing them may cure a recurring spongy pedal.

Brake hoses can fail because of improper installation. If a hose is twisted or installed in the wrong position, it can break. Or, if a car is driven on rough terrain, large wheel movements can strain a brake hose. Also, hoses can be damaged by rocks or road debris thrown up by the wheels. I've seen brake-hose damage caused by a car running over debris that gets tangled in a wheel and wrapped around the brake hose. Another cause of hose damage is letting a disc-brake caliper hang from its hose during maintenance: Use of correct maintenance procedures will avoid such potentially dangerous situations.

Check Clearances—If your car has wide wheels, fat tires, or other suspension modifications, look for chafe marks on the brake hoses. A tire touching a hose eventually can wear through it. Even something minor like a set of large-diameter heavy-duty shocks can cause similar problems.

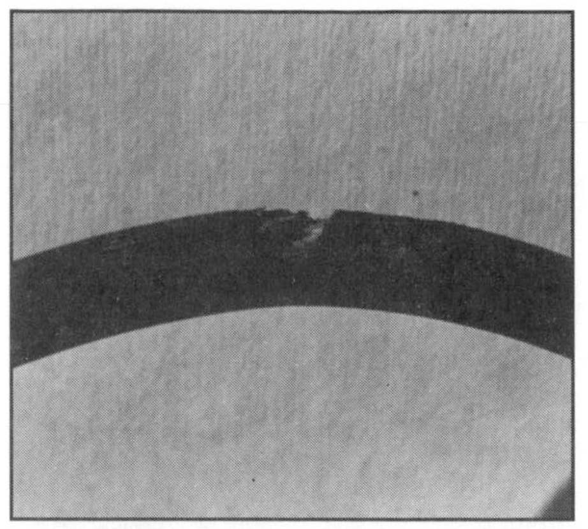

Someone previously installed wide wheels on my old Ford truck, but didn't check brake-hose clearance with front wheels at full lock. After some use, wheel-rim edge rubbed through line, causing a leak. Fortunately, damage was discovered during brake maintenance, rather than by "accident."

When brake line rusts, fittings can seize on steel tube. This can happen when seized fitting is turned with a wrench. Although such a line may not leak, it should be replaced.

Brake bleeding can be awkward and messy. Consequently, it's easy to do the job wrong or miss a bleeder. Take your time, plan the job right, and use correct tools. Start by jacking up the car and removing the wheels to improve access.

To check clearances, turn the front wheels to full right and left lock and check them. Check at both sides. Clearances may not be exactly the same. Jack up the car and look at clearance with the wheels at full droop. Next, jack up one rear wheel. This will compress the diagonally opposite front suspension. Check clearances, then jack up the opposite corner and recheck. Conditions and clearances change a great deal as the wheels travel up and down as they are steered. Try every possible combination to be sure. Although front hoses have the most clearance problems, particularly when turning, check the rear hoses as well.

If your car has been used off road, check the condition of the steel brake lines under the car. It's possible to pinch a line when going over a rough surface. If you find a damaged line, replace it. Look closely. A crack or a thin spot is very hard to see.

If your car is older, check the steel lines for corrosion. Rust can eat through steel lines, even though they are about as corrosion-proof as science can make them. The likelihood of having rusty brake lines is greater if you drive where salt is used on the roads in the winter. You can expect trouble from salt-water spray. Older cars from the Northeast and Midwest are likely to require brake-line replacement. Rusty brake lines is a usual problem on vintage cars. Consequently, most restorers replace them without bothering to check. Remember, the inside rusts from water in the brake fluid as well—something you can't check visually.

BRAKE BLEEDING

Bleeding is important to proper brake-system maintenance. It is usually done to remove air bubbles trapped in the system. Bleeding is a must every time the hydraulic system is opened up for any reason. It is impossible to prevent some air from entering the system if a fitting is loosened. Most systems have numerous high spots where air will be trapped.

Brake bleeding can be frustrating or easy, depending on the system design. Some bleeders are reached easily on the inboard side of the brakes, from underneath. On others, they may be tucked inside the suspension where they are real "knuckle busters" to get at. Some systems also have bleeders other than at the wheels, such as on the master cylinder. Consult your shop manual for any hidden bleeders. If you miss one, you will wonder why your pedal is spongy even after bleeding.

Bleeder screws can usually be turned with a small *six-point* box-end wrench. Avoid using an open-end wrench to prevent rounded-off bleeder screws and bloody knuckles. For difficult-to-get-to bleeders, special bleeder wrenches are available at auto-parts stores. These look like a box-end wrench with a huge offset.

Be aware that there is usually a bleeder at each brake, but some systems use a single bleeder to service both rear brakes.

Bleeders normally are found at:

- Each drum-brake wheel cylinder.
- Each disc-brake caliper—two or more bleeders may be used.
- Outlet port at some master cylinders, particularly those mounted at an angle.
- On some combination valves used with disc brakes.
- On accessories, such as trailer brakes.

Don't forget to look for bleeders on both sides of a disc-brake caliper. Some fixed calipers have bleeders on the inside where you would expect, plus another on the outside hidden by the wheel. If you don't remove the wheel, you'll miss the outboard bleeder.

Be sure to replace the rubber bleeder-screw caps if your brakes have them. They protect the bleeder-screw seat and threads from moisture and prevent corrosion seizure. These caps, which can be purchased from Volkswagen dealers, are a good addition to any road car.

Another frustration involves the metering valve used on some road cars. A special clip is used to hold this valve open so the brakes can be bled.

Note that caliper has a bleeder on each side. Forget to remove the wheel and you'll miss bleeding outboard side. This can result in a spongy pedal. Photo courtesy AP Racing.

Pressure bleeder is used in most repair shops because of its speed. Huge can of fluid is used to fill pressure bleeder. Because most racers fear possible use of old or contaminated fluid, pressure bleeders are not popular. For road cars, they offer speedy low-cost brake bleeding in a shop.

If it's not held open, the brakes can't be fully bled.

Consult the shop manual for the correct information on your car. If you don't have a shop manual, visit a car dealer and ask a mechanic. It's too dangerous for guesswork in this area. You may also have to buy a special tool from the dealer. This tool is often a spring designed to flex without damaging the valve as fluid flows through it.

Sequence of Bleeding—Bleeding should be done in a specific sequence to avoid the need to double back and rebleed. Follow this sequence if you're bleeding one wheel at a time:

- Master cylinder—if it is bleeder equipped.
- Combination valve.
- Bleeders on intermediate hardware in system.
- Wheel farthest from master cylinder, usually the right rear wheel. Note: If wheel cylinders have two bleeders, do upper one first; if a caliper has two bleeders, do inboard one first. If it has more, follow the manufacturer's recommended bleeding sequence.
- Bleed other brake at that end—usually left rear.
- Bleed next wheel farthest from master cylinder—usually right front.
- Bleed wheel remaining—usually left front.

After the bleeding process is complete, check the system by pressing hard on the pedal to be sure it stays firm. If the pedal is spongy, rebleed system or check for a leak.

To check for air in the brake system, have someone pump the pedal several times, then hold it down firmly while you remove the reservoir cap. Watch the fluid surface while your helper slides his foot off the pedal pad so the pedal snaps back and system pressure drops immediately. With a well-bled system, there will be a little turbulence at the surface of the fluid. If fluid boils up more than 1/4 in. above the surface, there's probably air in the system.

Bleeding Methods—Three commonly used techniques for brake bleeding are:

- Pressure bleeding.
- Manual bleeding.
- Special methods, such as gravity bleeding or vacuum bleeding.

These all work. People argue which method works better. All good arguments are so complex that nothing will ever be settled. So, I suggest that if you have trouble with one method, try another until you find the one that works best for your car.

I'll explain some of the problems with each method. Then I'll explain how each job is done.

Pressure bleeding requires a pressure-bleeder tool. The pressure bleeder pressurizes the fluid in the reservoir. Once pressurized, you open each bleeder in the sequence noted until no air is left in the system.

The advantage of pressure bleeding is that it is fast and one person can do the job. But, there are several disadvantages: The tool is expensive and a lot of fluid is wasted. If you don't use the pressure bleeder often, the fluid in it gets contaminated, even though it's in a sealed container.

Fluid flow in the lines is high if a bleeder is opened too far during bleeding. This can cause bubbles to form within the system from turbulence created as the fluid flows around sharp corners. This defeats the purpose of bleeding. Conversely, rapid fluid flow flushes out air trapped in high spots in the system. Bubbles clinging to the walls of rough components will be washed away by the flow.

With all the pros and cons, it is not clear if pressure bleeding is good or bad. Most auto-repair garages use it because pressure bleeding is fast—most racers, don't due to potential fluid-contamination problems.

Pressure bleeding has an advantage on an older car which has had little brake maintenance. The master cylinder does not have to be stroked to push fluid through the system. This is unlike the manual-bleeding method that requires full strokes of the brake pedal. On an older car, there may be corrosion and dirt inside the master cylinder. Stroking the pedal to the floor moves the piston deep into the master cylinder, where it would other-

Most racers prefer manual-bleeding method: One person opens and closes bleeders while helper operates brake pedal. Although optional, it's helpful for third person to watch reservoir and add fluid when needed. Method is reliable, but success depends on communication between those doing job.

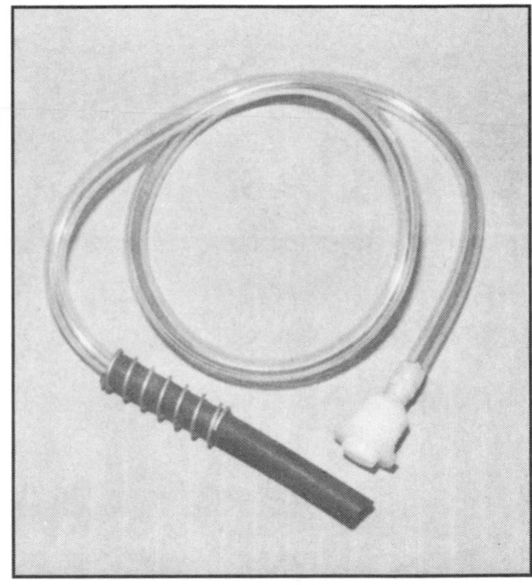

CTA Manufacturing's bleeder hose, available at many auto-parts stores, makes it possible for one person to manually bleed brakes. Hose is slipped over bleeder, bleeder is opened and free end is put in catch can or bottle. Fluid or air can only flow out of hose. When pedal comes up, one-way valve closes. This prevents air from entering hose. Using hose requires getting in and out of car to check for air bubbles in hose and fluid in reservoir. Photo by Ron Sessions.

wise never travel. Corrosion and dirt would likely damage the seals, causing them to leak. Pressure bleeding avoids this.

Manual bleeding is cheaper, but requires at least two people; while one operates the pedal, the other opens and closes the bleeders. The fluid reservoir has to be topped up during the bleeding process. Fluid flow can be fast or slow, depending on how fast the pedal is pushed and how much each bleeder is opened. Manual bleeding can be controlled better than pressure bleeding, so most racers prefer it. And, the job wastes little fluid and doesn't require expensive tools.

Gravity bleeding can be done only with certain brake systems. As you know, air in the brake fluid forms bubbles that float to high points in the system. If the high point is the master cylinder, air will come out automatically.

Gravity bleeding is not used often because it takes too long. Also, in almost every brake system, there are high spots that make it impossible. If you are designing a brake system from scratch, you may be able to make it possible to self-bleed by gravity, but most systems are not designed that way.

Vacuum-bleeding can be done by one person. A small hand-operated vacuum pump is attached to a jar—or plastic bottle—with a closed top. The jar has a hose that goes to the bleeder. Air is evacuated from the jar and the bleeder is opened. The vacuum pulls fluid and air bubbles from the hydraulic system and into the jar.

Vacuum bleeding sounds neat, but it has a potential problem. Brake-cylinder cup-type seals are meant to work with fluid pressure greater than the outside air pressure. With the vacuum pump, fluid pressure is below the outside air. This may draw in air past the seals, unless your hydraulic system has cup expanders. I don't

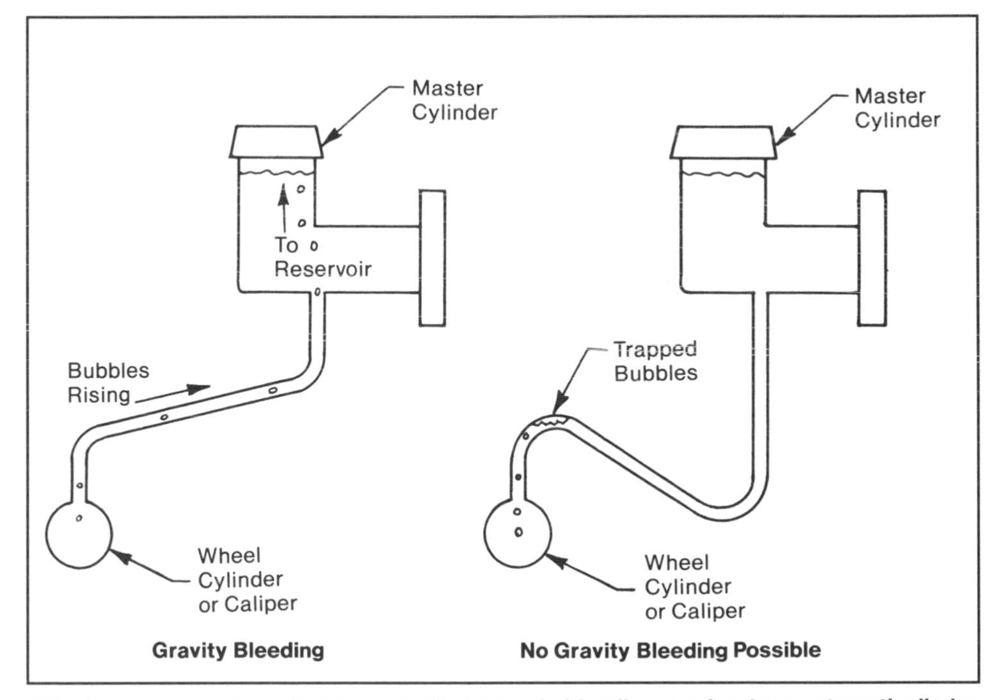

If brake system is free of high spots that trap air, bleeding can be done automatically by gravity. Air bubbles rise to reservoir. If system has high spots in lines, as shown at right, system will not gravity bleed.

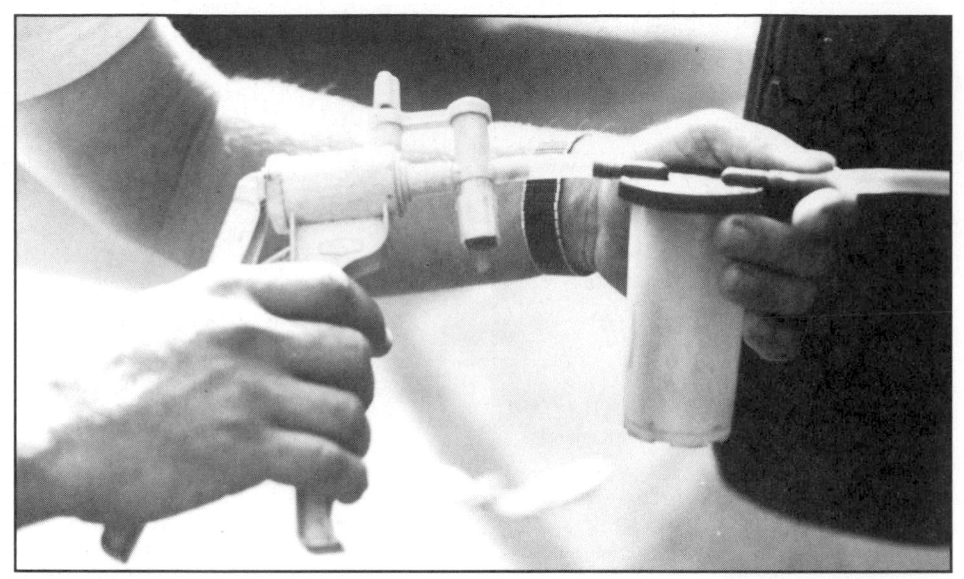

Vacuum bleeder sucks fluid or air into container through tube connected to opened bleeder. Although tool is easy to use and requires only one person, I don't recommend it. Seals in brakes can allow air to be drawn in if vacuum is applied to brake fluid.

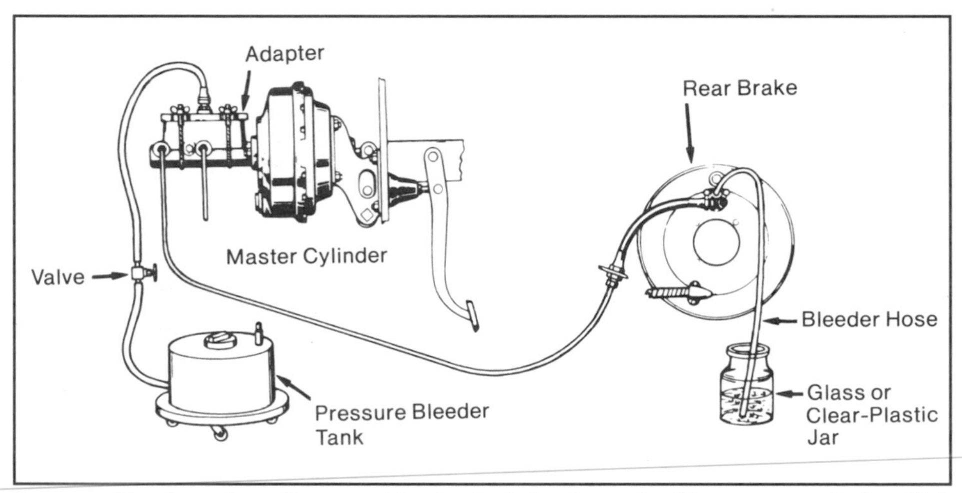

Pressure bleeder setup: Pressure-bleeder tank is charged with compressed air, which pushes fluid into brake system. Adapter clamps to reservoir, sealing fluid entering system. Drawing courtesy Bendix Corp.

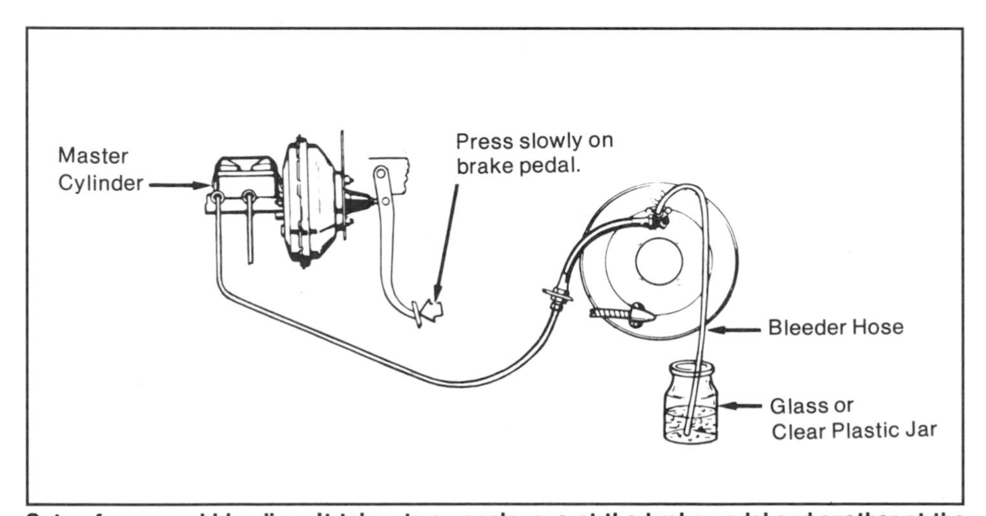

Setup for manual bleeding: It takes two people, one at the brake pedal and another at the bleeder. Person operating bleeder should give instructions to person operating the pedal. Bleeder is opened only when pedal is *being* pushed down; bleeder is closed *before* pedal is returned. Drawing courtesy Bendix Corp.

recommend the vacuum-bleeding method, although some people use it.

Many modern road cars are vacuum filled at the factory. The system is pumped down to a nearly perfect vacuum through the master cylinder by a very large vacuum pump. Then, fluid is forced in at about 100 psi. Because virtually all the air is removed before the fluid is put in, no bleeding is required. Systems filled like this must have cup expanders, but no residual-pressure valves.

Pressure and manual bleeding are detailed in the following pages. Before starting, consult your shop manual to see if special procedures must be followed for your car. I'll describe what is done for a typical car, but be aware that it may not be correct for your particular model.

How to Pressure Bleed—Pressure bleeders are used to save time by most professional mechanics. It is a one-man job. It is also the fastest way to bleed brakes. You must have a pressure bleeder to do the job, plus a source of compressed air. Pressure bleeding also requires a large amount of brake fluid to fill the pressure bleeder.

A pressure bleeder has a tank containing brake fluid, plus an air compartment to pressurize the fluid. The fluid and air are separated normally by a rubber diaphragm to prevent moisture in the air from contaminating the fluid. The fluid outlet connects and seals to the master-cylinder reservoir with an adapter that fits in place of the cap. This only works on a master cylinder designed with a sturdy integral reservoir. Some master cylinders with remote reservoirs cannot be pressure bled. Again, check the shop manual.

With the pressure bleeder connected to the master cylinder, pressure is applied to the hydraulic fluid from the air pressure in the tank. About 30 psi is used. Follow bleeding sequence described earlier. To do the bleeding, each bleeder is opened about 1/4 turn—no more. Otherwise, excess fluid flow results, allowing air and fluid to escape. When only fluid comes out, the bleeder is shut.

Each pressure bleeder comes with instructions. Also supplied are adapters that fit popular master-cylinder reservoirs. Bled fluid flows through a clear plastic hose pushed on the bleeder and into a container. Old fluid or any fluid bled through a hydraulic system should never be reused.

A pressure bleeder is convenient for changing all the fluid in a brake system. It allows the whole system to be refilled without the need to stop and top up the master cylinder. However, if you have the fluid changed by someone else, you may not get to choose the fluid that goes in your brake system. Whatever is in the tank is what you get, no matter what you think is best for your car. The fluid may be the best or it may be the cheapest the shop could buy. And, it may be new or old. This is why most racers won't use a pressure bleeder.

How to Manual Bleed—Manual bleeding is simple, but takes at least one helper, as mentioned earlier. The equipment needed includes some clear plastic tubing to fit the bleeder, and a glass or clear-plastic jar to catch the bled fluid. You must have a quart or two of fresh fluid, unless you are changing the fluid. Then, you may need about one gallon.

Bleeding sequence remains the same. Pedal pressure forces fluid and air bubbles out the bleeder. Push down on the pedal and hold it. Open the bleeder. Fluid should flow out the bleeder, through the tube, and into the jar. If the jar-end of the tube is covered with fluid, you'll be able to see air bubbles come out.

After the bleeder is opened, the pedal should be go down slowly until it stops against the floor or stop. Close the bleeder before the pedal is released. Usually, the pedal will have to be pumped several times to "bring it up" before opening the next bleeder in the sequence.

Watch the reservoir at all times. Plan to add fluid at least once while bleeding each wheel. If the reservoir runs out of fluid, air will be drawn into the master cylinder and you'll have to start bleeding all over again. It helps if a third person watches the fluid reservoir while one pumps the pedal and the other bleeds.

As you pull the hose off the bleeder, fluid in the hose will run out. When you bleed the next wheel, the first bleeding will push air out of the hose. This makes lots of bubbles in the jar. So don't assume this is air from the bleeder.

On a road car, don't pump the brake pedal with the bleeder open, even though the hose is filled with fluid and submerged in fluid in the jar. There's a risk of drawing contaminated fluid back into the system. Fluid must always be pushed out the bleeder, never drawn in. On a race car, the fluid is usually clean, so drawing in old fluid is less risky because it should be relatively fresh.

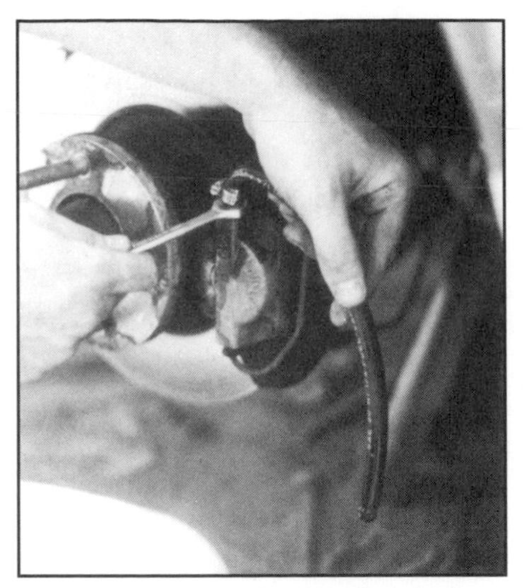

Not a good setup for bleeding brakes: Hose should be clear and free end immersed in fluid in clear bottle or jar. Bubbles can then be seen and mess is avoided.

Plastic bottle and clear plastic hose is handy for brake bleeding. Note that bottle is supported by wire looped over bleeder screw. Bottle at each wheel minimizes mess and brake-bleeding chore. Clear plastic bottle is better than one shown. Photo by Tom Monroe.

Simultaneous Bleeding—*Simultaneous bleeding* is another manual-bleeding method sometimes used on race cars with dual master cylinders. With this method, a front and a rear brake are bled at the same time. Simultaneous bleeding allows a full stroke of each master cylinder and a straight motion of the balance bar. To do this, you need another helper and tube and jar.

To perform simultaneous bleeding, partially fill the jars with fresh brake fluid. Connect one end of each tube to a bleeder and submerge the other end in a jar of fluid. Bleed the front and rear wheels farthest from the master cylinder first. If the calipers have two bleeders, bleed the inboard side first.

With gentle pressure on the pedal, *simultaneously* open both front and rear bleeders. Air will bubble from the hoses and some fluid will follow. With the bleeders open, slowly pump the pedal up and down until bubbles stop. Then, close the bleeders. Go to the other side of the car—inboard bleeders if you have them—and bleed those brakes. Remember, always keep the reservoir-fluid level high enough to prevent running out of fluid during the bleeding. If you suck air into the master cylinder, you must start over.

Once both sides are bled, test the action of the balance bar. With the tubes still in the jars, open one bleeder and push the pedal slowly to the floor. This simulates failure of one set of brakes. The balance bar shouldn't bind, nor the pedal bottom. If OK, close the bleeder and release the pedal. Test the other brake system by opening a bleeder at the other end of the car. This tests the balance bar at its maximum angle. One master cylinder is filled with fluid under pressure; the opposite one has no pressure at all. If everything is properly designed and built, the balance bar should be free at either extreme angle. If it binds, find out why and remedy the problem.

Simultaneous bleeding is not recommended for road cars. Unlike a race car, brake fluid in a road car gets bled infrequently, so it will be dirty and contaminated. Don't risk pulling old fluid back into the system by pumping the pedal with the bleeder open.

Eliminating Bubbles—Brakes are sometimes spongy after manual bleeding. Slow pedal movement avoids fluid turbulence that would otherwise form bubbles. But, slow fluid flow will not dislodge small bubbles that cling to rough surfaces. If you tap the calipers gently with a plastic mallet while bleeding, this will help dislodge those stubborn bubbles.

Bleeder Problems—Sometimes bleeders cause trouble. A common problem

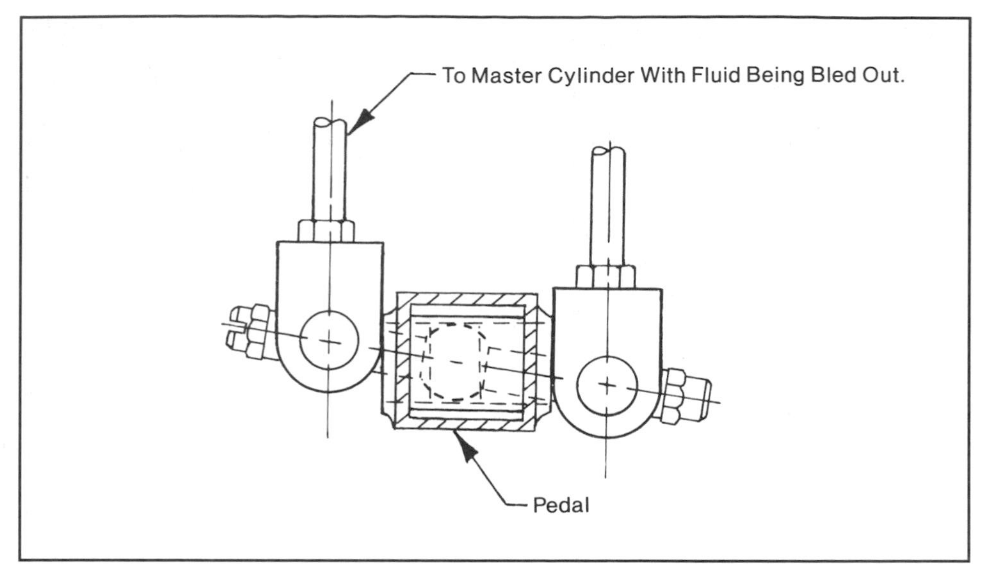

Master cylinder being bled strokes its piston forward until pedal hits stop. Other cylinder piston doesn't travel as far. As a result, balance bar assumes extreme angle when one master cylinder is bled. Balance bar must not bind at this angle, or breakage may result.

Penetrating oil is sprayed around studs before removing stuck brake drum. Before using penetrating oil, make sure adjuster is loose, parts are clean, and all attaching clips and screws are removed. Penetrating oil must be removed after use.

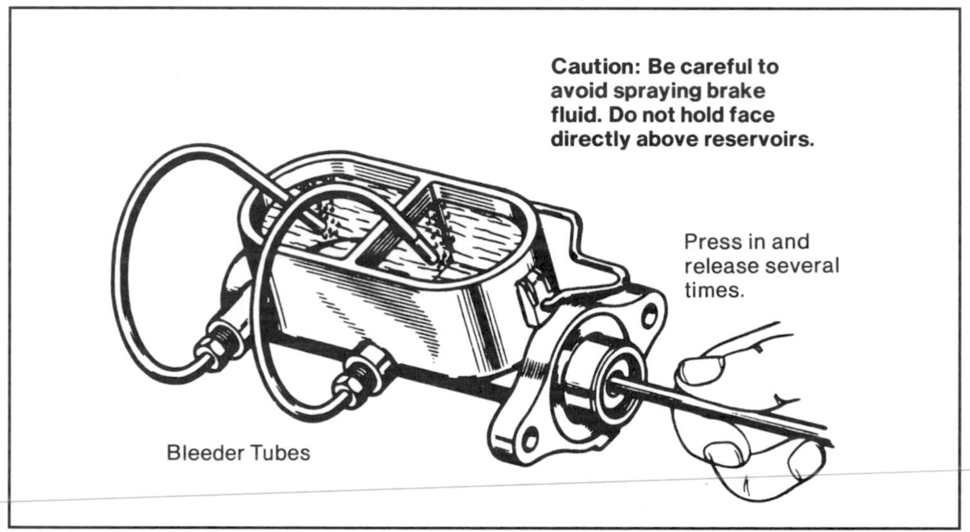

Dry master cylinder should be bled as shown. This can be done off the car. Bleeder tubes can be made of copper or steel brake tubing. When reinstalling master cylinder, transfer lines quickly to minimize fluid leakage. Drawing courtesy Bendix Corp.

is a rusted bleeder that seizes. It must be loosened with care with the correct wrench—no open ends. A broken-off bleeder screw means you'll have to replace the entire component—wheel cylinder, caliper, master cylinder or whatever—because it is almost impossible to remove the broken half without damaging the delicate seat.

Loosen a stuck bleeder screw by cleaning the area around it with a wire brush. Try turning it after cleaning. As a *last resort* try a good grade of penetrating oil. This is a last resort because oil can get inside and contaminate the brake fluid. After using penetrating oil, remove it by disassembling the component and cleaning it completely with brake-cleaning fluid or alcohol.

The best penetrating oil I have found is called *Kroil.* Available from Kano Laboratories in Nashville, Tennessee, it comes in a spray can with a plastic tube that attaches to the nozzle. This allows you to direct the oil exactly where you want it with less risk of contaminating other brake parts. Be careful! If penetrating oil gets on brake linings, they must be replaced.

To help prevent internal corrosion and sticking, install rubber bleeder-screw caps.

Filling Dry Tandem Master Cylinders—Fill a dry tandem master cylinder with care. It's not good practice to pour in fluid and pump the pedal. With air trapped inside, the piston will bottom and possibly cause internal damage.

If the master cylinder is off the car, fill it at the workbench. Remove internal bubbles by bleeding it as shown in the accompanying photographs. The tubes illustrated can be made of copper or steel. They install in the master-cylinder outlet ports.

After bleeding all the bubbles, install the master cylinder on the car. When connecting the brake lines, transfer them quickly to avoid losing all the fluid. This process makes bleeding easier, because no small bubbles get trapped inside the master cylinder.

A master cylinder used with disc brakes usually doesn't have a residual-pressure valve. Thus, *bench-bleeding* will only move fluid back and forth in the tube. To prevent this, hold a finger over the end of the tube during the return stroke of the piston. This prevents fluid from being drawn back into the tube on the return stroke. The drum-brake side of a tandem cylinder may have a residual pressure valve that prevents reverse flow of the fluid below the residual-pressure setting of the valve.

DRAINING BRAKE FLUID

When changing brake fluid, drain the hydraulic system completely. This *cannot* be done through the bleeders. This is due to the basic function of the bleeder. Bleeders are designed to let air out of a hydraulic system, so they are located at each high spot in the system. After draining from the bleeders, fluid remains in all low spots. Some people think that pumping the pedal will push all of the fluid out. Although this gets most out, some fluid will remain.

To get all new fluid, one method requires refilling and pumping out the system many times. This forces all of the old fluid from the low spots by mixing with the clean fluid. This process is better than nothing at all, but it is expensive to do it right because a lot of fluid is wasted. And, it never gives 100% clean fluid.

The correct way to drain old fluid is to open up the system at every low spot. This usually means disassembling wheel cylinders, removing or disassembling calipers, and unscrewing fittings. The master cylinder will also have to be removed and drained. Because this involves as much work as a hydraulic-system rebuilding job, it makes sense to replace all the seals at the same time.

Glycol to Silicone—When changing from a glycol-based fluid to silicone fluid, complete draining is a must. To take full advantage of the properties of silicone fluid, drain every drop of the old fluid, flush the system with alcohol, and rebuild all of the cylinders.

Cleaning brake parts should be done with special brake-cleaning fluid. CRC Brakleen is used at start of brake-maintenance job. Petroleum products should never be used on brake parts, as brake fluid is easily contaminated. Photo courtesy CRC Chemicals.

HYDRAULIC-CYLINDER REBUILDING

In time, the seals in a wheel cylinder, master cylinder or caliper will leak. Contamination or corrosion makes this happen sooner. Caliper seals that are old or have been subjected to high temperature also lose their elasticity and limit piston retraction. Usually, all brake-system cylinders need rebuilding.

Rebuilding a wheel or master cylinder should start with getting your car's factory shop manual. There are many different brake designs. Some require special rebuilding procedures. Most wheel or master cylinders can be rebuilt, but some are impossible because bore surfaces are usually pitted from corrosion. The cylinder then cannot be honed. Therefore, before buying rebuilding kits, disassemble the cylinders and inspect them. Even though your cylinders can theoretically be rebuilt, they may be too damaged to allow it.

To be safe, replace an entire cylinder if it leaks. For older cars for which new cylinders are not available, a machine shop can bore out the old cylinder and press in a stainless-steel or brass sleeve. White Post Restorations of White Post, Virginia, specializes in sleeving collector-car brake cylinders.

Brake-cylinder rebuild kits usually include rubber seals, dust boots and, sometimes, internal springs or seal expanders. Often, an older design will have newer-style seals in the kit. Many replacement cup-type seal kits include expanders. These add reliability to the seal. Rebuild kits are a lot cheaper than a new cylinder, particularly if you have to buy a caliper.

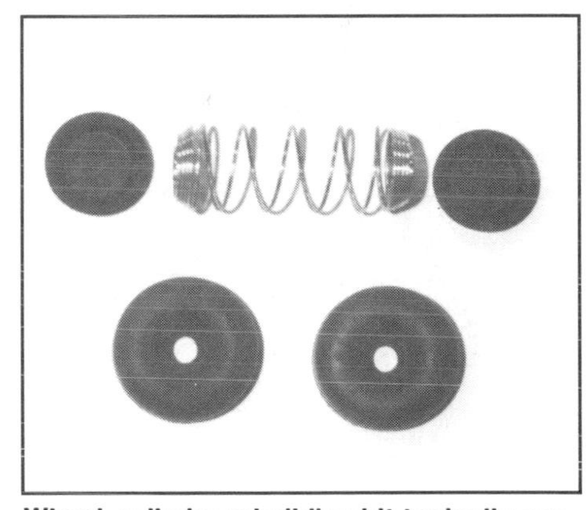
Wheel-cylinder rebuilding kit typically contains cup-type seals, boots and spring. Make sure wheel cylinder is rebuildable before buying kit.

Disc-brake calipers usually must be removed from the car for rebuilding. Some drum-brake wheel cylinders can be left in place on the backing plates for rebuilding, providing you can reach them with a hone.

Wheel Cylinders & Master Cylinders—The first task is disassembling the cylinder completely. If it is rebuildable, clean up the cylinder bore with a brake-cylinder hone. You can buy a hone at an auto-parts store. Get a hone with three small replaceable stones. They look like a miniature engine-cylinder hone. Use brake fluid—the type you'll be using in the brake system—as a lubricant while honing. *Never use solvent or any petroleum product on brake parts.*

If a cylinder can be rebuilt, a light honing is required to remove surface imperfections. After honing, check bore diameter with a feeler gage. If a narrow 0.006-in. feeler gage can be inserted between the piston and bore, you should replace the cylinder. You can also use an outside micrometer and telescoping gage to measure piston and bore diameters. If the difference between the two measurements is 0.005 in. or more, replace the cylinder.

If a cup-type seal is used in a cylinder that has excessive clearance between the piston and bore, a condition known as *heel drag* can occur. The seal gets pinched in the space between the

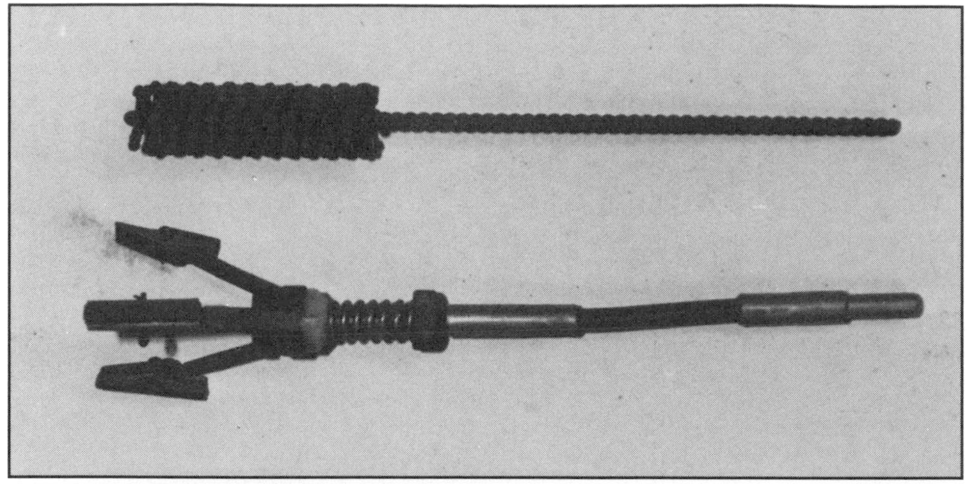
Two types of brake cylinder hones: Upper hone one is made for a specific cylinder diameter; lower one is more expensive, but will adjust to different bore sizes. Buy the adjustable hone, as it will be cheaper in the long run.

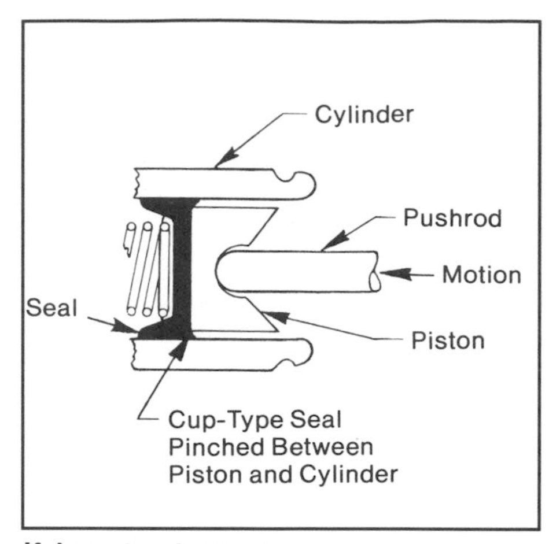

If bore-to-piston clearance is too great, rubber seal may wear excessively and can inhibit piston movement. The cause is heel drag—seal gets pinched between piston and bore. Honing beyond factory limits causes this condition.

Hone is driven by drill motor while brake fluid lubricates hone and keeps stones from loading up with metal particles. Never use oil or solvent for honing brake parts, as it will damage seals.

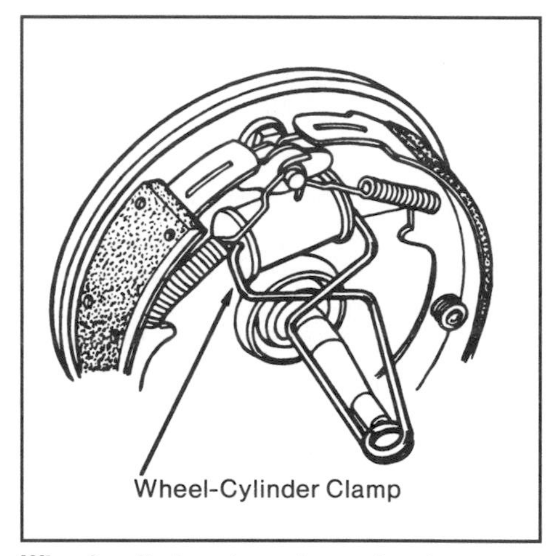

Wheel-cylinder clamp is used to keep pistons inside wheel cylinder while shoes are removed. Once shoes and return springs are reinstalled, clamp is removed. Drawing courtesy Bendix Corp.

piston and bore. This causes the piston to retract very slowly or stick. Avoiding heel drag is a must for proper braking.

When you replace the seals, wash all of the parts with clean brake fluid, blow out with clean dry air, and lubricate with clean brake fluid. Again, remember to use the same type as what you'll be using in the system. Use only brake-part cleaner for the first cleaning, but use only new brake fluid for lubricating. Don't assemble a cylinder dry. Grit and grease must be kept out of the brake system as well.

If you are rebuilding a master cylinder, fill it with the exact type of fluid you plan to use and bleed it before installing it on the car. This will make brake bleeding much easier later. Bench-bleeding a master cylinder is discussed on page 128.

Wheel cylinders don't have to be filled with fluid, only lubricated with it. They will fill easily during bleeding.

You may find a wheel-cylinder clamp handy to hold in the pistons while installing a wheel cylinder. The dust boots don't always hold in the pistons. The wheel-cylinder clamp, a simple tool available at most good auto-parts stores, keeps the internal spring from pushing out the pistons and seals.

Caliper Rebuilding—Disc-brake-caliper rebuilding is more complicated than rebuilding a master cylinder or wheel cylinder. You usually have to disassemble the caliper to remove the pistons. Don't get brake fluid on the pads if you plan to reuse them. On some calipers it is hard to remove the piston(s). Special caliper-piston removing tools are available. Another method is to use compressed air at the hole where the hose installs. If you use compressed air, the pistons will come flying out. Be careful! Flying parts can cause injury or damage! Wrap rags around the caliper and trap the parts. And don't use too much pressure. Be especially careful when removing plastic pistons. They are easily chipped and scratched.

Some caliper bores can be cleaned up with a hone; others cannot. Check your shop manual for specific instructions. For those that can be honed, the operation is similar to

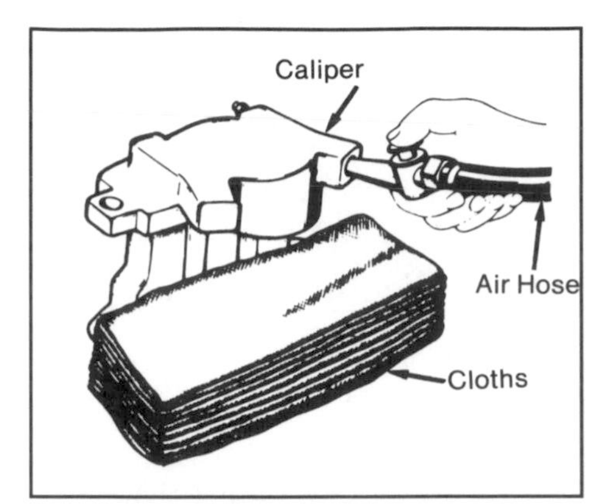

Usually, disc-brake piston can be removed with low air pressure applied through brake-hose hole. Make sure area in front of piston is well padded to prevent piston from flying out and bouncing into your face. Don't apply air pressure until you cover everything with a rag. When piston comes out, brake-fluid spray will follow! Job can be dangerous, so be careful. Drawing courtesy Bendix Corp.

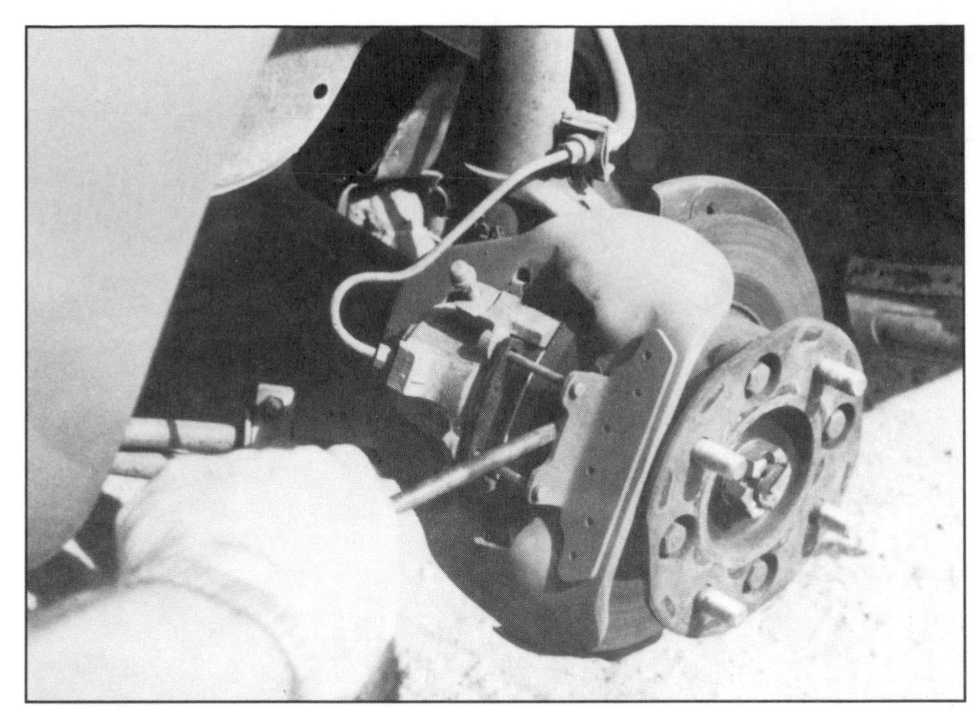

Retracting caliper piston on Honda with a screwdriver. Some manufacturers recommend using their special tool or a C-clamp. See your shop manual for specifics.

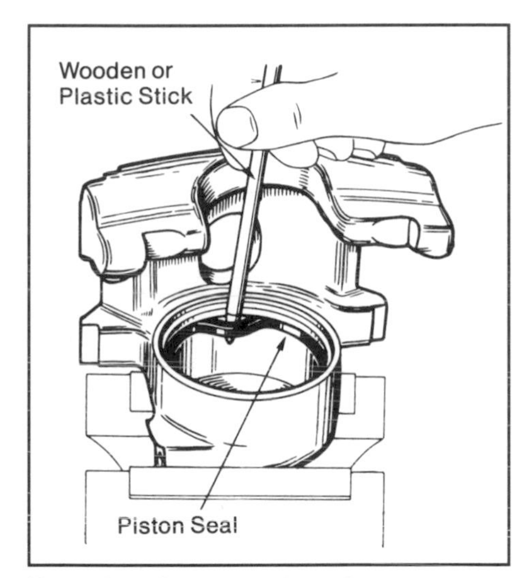

Removing piston seal from Chrysler floating caliper: Seal is installed in groove in caliper bore. Use a plastic or wooden tool to avoid damaging groove edges. Such a tool should also be used to install seals on pistons. Drawing courtesy Bendix Corp.

honing a wheel cylinder or master cylinder. Usually, caliper bores are limited to 0.002-in. oversize. Any larger, and the seals won't work. On a caliper that cannot be honed, the bore can be refinished with crocus cloth.

If its bore(s) is not scratched or pitted, a caliper can be rebuilt. When corrosion has pitted the bore, you must replace the caliper or sleeve it.

Caliper bores on old Corvettes are susceptible to corrosion damage. In fact, this has been such a problem that a company has come up with a solution. Stainless-steel Corvette calipers are available from Stainless Steel Brakes Corp., in Clarence, New York. These stainless-steel calipers minimize corrosion damage, particularly on cars using glycol brake fluid.

Some two-piece calipers can be disassembled for rebuilding. Others cannot, even though it may appear to be possible. Consult your shop manual before disassembling the caliper body. A two-piece caliper with an internal fluid passage has seals where the caliper splits. Make sure these seals are replaced before reassembling the caliper. The bolts holding the caliper body together are high-strength and require a specified torque. Be sure to tighten these bolts to the specified torque to avoid potential catastrophic problems later.

Removing or installing the seals on some calipers can be tricky. They fit into a groove, which must be clean and free of burrs. Be careful not to cut or damage the seal or groove when installing seals. Use plastic tools to avoid damaging critical parts. Also, don't ignore the rubber piston boot. It keeps moisture and dirt out of the caliper bore. If the boot is torn, water and dirt will ruin the caliper quickly, particularly on a road car.

Before installing calipers on a car, make sure they are perfectly clean and free of dirt and corrosion. This is particularly important on sliding calipers. There are contacting parts that must slide while under load. These sliding surfaces should be cleaned and lubricated with high-temperature brake grease to minimize friction and resist corrosion. Your shop manual will provide specific instructions.

Caliper-mounting bolts typically are high-strength and should be torqued to specification using a torque wrench. If the caliper-mounting bolts loosen, serious problems can result. So, make sure the correct bolts are used, they are in good condition, and they are torqued to the value specified in the shop manual.

LINING REPLACEMENT

The most frequent brake job is replacing friction material. Before starting this job, check the brakes for roughness by driving the car at highway speed and lightly applying the brakes. If the brakes shudder or the pedal pulsates, repeat the test using the parking brake—if the brakes are on the rear wheels. The presence or absence of shudder will tell you which pair of brakes is rough. Be careful while applying the rear brakes. Locked rear brakes make the car very unstable. Always have the release button or handle activated so you can release the parking brake quickly.

When replacing linings, you should also inspect the rotor or drum surfaces and other brake parts, and check for seal leakage.

Pad Replacement—Most disc-brake

Rotor is being surfaced with tool mounted on car. Rotor can be rotated by engine with this front-wheel-drive car. Tool machines both sides of rotor simultaneously, ensuring that surfaces will be parallel. Rotors that cannot be surfaced on car are removed and done on a special lathe.

Checking rotor runout is done with dial indicator clamped to a jack stand. Runout of 0.005 in. is considered upper limit for most cars.

pads are easy to change. They can be removed on many cars simply by removing a wheel, pulling a pin or locking device, and sliding the pads out. Sometimes this job takes only about five minutes per wheel. The only effort required is pushing the piston back into the caliper cylinder so the new thicker pad can be inserted between the piston and rotor. Your shop manual will tell you how to do this. Be careful not to damage a piston by prying on it.

With the pads out, inspect the rotor face. Check for cracks or scoring. Minor imperfections are OK, but large imperfections must be machined out. Rotors usually must be removed to do this. Take damaged rotors to a reputable brake shop for resurfacing. If you resurface one rotor, do the same to the opposite rotor. Otherwise, the brakes may pull until the newly surfaced rotor is worn smooth.

An easily made mistake that can occur when replacing disc-brake pads is that many pads can be installed backward. Although it seems obvious, make sure the friction material is installed so it faces the *rotor*. If you put a pad in backward, the brakes won't work correctly and the rotor will be destroyed.

When retracting caliper pistons to install the new pads, make sure the fluid reservoir isn't full. Otherwise, fluid will spill or squirt out the vent. This will be messy. You may find out how well glycol-based fluid works as a paint remover. Avoid this by opening the bleeder on the caliper before retracting the piston. Fluid will come out the bleeder rather than being pushed back to the reservoir. Run a hose into a jar from the bleeder to prevent spillage onto the floor.

As a policy, most brake shops surface rotors whenever pads are replaced. This results in good braking if surfacing is done properly. However, many times the job may not be required. To determine this, carefully inspect the rotors for damage and *runout.*

On most road cars, the pads wear out long before any rotor work is needed. Small surface scratches and scores of little concern are always present on a used rotor. And shops with poorly maintained lathes can ruin perfectly good rotors or drums. As a result, avoid machining unless the parts are scored heavily or your brakes are rough. Roughness is caused by rotor runout or thickness variation, and drum runout or ovality. Consult your factory manual for specific information.

Since 1971, all U.S.-made disc-brake rotors have a dimension cast into them. This is the *minimum-wear thickness,* sometimes called the *discard thickness.* A rotor is considered *unsafe* if it's machined thinner. So, never use a rotor that is machined to or less than the discard thickness—*discard* it!

Rotor Runout—Rotor runout—wobbling of the rotor friction surface—is critical, particularly on a race car. It should be checked every time the brakes are serviced. Runout can cause vibration or roughness in the pedal, excess pedal travel, or can even direct air into the caliper past the seals.

Rotor runout is usually caused by overheating. You can expect it if you use your brakes extremely hard. It can also be caused by overtorquing the wheel-mounting nuts. Many repair shops use an impact wrench to tighten wheels. Don't let them use one on your car. Used incorrectly, impact wrenches apply excessive torque to the wheel lugs, possibly warping the rotor-mounting flange.

Check rotor runout with the rotor mounted on the car. Before checking, make sure the wheel bearings are adjusted to proper preload. Set up a dial indicator so its tip is square to the rotor surface near the outer edge of the friction surface. Make sure the tip is on the friction surface.

Turn the rotor until you find the minimum indicated reading on the indicator. Mark it with a grease pencil,

Special puller is designed to remove brake drums. Thin jaws grasp back of drum. Handle is then turned carefully by hand—not with a wrench or lever. Excess force will break or distort drum or puller. Tapping drum lightly with a plastic mallet is OK to assist removing a stubborn drum.

Brake-spring-removal tool is a must when working on drum brakes. Bloody fingers or damaged springs can result from using pry bars or screwdrivers—springs can be dangerous.

machinist blueing, or masking tape. Zero the indicator dial. Now, turn the rotor and find the maximum reading. This is called *total indicated runout* (TIR), or simply *runout.* Mark it, too.

The tolerance for maximum runout should be specified in the factory shop manual. If not, make sure the runout is less than 0.005 in.

Rotor runout can be reduced by one of two methods—resurface the rotor or shim it at its mounting. Resurfacing should be used if the rotor surface is scored or deeply scratched. If the rotor surface is smooth, shimming can be used.

Before shimming a rotor, make sure wobble is not caused by out-of-parallel surfaces. This is commonly called *thickness variation.* Measure rotor thickness with a micrometer at the high and low spots you found with the dial indicator. The rotor should have a constant thickness at all points. If the wobble was caused by varying thickness, the rotor must be surfaced.

On road cars, thickness variation is the biggest cause of disc-brake roughness. Maximum thickness variation on new rotors is 0.0005 in. If a rotor has 0.0010 in. or more variation, it will probably cause roughness and must be machined or replaced. Rotors must be turned on a lathe with two cutting tools simultaneously—one on each side. *Never* cut rotors one side at a time. This will cause thickness variation. Most racing rotors with removable hats can be ground using a Blanchard grinder.

Shimming a rotor will work if:

- Rotor is flat and mounted on a tilt. You can see this on the indicator. A tilted rotor has one high spot and one low spot exactly 180° apart. Multiple high and low spots on a rotor require machining.
- Rotor is bolted to a machined surface. Some rotors are riveted on or are integral with the wheel-bearing housing or axle. Such rotors cannot be shimmed.

To shim a rotor, you must determine the location of the high and low spots. You should have already marked the two spots. Roughly calculate the shim thickness required to straighten the rotor.

Shims should be made from steel or brass shim stock, available in hardware stores in various thicknesses. Make sure you buy 0.001-in. or thinner material. Otherwise, you may not be able to adjust the rotor in fine enough increments.

Special tapered rotor shims are available for Corvette rotors from Stainless Steel Brakes Corp. This shim was developed to allow adjustment of rotors, which are made as an assembly with the mating part. Normally, Corvette rotors cannot be shimmed.

If you put shims only at the low spot, shim thickness should be about one-fourth of total runout. Install shim(s) on the proper bolt. Place shims one-eighth of total runout on the two adjacent bolts. Install rotor and torque bolts to specification. Remeasure rotor runout and adjust shim thickness as needed. More than one try may be required.

Don't worry about runout less than 0.002 in. It's all but impossible to make runout zero. And, surprisingly, a tiny amount of runout can actually help braking! It keeps the pads from continuously rubbing on the rotor surface by pushing them back a small amount from the rotor.

Brake-Lining Replacement—To replace drum-brake shoes, you must first remove the drum. Some drums come off easily after the wheel is removed—others are more difficult. You may have to use a special puller to remove a stubborn brake drum. Consult the factory shop manual for specific instructions before you start working.

If the drum should come off easily but doesn't, *don't* use a prybar between the drum and backing plate. If you pry against the backing plate,

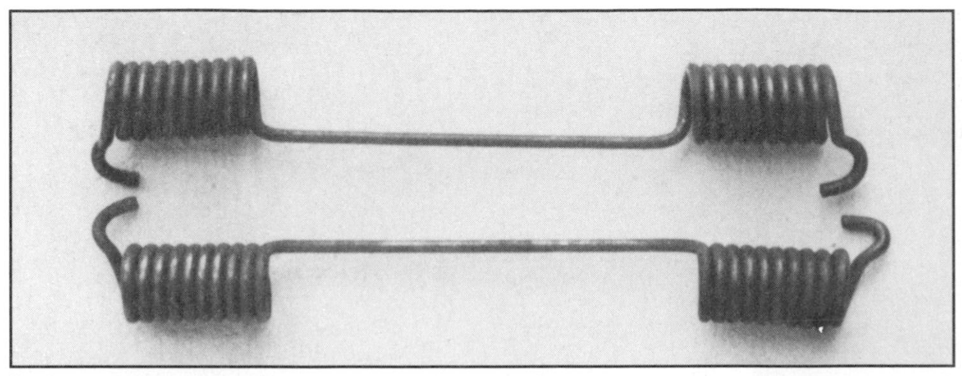
Springs are supposed to be the same, but obviously one is distorted. Longer spring should be replaced. Comparing an old spring with a new one is a good way of checking for stretched brake springs.

Brake drum is being machined to remove gouges and true its ID. Front-wheel drum is supported at its center while being turned on brake-drum lathe. Band wrapped around drum reduces chatter during machining.

you'll damage it. No backing plate is strong enough to be used as a prybar pivot for removing a stubborn drum. Instead, do the following:

- Make sure brake-shoe adjustment is loose. Check by rotating the drum. If tight, loosen adjustment manually.
- Clean area around the wheel studs and center hub with a wire brush.
- Apply penetrating oil to these areas.
- Tap edges of drum with a plastic mallet as you try to work the drum loose. Never use a metal hammer on a brake drum.
- As a last resort, use a brake-drum puller. This is a special puller with thin jaws. The puller grips the delicate edges of the drum. These drum pullers have a small handle so excessive force cannot be applied. Heavy force will break or bend the drum. Find and correct the problem before going any further.

A special tool is used to remove brake return springs. The springs have considerable tension on them, so be extremely careful. Don't use a screwdriver or other makeshift prybar. The accompanying photos illustrate how to use a brake-spring tool. This tool is inexpensive, makes the job go quicker, and avoids bloody fingers and bent springs.

If you are not familiar with the brake and you don't have a shop manual, make notes and sketch how the springs and adjuster are assembled. Note: Right- and left-side brakes may be different. So do both. Even better, take a photograph of both sides so you'll be sure to reassemble them properly. Many drum brakes function poorly after a rebuild because they were incorrectly reassembled.

Return Springs—Carefully inspect the return springs after removal. Check for distorted coils or bent ends. The springs from both sides of the car should be the same. Make sure the return springs are not stretched too far. Compare their lengths. Return springs can lose tension if they get too hot. Your shop manual should give you the needed specifications.

Some heavy-duty return springs have different specifications than the standard springs used on the same make of car. Carefully check specifications. Heavy-duty springs usually are identified by their color, but the color may be difficult to determine on an old set of springs. If in doubt, replace the springs.

WARNING: Never clean a brake with compressed air unless you are wearing proper safety breathing apparatus. Brake dust contains asbestos. If inhaled, it may cause lung disease. Don't risk blowing the dust in your eyes, either. The best method is to pour the dust out of the drums and wipe out what remains with a rag. Otherwise, wear a face mask and safety goggles.

Drum Inspection—Examine the drum rubbing surface. Check for cracks or scoring. Like a rotor, tiny scratches don't have to be machined out. Turning a drum makes it weaker, less stiff, and removes desired heat-absorbing weight. Only turn a drum if it's necessary. And, if one drum must be machined, also machine the opposite drum to the same diameter as well. If only one drum is turned, the brakes may cause pull to the side until the surface is worn smooth. Different diameter drums can also cause pull, regardless of surface smoothness.

If your drum brakes exhibit roughness or shudder, they should be turned or replaced. Sometimes, accessory wheels can cause drum distortion and roughness if the mounting surfaces are not perfectly flat.

Brake shoe is set up prior to being arced. Arcing shoes is highly desirable if drums are turned so shoes will have full contact with drum. Otherwise, shoes must wear in. Unfortunately, many shops have stopped arcing shoes because of laws limiting exposure to asbestos dust.

The usual maximum allowable increase in drum diameter is 0.06 in. Regardless, never turn a drum to its discard diameter.

Drum-Brake Shoes—Usually, shoes are sold on an exchange basis, so take the old ones along when buying new shoes. Don't touch the new lining surfaces.

If you had the drums turned oversize, find a shop that will also *arc*—machine—the shoes to fit the drums. This may be difficult. Many shops have stopped arcing brake linings due to the asbestos hazard. If you don't have the shoes arced, the linings will eventually wear to shape. Unfortunately, braking will not be as good until the shoes wear in.

Installing brake springs requires same tool used to remove them. End of tool is simply hooked over anchor pin and spring is popped on. Doing this any other way invites injury to yourself and the springs.

Parking brake is at output shaft of transmission on this front-engine, rear-drive car. Brake adjustment compensates for lining wear. Such parking brakes are rarely found on late-model cars.

Lubricate the shoes with a dab of high-temperature brake grease where they rub on the backing plate and at pivots. Always use brake grease, not chassis or wheel-bearing grease. Be careful to use an absolute minimum amount of grease. It's critical that you keep grease off the linings.

Assembly—Drum-brake assembly is the opposite of disassembly. Installing the automatic adjuster and putting the brake springs on are the only tricky parts. Here, you must use the special brake tool for safe installation. The springs sometimes can be mixed up and installed in the wrong place. Refer to the shop manual and your sketch, photos or notes to be sure. If an automatic adjuster was used, operate it manually to be sure it works properly before installing the drum.

To install the drum, you may have to run in the adjuster to allow the drum to fit over the new linings. Don't run it in more than necessary—just enough to get the drum on. If you go too far, you'll have to adjust the shoes back out from under the car.

Once both drums are on, you must adjust their clearances before driving the car. Even brakes with automatic adjusters must be set close or they won't work. Your shop manual will indicate the correct procedure.

PARKING-BRAKE-LINKAGE SERVICING

Usually, a parking brake requires little servicing. All that's needed is an occasional greasing of the linkage. On some cars, however, a separate parking brake needs adjusting or other servicing. Others have a parking brake on the drive shaft. Typically, this is a simple drum brake, often with an external-band drum brake similar to wheel brakes used on an antique car. The shop manual should provide specific instructions for servicing this type of brake.

Parking-brake linkages are adjustable, although generally this is not required. However, if the cable stretches, you may have to adjust it. Do not use the "Mickey Mouse©" U-bolt-type slack adjuster. It can break a cable. Instead, buy a new cable if it has stretched too far to be adjusted or if it has broken strands. And, only adjust a parking brake *after* the wheel brakes are operating perfectly.

One of the biggest problems with parking brakes on cars driven in winter salt is cable corrosion and sticking. A stuck cable can burn up brakes because it won't release, or makes the parking brake impossible to apply. If you have a sticking cable, replace it. Don't bother with lubricating them. Enclosed cables are difficult to lubricate and will probably stick again anyway.

Some cars with four-wheel disc brakes have a drum brake inside the rear-wheel disc brakes as a parking brake. This is adjusted and serviced the same as a normal drum brake.

PEDAL & LINKAGE MAINTENANCE

Except for periodic inspection, lubrication is about all a brake pedal and linkage needs. Make sure there is no slop or failure of the system. Examine the pedal pads and replace them if they are worn slick or coming apart. A foot slipping off a pedal can be as dangerous as a brake failure.

The brake pedal can be adjusted for free play. A brake pedal has a lot less free play than a clutch pedal. There should be only a slight amount of pedal movement between its fully retracted position and the beginning of master-cylinder-piston movement. Consult your shop manual for exact specifications. Usually, free play never needs adjusting, unless the master cylinder has been replaced.

BRAKE-BOOSTER CHECK

If power brakes are causing problems, first check the hydraulic system. This includes the master cylinder and wheel cylinders. If the basic brake system operates properly, check the booster.

The following trouble-shooting procedure is for a vacuum booster:

CRC Disc Brake Quiet compound is applied to backside of brake pads to reduce brake noise. CRC recommends using compound on outboard pads. They can be reassembled on the car following a 10-minute curing time. Photo courtesy CRC Chemicals.

- With the engine off, press the brake pedal several times. This bleeds off all the vacuum in the booster and the resulting power assist.
- Press lightly on the brake pedal and start the engine. The pedal should drop slightly under your foot if the booster is working properly. If there is no drop, the booster is not working. Without the booster, pedal effort will be much higher than normal.
- Before replacing the booster, make sure it's getting the correct supply of vacuum. Check all vacuum lines for leaks and kinks. A vacuum leak often sounds like a hiss as the engine idles. Check intake-manifold vacuum with a vacuum gage and compare it to the specs in your shop manual. Poor engine vacuum may mean burned valves, hose or intake-manifold-gasket leak or some other engine problems.
- If the vacuum supply is OK, correct the booster. Most boosters are replaced rather than overhauled due to their low cost and critical construction.
- Make a final check, if you found the vacuum booster to be at fault. A common cause of vacuum-booster failure is automatic-transmission-modulator failure. This allows automatic-transmission fluid (ATF) to be drawn into the engine and, subsequently, the vacuum booster. The ATF attacks the booster diaphragm, causing it to fail eventually. Transmission-modulator failure shows up as fluid in the vacuum hose from the manifold to the booster, white smoke out the exhaust, or a malfunctioning transmission. Replace the modulator if it is at fault.

If a hydraulic booster is used, the troubleshooting process is similar:

- With the engine off, pump the brake several times to bleed off all hydraulic pressure in the accumulator.
- Push on the brake pedal and start the engine. The pedal should drop under your foot, usually followed by a slight surge. This confirms that the booster is working.
- If the booster doesn't pass this test, check the power-steering pump. Consult your shop manual for specific checks. A slipping belt or low power-steering fluid may be the cause.
- If there's pressure from the power-steering pump, consult your shop manual for instructions on servicing your booster. Be sure to read the instructions carefully. High pressure stored in an accumulator can be extremely dangerous if the unit is not disassembled correctly. Often, a pressure of 1500 psi is involved. There's also a very strong spring behind it! If you don't have the proper instructions, leave the unit alone. Take the car to a mechanic knowledgeable of your type of car.

NOISY BRAKES

Disc-brake squeal can be very annoying on a road car, particularly if the rest of the car is quiet-running. Usually, squeal is caused by vibration of the front brake pads against the rotor. Some manufacturers use anti-squeal shims or other devices designed to damp these vibrations. When a car gets old, anti-squeal devices can become ineffective. Replacement of parts may eliminate noise. Consult your factory shop manual for this maintenance.

High-friction linings tend to be noisier than low-friction linings. And, asbestos-filled linings generally are quieter than those without asbestos. This is the main reason why manufacturers have taken so long to develop asbestos-free linings.

There are anti-noise compounds designed to be applied the pad backing plate as shown at left. One of these is CRC Chemicals' Disc Brake Quiet. It is a thick paste, similar in appearance to RTV sealer or bathtub calk. When correctly applied, it can quiet squealing disc brakes. Before using this type of material, make sure the brakes are clean and in good operating condition. Noise can be caused by serious malfunctions or wear in a brake.

Sanding lining and rotor surfaces will quiet noise temporarily. Usually, the ends of the linings vibrate, so grinding chamfers on the leading and trailing lining edges will sometimes eliminate noise.

If drum brakes squeak as you apply and release the pedal, it's probably the edge of the shoes rubbing the backing-plate shoe ledges. Apply a thin film of high-temperature brake grease to the shoe ledges to quiet the noise. Be careful not to get grease on the lining.

There is no sure cure for brake noise. If one existed, the manufacturers would be using it.

Modifications 12

Grand Prix cars have some of the best-designed brakes. Mechanics modify brakes to suit course and temperature conditions on race day. Lightest brakes that work are used to avoid excess weight. Smallest-possible cooling scoops are used to reduce aerodynamic drag. As with engines in highly-competitive racing, brakes must be pushed to their performance limits, or you don't win!

If your brake system is not working as desired, consider modifying it for better performance or reliability. Modifications are often the result of testing, either on the road or the track. However, don't attempt any modification until you:

- Inspect and repair the brakes so they operate properly.
- Test the brakes and record the results so you'll know what aspects of their performance need improving.
- Analyze your car to determine the forces and pressures in the brake system. Also, test the car to determine brake temperatures. You may have done this already.

The first step is necessary before you can justify spending time and money on modifications. Most new cars have well-engineered brakes. They will work fine if properly maintained. However, if the car is

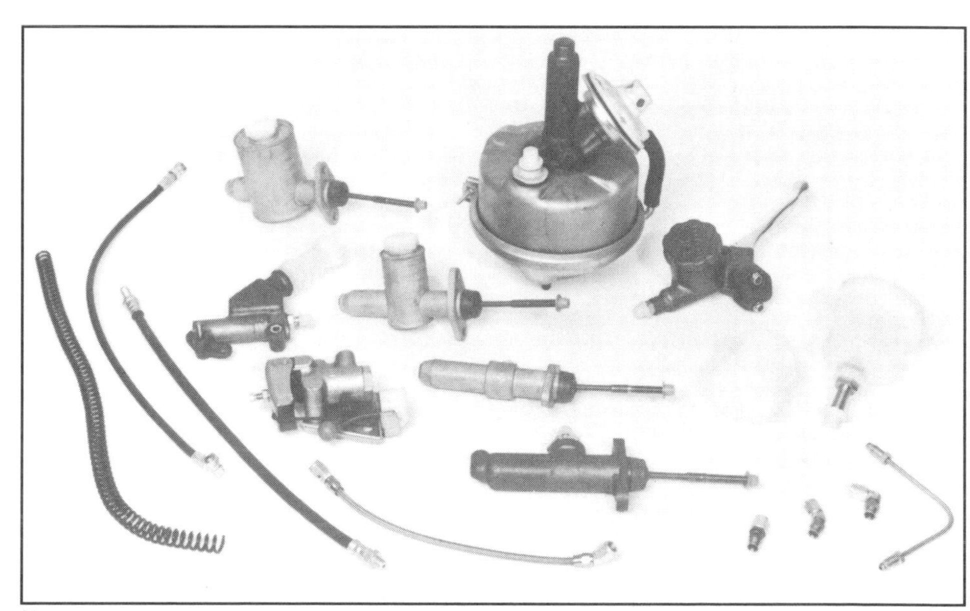
Numerous components are available to help improve a brake system. Trick is learning how to choose and apply them. Some products available from AP Racing include: steel-braided hoses, various size master cylinders, vacuum boosters, and fluid reservoirs. Photo courtesy AP Racing.

Before you begin modifying a car's existing brakes, first make sure they are in top condition and test them. Modification may be unnecessary. Don't fix something that doesn't need fixing. A race track is ideal for testing.

Formula Ford brake is small, light and simple. It works great on a light race car with 110 HP. If allowed, bigger brakes may stop car faster, but lap times might be slower due to added weight. Always use a brake that's big enough, but not too big. Extra weight hurts in racing.

driven beyond its design limits or if it is highly modified to different specifications, the brakes may need modifying.

To do an intelligent job of modification, first identify what needs modifying. The next step is to find out exactly what's wrong with brake performance. If you list these problems, you are less likely to lose track of them during the modification process.

Be specific when listing the problems. Don't just say, "The brakes are lousy." For example, say "The brakes fade after two stops from 100 mph," or "Temperature reaches 600F after three stops from 100 mph." The more specific you are, the easier, cheaper and more effective modifications will be.

Use the formulas in Chapter 9 to calculate internal brake-system forces and pressures. These can be found by measurement, if you prefer. You must understand clearly how and why you're modifying the brakes before you can do it properly. Temperatures must be measured by testing modifications, including cooling changes.

Following are some brake modifications and objectives to consider:

- Install bigger brakes to increase brake torque and reduce brake temperatures.
- Improve cooling.
- Eliminate fade.
- Reduce pedal effort.
- Increase lining life.
- Add brake-balance-adjustment features.
- Reduce deflections.
- Improve stability.

BIGGER BRAKES

Bigger brakes almost always improve brake performance. The additional weight of a bigger disc or rotor reduces temperatures. If drum or rotor diameter is increased, pedal effort will also be reduced. And, if swept area is increased, the linings probably will wear longer.

The disadvantage of bigger brakes is the adverse effects added weight have on acceleration and handling. More weight slows acceleration. Additional rotating weight makes it even worse. Also, added unsprung weight bouncing up and down with the wheels reduces traction in bumpy corners and when accelerating or braking on bumpy surfaces. So, don't install the heaviest brakes you can find. Select brake size to suit the performance needs of the car.

On Grand Prix cars, the brakes are sized closely to the performance of the car. They are no bigger than what's needed to do their job. On tracks that are hard on brakes, either bigger rotors or two calipers per rotor are used. On tracks with less braking requirements, brake weight is reduced with lighter components. This may be carrying things too far for most applications, but serious racers with unrestricted funds do it to win in tough competition.

Regardless of the application, always use the largest-*diameter* brakes possible. A large-diameter drum or rotor reduces pedal effort and makes braking easier. You can perhaps eliminate power assist by installing larger-diameter brakes. Thicker rotors are not necessarily required when rotor diameter is increased. Remember: Add brake weight only if required to cure a severe cooling problem. If a light, solid rotor cools well enough, don't use a heavier, vented rotor. When changing to larger drums, consider reducing weight by switching from iron to aluminum. Aluminum drums cool better than iron drums of the same size. And they are lighter. They are well worth the expense.

Separate cooling problems from pedal-effort problems. If the brakes have both high pedal effort and excessive temperatures, the needed changes are obvious. Your car will benefit from both larger-diameter and heavier brakes to cure both problems. However, if pedal effort is the only problem, stay with lighter rotors, calipers or drums.

It's easy to predict the effect larger rotor or drum diameters will have on pedal effort. Because pedal effort is related directly to brake torque, use the formulas given in Chapter 9. Calculate the new brake torque and compare it to the old brake torque. The new pedal effort will be the *indirect* ratio of the two numbers. Pedal effort always decreases with increased brake diameter.

Temperature rise per stop can be calculated if you know the weight of the old and new rotors. Because temperature rise per stop is related directly to brake weight, doubling rotor or drum weight should reduce temperature rise by one-half. The weight of an iron drum affects the temperature rise the same as rotor weight. You can pre-

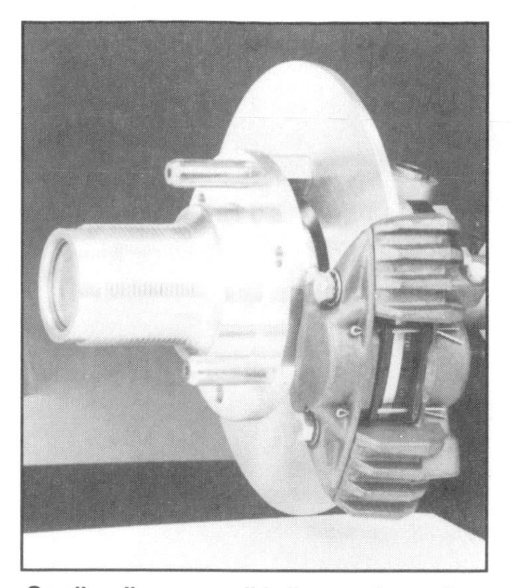

Small caliper on solid disc works well on front of sprint car run on dirt. Photo courtesy AP Racing.

Porsche Can-Am car under hard braking: Because nose is so low to track surface, underbody airflow is blocked. Ducts route cooling air to brakes, page 141.

dict how many stops from a certain speed it will take to reach a certain temperature from the formula for temperature rise, page 11. It doesn't indicate anything about cooling, however.

Cooling is nearly unchanged with brake-size changes. Improve cooling with air ducting or aerodynamic changes to the car.

Installing Bigger Brakes—Finding bigger brakes for a road car can be a problem. You may find a bolt-on conversion that increases brake size if you have a popular performance car. Or, you may find bigger brakes on a station wagon that will bolt on. Often these brakes will have wider drums or heavier rotors, but sometimes diameter may also be larger. Look for larger-diameter wheels to go with larger-diameter brakes.

By studying shop manuals for different model years for your car, you may discover that the brake-drum or rotor size is larger for a certain year. If you discover this situation, go to a dealer or junk yard and check whether the brake will bolt on. It may be that the size or design of the brake mounting also changed, eliminating the interchange possibility. Another possibility is that some cars could be ordered with optional heavy-duty brakes. Sometimes these brakes are larger. Study the parts books for options. Your dealer can help with this. Look for *Police, Taxi,* or *Heavy Duty* listings in the parts book.

On a race car, changing to bigger brakes usually is easier than on a road car. First off, special brakes typically are used on a race car, and they generally are available in various sizes. The mounting pattern for a rotor is usually specified when ordering it. A rotor 1-in. or so larger in diameter may accept the same caliper. Check with the manufacturer or dealer before assuming they will work. If you change rotor thickness, either the caliper will have to be changed or the linings machined thinner. Companies like Hurst, Tilton Engineering and JFZ can help with either approach.

COOLING IMPROVEMENTS

Modifications to improve cooling are common on race cars. Some road cars can also benefit from cooling improvements, but your choices of what to do are limited.

There are several ways to improve brake cooling:

- Add or enlarge air ducts to brakes.
- Change or modify wheels for increased airflow to brakes.
- Change from solid to vented rotors on disc brakes.
- Change to finned or aluminum drums.
- Remove or modify dust shields on disc brakes.
- Vent backing plates on drum brakes.
- Install water-cooling system.
- Change car-body aerodynamics for increased airflow to brakes.

Air Ducts—The easiest and most effective way to improve brake cooling is to add air ducts. These start with scoops or holes in the body on most full-bodied race cars. On open-wheel cars, the ducts can be scoops attached to the brakes, uprights or axle.

Vented rotors in place of solid rotors will improve brake performance if temperatures are too high. Scoops or modified ducting should be tried first.

A brake duct must be designed carefully to get the most air for the least amount of added drag. The most critical parts of a duct are its entrance location and where air exits the duct at the brakes. A constant-diameter tube to carry the air is often used between the duct entrance and exit.

When installing a brake duct, be sure its entrance faces forward and is in a high-pressure area. The vertical

Dual master cylinders are used in place of stock tandem master cylinder on this dual-purpose car.

Aftermarket spoiler has holes for brake-cooling ducts. High pressure exists in front of spoiler, making ideal spot for brake ducts. Spoiler without ducts adversely affects brake cooling. On a race car, tubes should be used to route air directly to brakes.

This may help front-brake cooling, but not looks. Neat brake scoops under bumper or behind grille would do same job and not require cutting torch.

Brake ducts on McLaren Can-Am car are simple holes in body with tubes attached. Forward surfaces of sloping front fenders are in high-pressure areas.

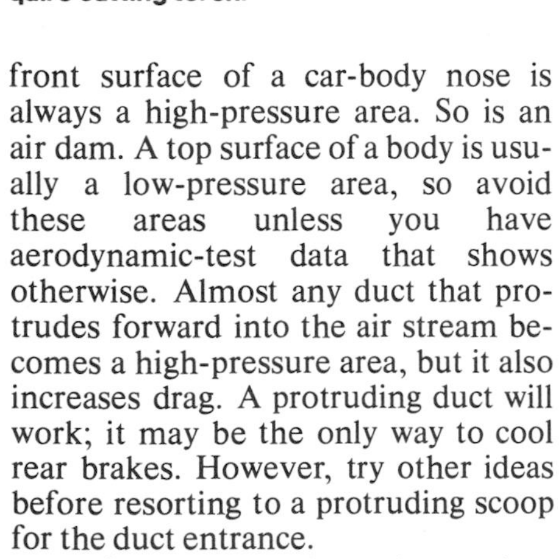

front surface of a car-body nose is always a high-pressure area. So is an air dam. A top surface of a body is usually a low-pressure area, so avoid these areas unless you have aerodynamic-test data that shows otherwise. Almost any duct that protrudes forward into the air stream becomes a high-pressure area, but it also increases drag. A protruding duct will work; it may be the only way to cool rear brakes. However, try other ideas before resorting to a protruding scoop for the duct entrance.

A front-mounted engine-coolant radiator is always in a high-pressure area. Tapping into it with a brake duct for the front brakes will not add drag, but will reduce engine cooling slightly. Engine-induction plenums on Can Am and sports-racing cars with large overhead air scoops are always at high pressure. They can be tapped for rear-brake cooling without increasing drag. Testing should be done to check that reduced cooling or plenum pressure does not cause engine overheating or reduce top speed, respectively.

NACA Duct—A flush duct entrance, known as a *NACA duct,* is often used for rear-brake-duct entrance. If you have a wind tunnel or do serious aerodynamic track testing, you can find a suitable high-pressure area in which to locate such a duct. A NACA duct may flow some air in a medium-pressure area, but it won't flow much, if any, in a low-pressure area. If you must guess at a spot for a duct, at least try a *wool-tuft test* first.

Wool-Tuft Testing—A wool-tuft test

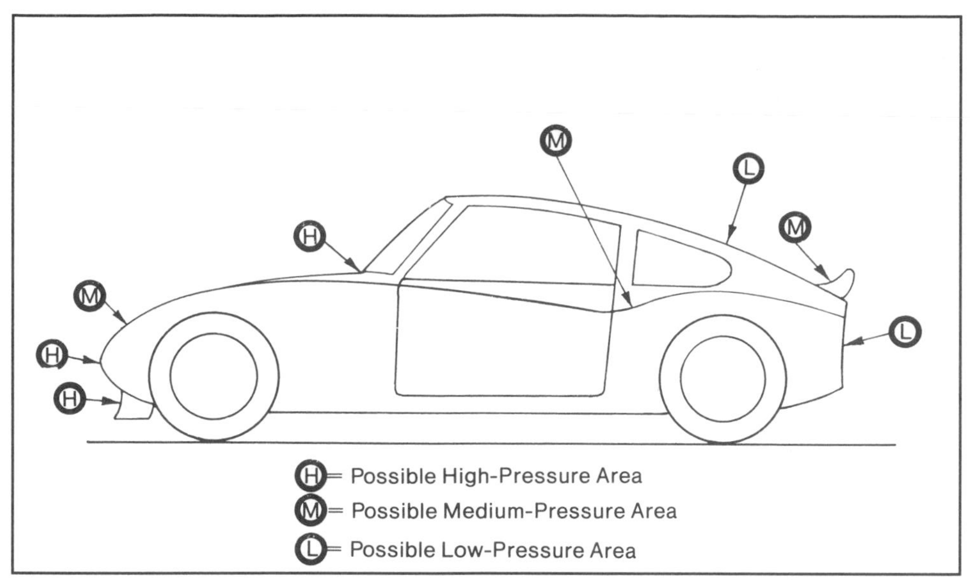

High-pressure areas on car body exist at nose leading edge and base of windshield. Low pressure always exists behind tail. Accurate wind-tunnel tests or pressure measurements must be used to find other areas suitable for duct entrances.

Cooling duct on March Indy Car maximizes airflow and minimizes drag. Rather than projecting duct intake into airstream, which would also increase drag, advantage is taken of existing high-pressure area at front of tire. Duct directs air from intake at front of tire, through duct, and into center of suspension upright and brake rotor. Photo by Tom Monroe.

Wool tufts are installed on car body before aerodynamic testing. Bits of yarn are taped in even rows. Assuming body is symmetrical, tufts on one side of car are sufficient.

requires driving the car. Tape short lengths of yarn—tufts—on the body a few inches apart with transparent tape. The yarn strips should contrast with the body color so you can see them. Cut the yarn about 5-in. long and tape each piece across the car with the tape in the middle. When the car is in motion, the yarn will form a U-shape pointing in the direction of airflow. If the flow is *attached* to the body—it is not turbulent—the yarn will lie flat and straight with little wiggling.

Place the duct in an area on the most forward-facing surface where smooth airflow is indicated. If the yarn points rearward and wiggles slightly, such an area is a second choice. This is *turbulent attached flow,* and usually is found near the rear of a body. If the yarn wiggles violently and points in all directions, this is *separated air.* It is low pressure and *not* the place for a duct entrance. You'll always find separated airflow behind a body, behind blunt protrusions, and often, all over the tail end. Regardless of the indicated flow, never install a duct entrance on any rearward-facing surface, even if it is only slightly rearward-facing.

After installing the ducts, retest with wool tufts. Put tufts in and around the brake-duct entrance and check that air actually flows into the opening. Some duct entrances have air coming out of them because pressure at the exit is higher than at the entrance! So, test the duct entrance alone and with the entire brake duct installed. There may be critical differences.

Brake ducts on Porsche Can-Am car use different designs at front and rear of body. Front ducts face direct airflow; rear ducts use flush NACA-type ducts. Effectiveness of rear ducts should be confirmed by testing—pressure on top of body can't be determined by simple observation.

Tuft test was to confirm that air flows smoothly over body at radiator-duct exit. High-pressure areas are indicated by tufts laying flat on body surface. Flow is somewhat disturbed, but in proper direction.

Brake ducts on this stock car are mounted under front bumper. A flexible hose leads cool air to front brakes. This type duct is easily modified to change brake cooling.

A well designed duct is simple and neat. Ducts sealed to center of rotor such as this are much more efficient than those that just point at center of rotor. Flex hose used for air ducts on cars is available in auto parts or surplus stores. Notice both ends of hose are securely attached with hose clamps.

Air-Pressure Gage—An air-pressure gage that measures in inches of water (in.H_2O), page 113, can be used to find high-pressure areas and to improve ducting. Mount the gage so the driver can see it. Run rubber or vinyl tubing from the area you want to test to the high-pressure port on the gage. Tape the tubing to the body. Position the tube on the surface of the body so air flows across the ends, not into it.

If you leave the low-pressure port disconnected, the gage will measure the pressure difference between the test area and where the gage is located. However, you can run a tube from the low-pressure port to the area of the car you want the duct to pressurize—the brake rotor. The gage will then measure the pressure difference between the test area and brake-rotor area. Before running the test, blow *very* lightly at the end of the tube in the test area. If the gage reading increases, it is hooked up correctly. Run the car at constant speed, say 80 mph, and read pressure. If the gage reads below zero, the duct will not work. Air flows from high to low pressure only. Tape the high-pressure tube to different locations around the body until you find the highest-pressure area that is practical to accept a duct inlet.

If you have a duct exit that seals to the rotor and you want to improve it with larger tubing, fewer bends or a different inlet location, drill a hole in the rotor plenum close to the rotor. Connect it to the high-pressure port on the gage. Disconnect the low-pressure gage port. Run a constant-speed test before and after modifying the duct. If pressure increases after the modification, the duct will flow more air. The same procedure can be used for optimizing engine-induction ram-air-scoop and radiator-inlet sizes. Just be sure the high-pressure gage pickup is in front of and close to the item being pressurized.

There are some complex methods to determine the exact dimensions of a NACA duct. The easiest way to design a duct is to copy one from another car. A lot of wind-tunnel testing and track testing has been done on factory-built race cars. There's no reason to repeat their work. They can't hide the duct shape, so go into the pits with a camera and photograph ducts on winning factory-designed cars. Enhance the information you'll get with the photo by measuring the duct, if the car owner will allow it.

To repeat, the front of a car body or nose spoiler—air dam—is an ideal place to install a duct entrance for the front brakes. Make a round, smooth-edged hole the size of the duct. Rounded edges at the entrance minimize turbulence inside the duct. Use a round tube or duct to route air to the brake. Use smooth tube for straight ducts and corrugated heater-type ducting for curves. Mount the duct securely so it won't fall off. Check wheel and suspension clearances in all turn and wheel-travel extremes: bump, droop and steering lock.

Duct Exit—The duct exit should be at a specific location. On a drum brake, direct the exit at the drum so air hits both the drum edge and backing plate. If the backing plate is vented, make sure air blows into the vents. All race cars that use drum brakes should have vented backing plates—assuming rules allow them—but road cars may not. If there are no vents, point the duct more at the drum and less at the backing plate. The accompanying drawing shows how to locate a cooling-duct exit for drum brakes.

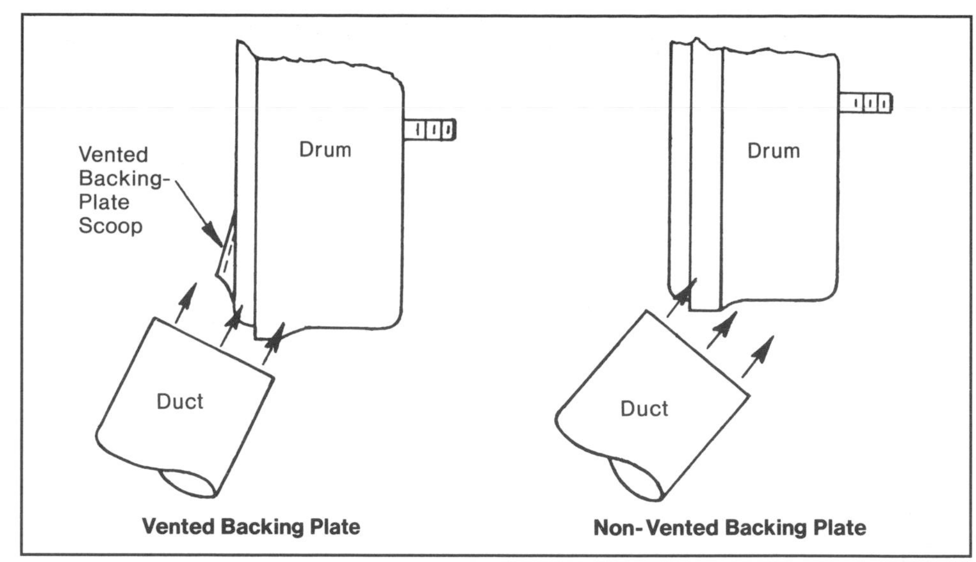

Cooling duct should be pointed as shown depending on backing-plate design. If there are cooling holes or scoops in backing plate, duct should point toward backing plate and drum. If there are no holes in backing plate, point duct entirely toward drum.

A disc brake with a solid rotor should have the duct pointing at the forward edge of the rotor so both sides of the rotor are cooled. This is not a suitable duct-exit location for a vented rotor. I discuss ducting vented rotors later.

It's difficult to cool both sides of a solid rotor equally. A duct pointed to the inboard side of the rotor will take care of that side very well. However, the outboard side of the rotor gets little or no air if it is close to the wheel. Consequently, the duct should be designed to force air between the wheel and the outboard side of the rotor. If the rotor has one hot side and one cool side, it may warp. A warped rotor will adversely affect braking. The nearby drawing shows a duct designed to cool both sides of a disc-brake rotor. This may be easy or impossible, depending on the specific configuration.

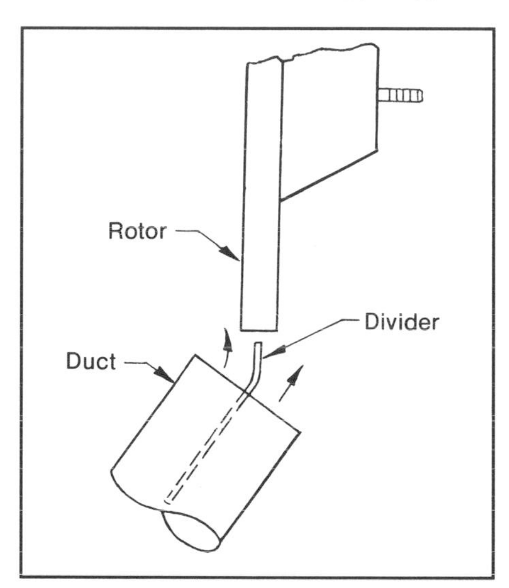

It is essential to get cooling air to both sides of a solid rotor. Divider in duct will split airflow so it can be directed to each side of rotor. Some experimenting with dividers and duct locations may be necessary to get it right.

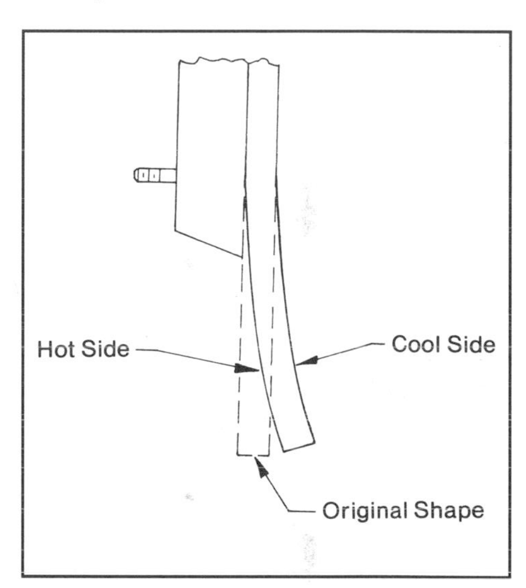

If a rotor is cooled more on one side than the other, it can warp as shown here. Hot side expands more than cold side and rotor takes the shape of a saucer. Braking is not good on a warped rotor. Equal cooling on each side prevents this situation.

A duct design must be proven by testing. This is because brake cooling is complicated by other factors. For example, wheel design also affects cooling of the outboard side of the rotor. Measuring temperature on both sides of the rotor and caliper will confirm how successful a duct design is.

A disc brake with a vented rotor should have the cooling air directed to the center of the rotor. As it turns, the rotor acts as an air pump, although not an efficient one. Air is drawn into the center of the rotor and is thrown out the edge. Vented-rotor ducts can be sealed to the center of the rotor. This is much more efficient than merely pointing the duct at the rotor center because air is forced through the rotor. See the drawing for how a vented rotor should be ducted. If more cooling is required, seal the duct to the center of the rotor.

Duct Area—The ideal air-duct cross-sectional area should be at least as large as the rotor cooling-passage-inlet area. To determine this area, measure the distance between the inboard and outboard rotor side plates. Also measure the inner diameter of the inboard rotor side plate. Calculate the area using this formula:

Inlet area = (3.14)(inboard-plate ID)(distance between plates)

For a rotor with a 6.25-in. inboard-plate ID and 0.63 in. between plates:

Inlet area = (3.14)(6.25 in.)(0.63 in.)
= 12.36 sq in.

Fiberglass scoop on this Indy Car brake is sealed to front upright. Air is ducted directly to center of vented rotor. This duct was obviously designed with rest of the car, and not merely added on after brakes overheated. Photo by Tom Monroe.

This brake duct directs air to center of vented rotor. Rotor motion sucks cool air in and blows it out the vents. This duct also puts cold air on rubbing surface of rotor, but only on inboard side. This could lead to uneven cooling of the two rubbing surfaces, and may cause rotor to warp. A baffle should be added to upright to seal gap between inboard braking surface and upright. This would force all air collected by scoop through rotor-cooling passages.

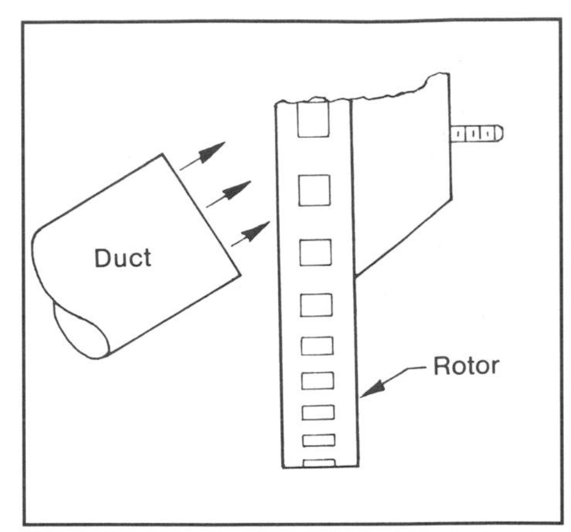

Vented rotor needs cool air directed at its center. Air from duct is picked up by rotor and sucked through vents. Best cooling ducts are designed with duct exit sealed against rotor center.

Common duct diameters and areas are as follows:

5-in. Diameter/Area 19.6 sq in.
4-in. Diameter/Area 12.6 sq in.
3-in. Diameter/Area 7.1 sq in.

Ideal duct size for this rotor is 4 in. If there isn't enough room to package the correct duct in your car, get as close as you can. The exit adaptor that seals the duct to the center of the rotor should almost touch the rotor. Allow about 1/16-in. clearance. If it rubs, the adaptor will make noise and wear the excess away, but it won't hurt the rotor. Be sure the adaptor does not restrict airflow.

The cross-sectional area of the adaptor should be as large as the duct. Squeezing a tube oblong reduces area. Start with a cone and form the exit end to the proper shape. It should seal the entire center of the rotor. The adaptor should blend to a round section to which the duct will connect. Make provisions for mounting the adaptor to the spindle or suspension upright. It must be secure and move with the wheel.

If your car already has ducts, but brake temperatures are too high, duct modifications may help. Perhaps the duct entrances or exits are in the wrong place. Tuft test them to see if air is really flowing into the duct. If everything looks good, try installing larger-diameter ducts. Test the brakes with the new ducts under the same conditions as the old ones and see if they made any difference. You can always flow more cooling air through a bigger hole, but it always adds drag. The car will lose top speed with larger ducts.

Changing to Vented Rotors—Solid rotors are cheaper and lighter than vented rotors, but they can get too hot. Vented rotors are better for several reasons. They are heavier, thus have less temperature rise per stop. They also have four surfaces instead of two to provide cooling. Finally, vented rotors also pump air through the brake. Vented rotors usually are necessary on a heavy race car such as a stock car.

The best vented rotors have curved vents. Curved-vent rotors pump more air than straight-vent rotors, plus curved-vent rotors are more resistant to cracking. However, this makes rotors used on the right and left sides of the car different. They are designed to pump air as they rotate in one direction only.

If you can't afford curved-vent rotors, use straight-vent rotors. They will run considerably cooler than solid rotors. Regardless of which way you go, remember that vented rotors are much thicker than solid rotors. The calipers will have to be replaced or modified if you change to vented rotors. But, most of the time, you'll need a different caliper. On some two-piece calipers, spacers can be installed to allow the thicker rotors to be used. Longer bolts are required to make this modification.

If you change to vented rotors, change the brake ducts to supply air to the center of the rotor. This will provide a constant supply of cool air to be pumped through the interior of the rotor.

Modified Wheels—Some wheels help brake cooling by providing a place for cooling air to exit the wheel well. This can happen by the wheel acting as an air pump, or the wheel merely having enough vent holes in it to allow hot air to escape.

When I was developing the first monocoque split-rim racing wheels, I found some interesting effects from wheel centers. On some cars, the brakes ran much cooler if vent holes were drilled in the wheel centers. The outside of the wheel would be covered with black dust from the brakes. This indicated that air really flowed through the vent holes. On other cars, vent holes did not affect brake cooling.

After some study, we learned that the shape of the corners of the body nose made the difference. As air flows around the corners of a rounded nose, a low-pressure area is created at the front wheel wells. This draws air through the wheels and helps cool the brakes. On cars with flat sides and a square-edged nose, this low pressure does not exist. Airflow through the wheels had to be supplied entirely by the brake-cooling ducts.

This brake duct suffers from too many bends and tight clearances. Designing brake cooling ducts with rest of car can help avoid these problems. Best duct is nearly straight with large, smooth bends.

Angled vent and grooved rotors are directional. Left-side rotor must turn counterclockwise for maximum cooling.

Because of the large area of a wheel well compared to an air duct, suction at a low-pressure wheel well can move much more air than an air duct. When low-pressure wheel wells with vented wheels and sealed air ducts are combined, the results are dramatic for moving air through a vented rotor.

Before you try vented wheels, first test the nose of the car with wool tufts. If air flows out of the front wheel wells, vented wheels will probably improve brake cooling. This will occur if the nose of the car has rounded corners, allowing air to flow easily around the sides of the car. The air speeding up as it rounds the corners of the body causes a suction at the front wheel wells. This flow can be observed when watching cars run in heavy rain. If air flows out of the wheel wells, a great deal of spray will come out too; you can see it.

Many racing and street wheels come with vent holes. If the wheels are also designed to act as an air pump, the effect will be greater. Hubcap-like lightweight fans have been fitted to wheels to help pump air out. Chevrolet Corvair cooling fans have been used for this.

The most important aspect of venting air through a wheel is the total amount of cut-out area. As this area increases, airflow increases. A few large holes are better than many smaller ones, but you may not have much choice. Most wheels must have small holes for structural reasons.

Another item that helps brake cooling is wheel material. An aluminum or magnesium wheel conducts heat

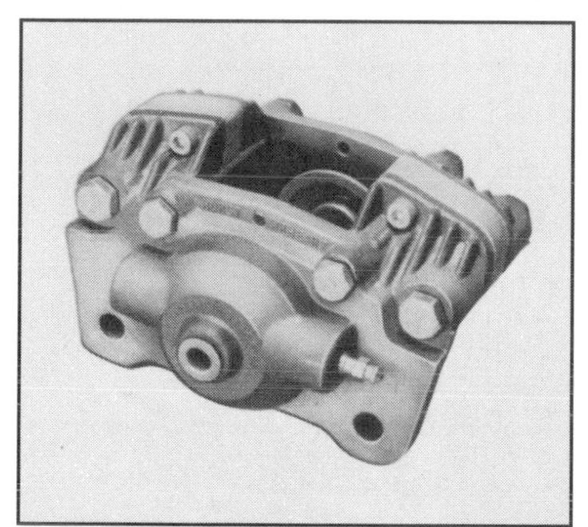

Hurst/Airheart caliper is available with or without a spacer between caliper halves. With spacer, caliper can be used on a 0.810-in.-thick rotor. Without spacer, caliper is suited for a 0.375-in. rotor thickness. If converting to thicker rotors and you have unspaced calipers, merely add spacers. Photo courtesy Hurst Performance.

Centerline wheels were originally a solid-center construction. As need to direct cooling air through wheels became evident, these vented versions were created. Because this wheel is so deep, duct should be used to take advantage of cooling holes.

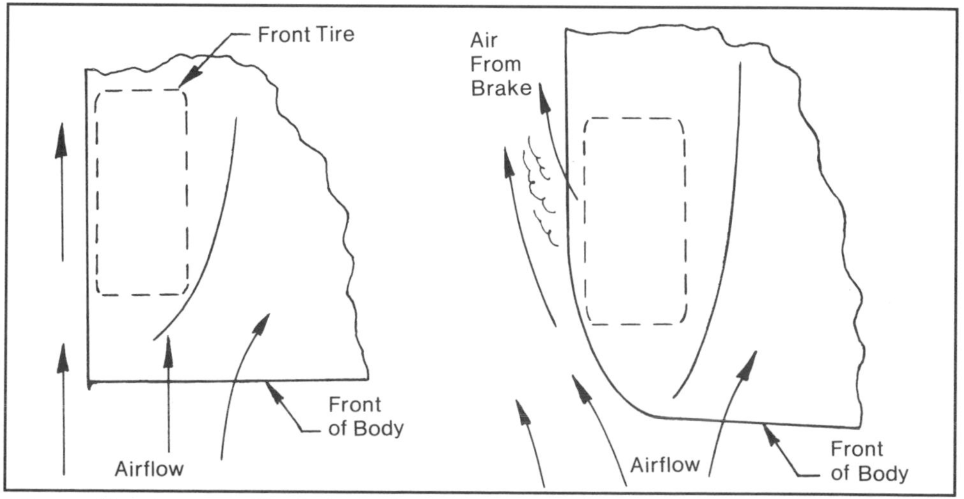

Shape of front corner of body determines how much air is pulled out through holes in wheel. This drawing shows two different body shapes, looking down on top of body. Car on left has a square nose, with little suction on front wheel well. Rounded-body on right drawing tends to pull a lot of air through front wheel.

This Porsche race car has cooling fans mounted outside of each wheel. At high speed, fan sucks cool air out through wheel. Cool air is ducted to brake through car nose.

Although this optional finned Corvette drum is same diameter as standard drum, it is 1/2-in. wider. Thus, it is not possible to merely slip finned drums on in place of non-finned Corvette drums. Entire brake assembly must be replaced. Be careful to take all dimensions when converting to heavy-duty brake drums.

Aluminum brake drums were used on large Buicks before they changed to disc brakes. Aluminum drums are great for cars requiring drum brakes. Often, cooling problems can be eliminated by changing from iron to aluminum drums.

Wally Edwards improved braking on his vintage sports car by adapting Buick aluminum drums to early Ford brakes. Many old street rods can be improved by this modification. Finned aluminum drums look good behind wire wheels on a show car, too.

better than a steel wheel. The wheel conducts heat from the brake and wheel hub. Although the wheel and tire will run hotter, the brake runs cooler.

Aluminum or magnesium wheels generally are used in racing because of their lower weight. However, improved cooling and better handling result with racing wheels. The lighter weight also improves acceleration.

Changing Brake Drums—Finned brake drums improve cooling over drums that don't have fins. If your brake drums are not finned, try to obtain some. It is not a simple modification, generally because few cars offer finned drums as a modification or as optional equipment. On a road car, check with your dealer to see if special drums are offered. On early Corvettes, for example, optional heavy-duty brakes used finned drums. However, the entire drum was wider. This required an entire brake-assembly change.

If you change drums, use aluminum drums in place of iron ones, if possible. They are lighter and cool better. Most aluminum drums have fins for maximum cooling area. For years, Buicks and some Lincolns used large, finned aluminum brake drums. You may find some in a wrecking yard that can be adapted to your car. Also, many foreign cars have aluminum drums on the rear.

On some sports cars equipped with drum brakes, optional aluminum drums were offered by the factory or from specialty houses. If you have a car that had *Al-Fin* drums as an option and are lucky enough to find some, they are a worthwhile modification. Not only do they improve cooling, they look great behind wire wheels!

Modifying Dust Shields—Disc-brake dust shields block cooling air. If a road car is modified for racing, the dust shields should be removed to allow cooling air to hit the brake rotor. This is not a good modification if the car is driven on the street. The rotor will suffer from road-dirt and grime contamination, and wear will increase. The manufacturer wouldn't install them if they weren't necessary. I'll qualify this:

If you remove the dust shields, restrict use to racing only, *unless you live in an area with no dirt roads and no snow*. Then, if it is a dual-purpose car, you'll have to make a choice between improved brake cooling and more brake maintenance from street use. This depends on the specific car and the conditions under which it is driven. You'll have to test and keep records to make sure you made the right choice.

Ultimate in vented backing plates are found on this old oval-track racer. Entire backing plate is cut away except for small tab holding wheel cylinder. On most cars, however, the brake shoes will not be retained and will possibly become misaligned sideways. A more normal backing-plate modification is drilling cooling holes in it or adding air scoops.

Backing plate on this car is modified by drilling holes in low-stress areas. Scoops are added to direct cool air to interior of brake drum. If racing rules allow it, this is a desirable modification for drum brakes For street use, scoops and drilled backing plates will make brakes more sensitive to wet-weather driving.

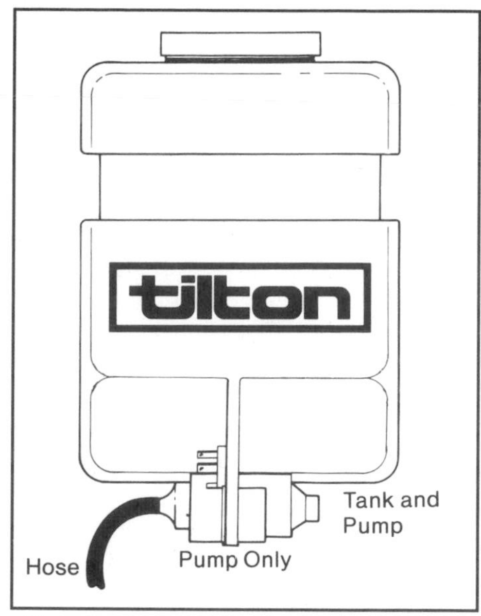

Tilton water-cooling system includes a reservoir with a small electric pump. Spray nozzles and fittings complete kit. System is designed to spray cooling water into brake ducts to lower temperature of cooling air directed at brake. This system works best on hot, dry days. Drawing courtesy Tilton Engineering.

Vented Backing Plates—Drum-brake backing plates should be vented for racing. This involves cutting a hole in the front and rear of the backing plate to allow air to flow in and out of the drum. Sometimes, a scoop is added to the front hole to direct air into the drum. Venting can be a large cut-out or a number of drilled holes. A large cut-out works best, but is structurally weaker. A screen can be used to keep rocks out. However, if the track is clean, the screen will not be necessary. Before disc brakes became popular, for maximum cooling, stock cars ran 12 X 3-in. drum brakes with brackets in place of backing plates.

When modifying backing plates, don't weaken the backing-plate structure. Brake shoes mount to the backing plate, so all the braking torque is transmitted to the suspension through the shoe anchor points and, consequently, the backing plate. Cut the backing plate only in areas far from where these highly stressed anchor points attach. If you are not sure how to do this, turn the job over to an engineer or experienced race-car technician.

Water Cooling—One of the latest tricks for cooling brakes is water cooling. A device is used to spray water into the brake-cooling ducts. The water vaporizes in the duct and cools the incoming air. Squirting water directly on the rotor has been used for years with varying degrees of success. The idea with this new system is not to squirt water directly on the brake. Steam formed will not cool the brake as efficiently as air. Also, it may cause the rotors to crack. Instead, the water is strictly an air cooler.

Brake Evaporative Cooling is an accessory water-cooling system available from Tilton Engineering. It is similar to a windshield-washer system, with a water reservoir and an electric pump. Water is fed to the spray nozzles with small hoses.

To operate the brake water-cooling system, the driver turns on a switch mounted on the instrument panel. Because the driver can't measure brake temperature, he must run the system throughout the race or just when he thinks he needs it. In a long race, water supply may be critical, so planning should be done. Huge amounts of extra water shouldn't be carried unless necessary.

Use water cooling as a last resort. Water is heavy and the system requires constant maintenance. First, make sure the air ducting is sized and sealed correctly, and the inlets are in a high-pressure area. Also, be sure the wheels are vented correctly. If all this is not enough, then try water cooling.

Because the water-cooling system is meant to cool the air in the brake-cooling duct, the system only works well on hot, dry days. If the air is cool and damp, little extra cooling is done. On a hot dry day, water cooling can make a considerable difference.

Modifying Body Aerodynamics—The subject of brake ducts has been discussed previously. If you've done all of the modifications previously mentioned to improve duct performance and you still find that the ducts are not accomplishing the job, or airflow from the front wheel wells is low, consider changing the body. Before you attempt this potentially difficult job, test the car with wool tufts to establish a baseline. Take photos, movies or video tapes of the tufts so you can review what happens as changes are made. First check the duct entrances for turbulent or reverse airflow.

Frequently, rear brake ducts are located in turbulent air. Typically, turbulence on an open-cockpit car is caused by objects, such as the windscreen, rollbar, or the driver's head, in front of the duct intake. Some fairings or reshaping of parts can greatly affect airflow over the car. Consider adding a streamlined, tapered fairing behind the driver's head to reduce turbulence over the rear body. This will also reduce total drag on the body.

If you have a Chevrolet sedan, you may wish to install 1984 Corvette disc brakes. Problem with this conversion is finding a way to mount calipers on your suspension, and resulting caliper clearance problems. If pedal effort gets too high, you could possibly use master cylinder and booster from the Corvette. A modification such as this can be difficult and will take a lot of planning. Photo courtesy Girlock Ltd.

Rotor, below, was drilled with a few large holes. Above rotor is typical; it has many smaller holes. Most drilled rotors have more holes in them and smaller ones. A 3/8-in.-diameter hole is probably an average size, but I have seen them from 1/4-in. to 1/2-in. diameter.

FADE REDUCTION

Brake fade is usually the direct result of high temperature, so first consider cooling modifications and harder linings. Other modifications to help reduce fade are also possible.

Converting from Drum Brakes to Disc Brakes—There will be a reduction in fade by converting from drum brakes to disc brakes. Unless you have an old car that uses four-wheel drum brakes, drums will be used on the rear wheels only. On a road car, the biggest problem with converting rear brakes to discs is retaining the parking brake.

Although disc brakes reduce fade, there is usually only a small gain to be made in converting *only* the rear brakes. Because front brakes do most of the work, rear-brake fade is not a serious problem. Consequently, you should also consider modifying the front brakes with larger rotors, high-performance linings, cooling improvements or other changes.

Converting from drums to discs can be difficult. Major problems include maintaining clearances and mounting the calipers. A caliper is often a difficult item to place in the wheel and suspension. Before throwing up your hands in despair, study brakes on other cars made by the manufacturer of your car. On some road cars, there are high-performance versions using disc brakes on the rear. For example, the Thunderbird, Lincoln and Corvette used disc brakes on the rear, while other models of the same year used drums. Because the wheel-stud pattern is the same, rotor mounting problems may be solved easily. However, caliper mounting may not.

Mounting calipers usually requires a special bracket, available from racing-brake manufacturers. The bracket usually is for drag- or stock-car-racing applications using racing brakes. This bracket may work on your car, even though your car may use passenger-car rotors. Contact racing-brake suppliers for information on your car.

If you have a mount for the rear calipers, the job isn't finished yet. Because drum brakes have servo action and disc brakes don't, brake balance no doubt will be wrong. Refer to Chapter 9 for how to calculate the forces in an all-disc-brake system.

I suggest you use a balance bar or proportioning valve to balance your new system. Also, if you don't change the master cylinder, you can expect pedal effort to increase, because drum-brake servo action requires less pedal effort. Make sure you can live with the increased pedal effort before starting this modification. If your car already has excessive pedal effort, changing the master cylinder or going to power brakes may be the only answer.

Rotor Drilling or Grooving—Drilling holes or cutting grooves in rotor sur-

Grooves on this rotor run in same direction as cooling vents. This rotor turns clockwise as viewed in photo. Notice how each groove lines up with metal web between cooling vents. This reduces chance of cracks forming at a groove. Trend in racing-brake development is toward grooves and away from cross-drilled holes.

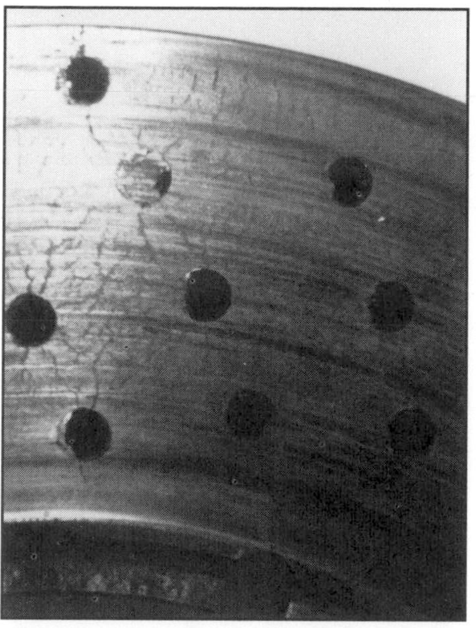

This rotor has many cracks due to thermal stress. Notice how holes seem to attract cracks. If you drill the rotor, you can expect cracking to be worse than if left undrilled. This rotor would have cracked anyway, but cross-drilling makes it worse. Cross-drilling removes weight, so rotor can't absorb as much heat.

If removing a solid rotor for modification, you may want to machine a groove in the edge. This removes some weight from a solid rotor, slightly increases its surface area, and does not change swept area. Do not try this if rotor temperatures are too high. You may need more rotor weight, not less.

faces *can* help reduce fade. Holes or grooves allow gas and dust from the linings someplace to go instead of wedging between the linings and the rotor surfaces.

However, rotors tend to crack at the holes or grooves because of increased stress at these points. Consequently, frequently inspect drilled or grooved rotors for cracks. Replace them if they start to fail. Improved fade resistance comes at the price of more-frequent rotor replacement.

Cross-drilling a brake rotor appears to be a simple job. It isn't. The problem is maintaining rotor balance. It's almost impossible to come up with a hole pattern that doesn't affect rotor balance.

A disc-brake rotor is heavy and rotates at moderately high speed. Rotor unbalance will be noticeable, particularly on a lightweight high-speed car. Therefore, if you wish to cross-drill the rotors, have a machine shop do it. They can lay out a symmetrical hole pattern on a rotary table and drill the holes concentric with the center of the rotor.

It is also possible to mill grooves in a rotor to produce the same effect as holes. Grooving doesn't remove as much weight, so balance is less of a problem. However, to do the job properly, use a milling machine. It's back to the machine shop, unless you can do it in your home workshop. Grooves should not run radially on a rotor—they should have a diagonal slant to them. Such a design is used for the rotors on many new motorcycles. The exact dimensions of the grooves aren't critical—0.06-in. wide by 0.03-in. deep will work.

As with cooling holes in a vented rotor, grooves should be cut differently from one side of the car to the other. According to rotor rotation, the grooves should slope away from rotation as they extend outward from the rotor friction-surface ID. The purpose is the same as rotor vents. Gas should be pumped outward through the grooves to the rotor OD as the rotor rotates.

After drilling holes or milling grooves, deburr the edges of each cut. A sharp edge acts like a cutting tool and will wear away the linings quickly. Use an abrasive stone or fine file to smooth all of the edges. Go over them twice to be sure.

Drilling holes in a brake drum has more effect on reducing fade than drilling a disc-brake rotor. Perhaps this is because drum brakes fade more easily than disc brakes. Therefore, any improvement is more effective. Few vehicles come with drilled brake drums, so some experimentation is required. You probably won't find anything to use as an example.

Drilling holes in a brake drum is tricky. Never drill through reinforcing lip at edge of drum. Hole location chosen here is near center of lining, where it will do the most good. Holes in drums should be experimented with carefully to avoid dangerous drum failure. Debur holes after drilling.

Drilling a brake drum also has its negative side. Drilling weakens a drum. A drum is subjected to high tension loads each time the brakes are applied, resulting from the outward pressure of the brake shoes. The shoes expanding in a drum tend to burst it. Holes at the inboard edge of a drum can cause cracks to start.

Because correctly drilling a brake drum involves a knowledge of what the stresses are, it's very difficult to determine the proper hole size and

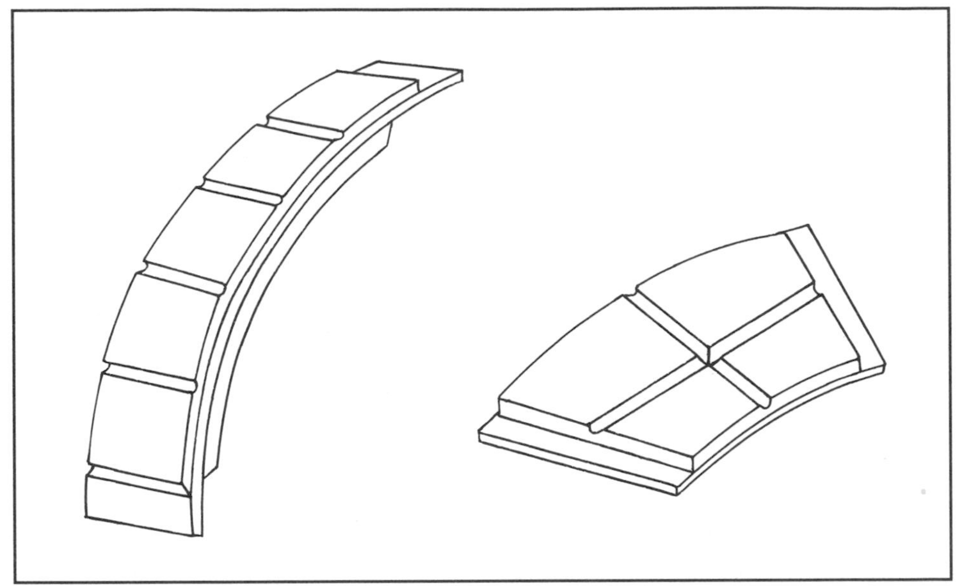

Disc-brake pads often come with one groove cut in radial direction. Cut second groove for improved fade resistance.

pattern. The stress on a brake drum is determined by the design of that particular drum. Therefore, start by drilling a *few* 1/4-in. holes. Drill the holes in the center of the friction surface. Do not drill holes at the inboard lip. The inboard edge is the weakest area of a drum. After drilling the drums, test to see if the holes help reduce fade. If there is still a fade problem, drill more holes. Run another test. By doing a number of identical back-to-back tests, you'll be able to determine how many holes are required. Don't drill more than necessary.

Use drilled brake drums only for racing. Inspect them for cracks after each race. If you find cracks starting at the edge of the holes, replace the drum. The replacement drum should have the holes drilled in a different pattern unless the other drum lasted a long time before cracking. Find areas of low stress to drill the holes. Experimentation is the only way to determine this.

Grooved Linings—Grooved linings have the same purpose as drilled rotors or drums. Gas and dust from the linings will enter the grooves instead of building up a lubricating layer between the friction surfaces.

Grooving linings has no affect on balancing. Also, it is easier than cutting an iron drum or rotor—you can do it with a hand saw grinder.

A disadvantage of grooved linings is that they wear faster. Reduced lining surface area increases wear. So don't get carried away with the amount of lining you cut. Again, testing is needed to determine how much grooving is needed.

On drum brakes, cut the slots across the shoe. Don't cut through the lining. A small slot is all that's needed, or 1/8-in. wide, about half way through the lining. Be careful not to cut the lining so it will come off if it cracks through a groove. Leave a pattern of at least four rivets in each uncut segment. For bonded linings, there should be no less than 2 in. between grooves.

WARNING
When cutting linings, wear a respirator to avoid inhaling potentially dangerous asbestos dust. Also, don't get the dust on your clothing or hands. Finally, clean up the dust when finished. If you have a shop vacuum, use it while cutting. Then you won't have much of a mess to contend with later.

On disc-brake pads, most linings already have a groove running across the center of the pad in a radial direction relative to the rotor. If yours are grooved, cut another groove 90° to it through the center of each pad. However, if your pads don't have a radial groove, put one in. Test the pads before cutting the second groove. If there's enough fade reduction with a radial groove, don't cut another groove.

REDUCING PEDAL EFFORT

If your car has excessive pedal effort during hard braking, it must be reduced. I'm not talking about fade here—fade only occurs when the brakes are hot. If the pedal effort is excessive when the brakes are cool, other modifications are needed. I discuss here how to make modifications to reduce pedal effort.

To reduce pedal effort, use:

- Softer brake-lining material.
- Smaller-diameter master cylinder.
- Power-brake booster.
- On drum brakes, larger-diameter wheel cylinders.
- On disc brakes, more total piston area.
- Smaller-diameter tires.
- Larger-diameter brakes.
- On drum brakes, more servo action.
- Increased pedal ratio.

Some of these modifications are easy to figure out and do. Others are expensive and require testing. I listed them in the order you should try them, with the easiest and less-costly methods first.

Softer Brake Linings—As discussed in Chapter 4, different friction materials have different friction coefficients. To review, linings with the highest friction coefficient are called *soft linings*. You can reduce pedal effort merely by changing to a softer lining. It is inexpensive and simple to make the change. If it doesn't work because of fade, excessive wear or whatever, you can easily change back to the original linings.

Of course, the problems with going to soft linings are that fade usually occurs at lower temperature and wear increases. So, if you already have fade problems, don't use softer linings. In fact, you may have to use harder linings, which will *increase* pedal effort. The same goes for wear. If wear is high already, don't use softer linings. However, if it's acceptable, try the softer linings, but check wear. There's no way to predict wear without trying the linings in actual use.

Although softer linings may give reduced pedal effort, it will be difficult to predict the change. The friction coefficient often depends on temperature. And, you can't predict the temperature. Softer linings will increase temperature because the front brakes will do even more work and the rear brakes less.

If high pedal effort was preventing

you from stopping at maximum deceleration, softer linings will increase deceleration. On a race car, this means higher friction-surface temperature. The temperature at the friction surface determines the fade characteristics and friction coefficient of the lining. If your brakes were balanced properly before you changed to higher-friction pads, you may have to rebalance your brakes.

To repeat what I said in Chapter 4, people talk about hard and soft linings without knowing the designations to describe how hard or how soft a lining is. Only brake-lining manufacturers and, perhaps, race-car engineers know these numbers—and they won't tell. It would be ideal to have curves showing coefficient of friction versus temperature for the various lining materials. If you could measure the temperature of your brakes, you could then predict what pedal effort would be. Because you don't have this information, you must test to see how the linings perform.

Soft linings are a practical modification for street use. However, on a race car, softer linings usually result in fade or excessive wear. If soft linings don't solve your high pedal-effort problems, let's look further.

Changing a Single or a Tandem Master Cylinder—This modification is usually inexpensive and easy. Also, the results can be calculated quickly. Pedal effort is reduced if you use a smaller-diameter master cylinder, and balance will not be changed. Use the following formula to calculate new pedal effort:

$$\text{New pedal effort} = \frac{P_{EO} D_N^2}{D_O^2} \text{ in pounds}$$

P_{EO} = Old pedal effort in pounds
D_N = New master-cylinder diameter in inches
D_O = Old master-cylinder diameter in inches

As an example, assume you are changing a 1.00-in.-diameter master cylinder to a 0.75-in.-diameter cylinder. Assume the old pedal effort with the 1.00-in. master cylinder was 100 lb at maximum deceleration. Putting these numbers into the formula gives:

$$\text{New pedal effort} = (100\text{ lb})\frac{(0.75\text{ in.})^2}{(1.00\text{ in.})^2}$$

$$\text{New pedal effort} = 56\text{ lb}$$

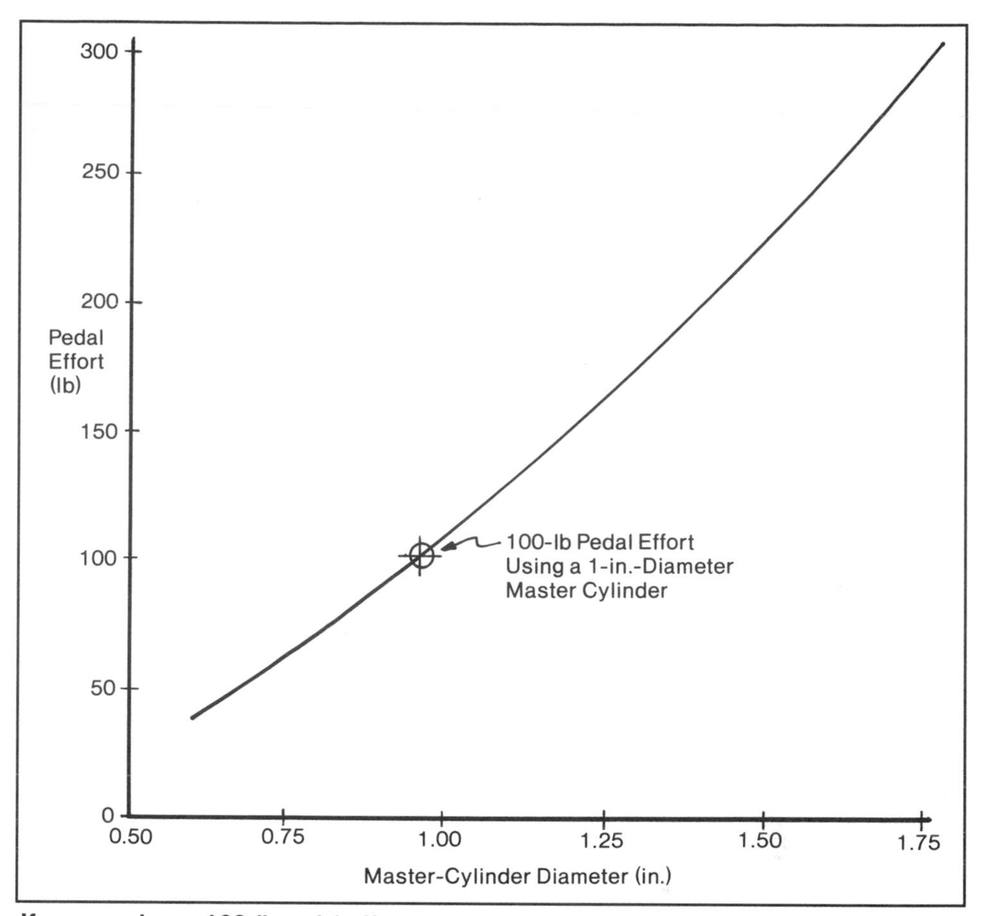

If your car has a 100-lb pedal effort using a 1.0-in.-diameter master cylinder, this curve shows how master-cylinder diameter can change pedal effort. Plot a curve like this using formula in text relating pedal effort to master-cylinder diameter. This curve is handy for picking a new single master-cylinder or tandem master-cylinder size. Remember, as pedal effort is reduced, pedal travel increases.

At the same deceleration, pedal effort dropped from 100 lb to 56 lb. This would be true for any car changing from a 1.00-in. master cylinder to a 0.75-in. master cylinder. It does not depend on any other factors. This formula does not work for a brake system with dual master cylinders unless you are making the same modification to both master cylinders.

Before changing master-cylinder diameter, find out if a smaller one exists that is interchangeable. Some companies manufacture different-diameter master cylinders that have the same mounting-bolt patterns. Reread what to look for when buying a new master cylinder beginning on page 46. Make sure you get a master cylinder with the correct features, such as with or without a residual pressure valve.

When master-cylinder diameter is reduced, pedal stroke increases. The amount of fluid displaced to apply the brakes—move friction material in contact with rubbing surfaces—depends on clearances in the brake system. Displacement is not affected by changing the master cylinder. If 0.5 cu in. of fluid is displaced to take up all the clearances, 0.5 cu in. of fluid will have to be displaced with either the old master cylinder or the new one.

Because the volume of fluid displaced equals the piston area times its stroke, there will be a larger stroke with a smaller-diameter master cylinder. The new movement can be calculated as follows for a single master cylinder or tandem master cylinder with equal-diameter pistons:

$$\text{New piston movement} = M_O \frac{D_O^2}{D_N^2} \text{ in inches}$$

M_O = Old piston movement in inches
D_O = Old master-cylinder diameter in inches
D_N = New master-cylinder diameter in inches

If you measure movement at the master-cylinder pushrod, the answer will be movement at the pushrod.

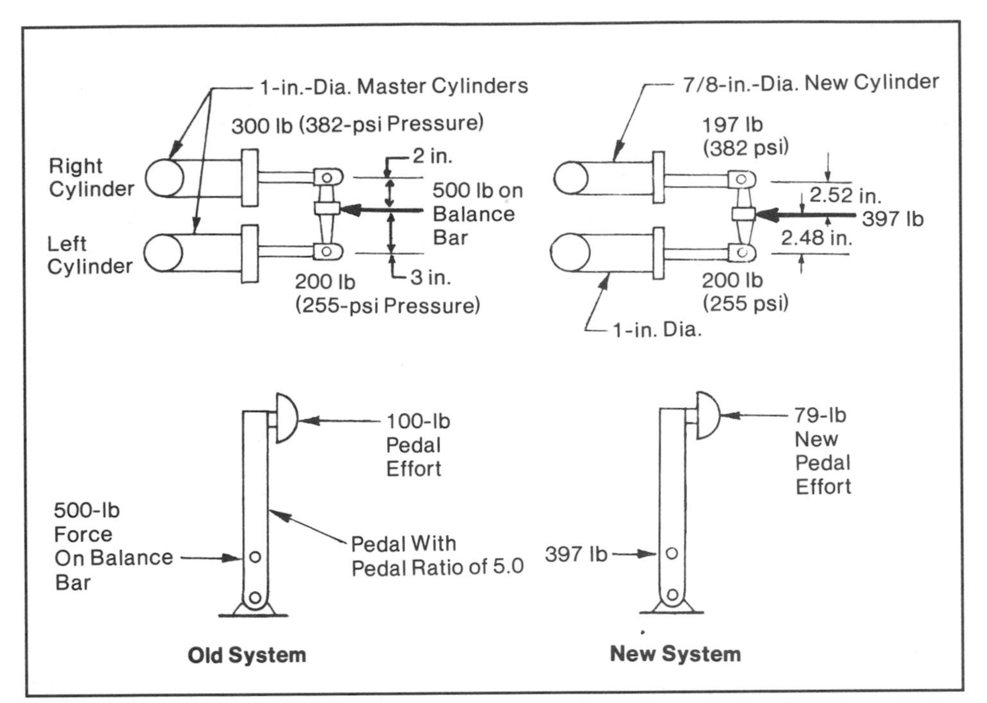

Drawing shows example problem described in text. Changing right-hand master cylinder from 1-in. to 7/8-in. diameter reduced pedal effort from 100 to 79 lb. In addition, balance bar is nearly centered in its adjustment with new master cylinder.

Make sure the new master cylinder can supply the required fluid displacement without bottoming.

First, be sure that the pedal will fully stroke the master cylinder without bottoming on the floor. Do this before you connect the lines. If the pedal bottoms, increase pushrod length until the master cylinder bottoms the same time as the pedal. Be sure to adjust the return stop so the master cylinder will return fully.

After the lines are connected and the system is bled, drive the car. Heat the brakes by making a few hard stops, then try to lock the brakes. If the pedal bottoms before wheel lockup, the new cylinder doesn't have enough travel. If this is done, the pedal may be out of position when retracted, requiring pedal, pedal bracket and cylinder modification.

If pedal movement is already excessive or the pedal is too high, reducing the master-cylinder diameter is not a good idea. However, if pedal travel is excessive because there's too much deflection in the brake system, that can be corrected. If you are troubled with both long pedal travel and high pedal effort, first try modifying the system to eliminate excessive deflection. Then, you may find that reducing master-cylinder diameter will now work.

There's another thing to consider when reducing master-cylinder diameter: Fluid pressure is higher with a smaller cylinder. Usually, brake-system components have a maximum safe operating pressure of 1000—1500 psi. If the safe operating pressure is exceeded, the system can be dangerous and unreliable. The maximum operating pressure is troublesome, particularly on special cars with lightweight calipers. Some racing calipers cannot take as high a pressure as road-car calipers. If calculated operating pressure is too high, *do not* reduce master-cylinder diameter. Instead, use other methods to reduce pedal effort, such as larger caliper-piston diameters.

Changing Dual Master Cylinders—There are two reasons for changing master-cylinder diameter on a car equipped with dual cylinders: to change pedal effort or brake balance. Changing one master cylinder affects both simultaneously.

Calculating new pedal effort resulting from changing one master cylinder in a dual system is more complex than with a single-cylinder system. You must also consider the brake-balance change that occurs when only one cylinder is changed. If your brakes were balanced before the change, you *must* change the brake balance afterward to obtain maximum braking.

Use the formulas in Chapters 6 and 9 to calculate the new pedal effort. If you are changing one master cylinder to reduce pedal effort, and brake balance is correct, follow these steps:

- Find the balance-bar setting. Refer to the formulas on page 75.
- Find approximate pedal ratio.
- Find the diameter of each present master cylinder.
- Estimate pedal effort before changing the master cylinder.

Calculate new pedal effort by beginning with the following steps:

- Calculate present force on balance bar: Multiply pedal effort by pedal ratio.
- Calculate force on each master cylinder from the formulas on page 75.
- Calculate area of each master cylinder.
- Calculate pressure in each master cylinder: Divide force on each master cylinder by its area.
- Choose a new diameter for the master cylinder.
- Calculate piston area of the new master cylinder. Pressure in new master cylinder will not change *if brakes were properly balanced before modification.*
- Calculate force on the new master cylinder by multiplying pressure by piston area. Because the *area* changed and not *pressure,* only the force changes.
- Add new cylinder force to force on the other master cylinder. Total force is the new force on the balance bar. Notice that the balance bar *must* be adjusted to give the correct forces.
- Calculate new pedal effort: Divide the new force on the balance bar by pedal ratio.
- Calculate position of the balance bar for new forces using the formulas.

To illustrate, let's run through a typical problem. Assume the brake system is as shown in the accompanying drawing. Original master cylinders are 1.00-in. diameter.

With original pedal effort of 100 lb, I'll try a 0.875-in. master cylinder to reduce this effort. Assuming that the brakes were balanced before the modification, adjustment to the balance bar will be made after the modification. Pedal ratio is 5.0. Using this information:

Force on balance bar = $R_P P_E$
R_P = Pedal ratio
P_E = Pedal effort in pounds

Force on balance bar = (5.0)(100 lb)
Force on balance bar = 500 lb

From formulas, the force on each master cylinder is:

$$F_L = (2\text{ in.})\frac{(500\text{ lb})}{(5\text{ in.})} = 200\text{ lb}$$

$$F_R = (3\text{ in.})\frac{(500\text{ lb})}{(5\text{ in.})} = 300\text{ lb}$$

F_L = Force on left master cylinder in pounds
F_R = Force on right master cylinder ir pounds

Notice that the sum of master-cylinder forces must equal total force on balance bar. That is a check on the previous calculation.

Using piston diameter, old master-cylinder area is:

Area = $0.7854D_M^2$
D_M = Master-cylinder diameter in inches
Area = $(0.7854)(1.00\text{ in.})^2$
Area = 0.785 sq in.

Instead of calculating area, you could use the table on page 101 to find master-cylinder area.

Hydraulic pressure in each master cylinder is:

Pressure = Force/Area in pounds per square inch

$$P_L = \frac{200\text{ lb}}{0.785\text{ sq in.}} = 255\text{ psi}$$

$$P_R = \frac{300\text{ lb}}{0.785\text{ sq in.}} = 382\text{ psi}$$

P_L = Pressure in left master cylinder
P_R = Pressure in right master cylinder

Now, change the right cylinder to 0.875-in. diameter. I selected the right one to change because the balance bar is already adjusted for more force on the right cylinder. With the new smaller cylinder, less force will be required to give the same pressure. Thus, the balance bar will be adjusted away from the new cylinder.

Find the area of the new master cylinder from the table or as follows:

Area = $0.7854D_M^2$ in square inches
Area = $(0.7854)(0.875\text{ in.})^2$
Area = 0.601 sq in.

Remember that pressure in the new master cylinder remains the same. However, the force changes. The new force on the right master cylinder is:

Force = (Pressure)(Area) in pounds
F_R = (382 psi)(0.601 sq in.)
F_R = 230 lb

Conventional power booster will not fit this Triumph Spitfire. There is not enough room around the master cylinder. For this application, ATE makes a booster that can be mounted in any location. It uses hydraulic pressure from the existing master cylinder to operate the booster rather than using force directly from pedal linkage. This remote booster is called the ATE T-50 series. Bendix makes as similar unit called the *Hydro-Vac*.

New force on balance bar is found by adding left-cylinder force to right-cylinder force. Left-cylinder force is the same as before—we didn't change it.

Force on balance bar = $F_L + F_R$ in pounds where:

F_L = Force on left master cylinder in pounds
F_R = Force on right master cylinder in pounds

Force on balance bar = 200 lb + 230 lb
Force on balance bar = 430 lb

New pedal effort can now be calculated:

$$\text{Pedal effort} = \frac{F_B}{R_P}$$

F_B = Force on balance bar in pounds
R_P = Pedal ratio

$$\text{New pedal effort} = \frac{430\text{ lb}}{5.0}$$

New pedal effort = 86 lb

With an original pedal effort of 100 lb, pedal effort is reduced by 14 lb. Brake balance must now be readjusted using the new force on the right master cylinder. Because the right and left forces are close to being equal, the balance bar can be adjusted close to its center. Changing the right cylinder made the balance change in the desired direction. If we had changed the left cylinder, balance-bar adjustment would have gone in the opposite direction—way off center.

Also, if the left master cylinder had been changed instead of the right one, pedal effort wouldn't have been reduced as much. Why? It's more effective to change the master cylinder with the highest force when doing it to reduce pedal effort.

On a dual master-cylinder system, it's relatively easy to find master cylinders with different diameters that have the same bolt pattern. Racing-brake-manufacturer catalogs usually supply all the necessary information.

Finally, remember that piston movement increases when master-cylinder diameter is reduced. Make sure the new master cylinder provides the required fluid displacement before bottoming.

Also, if you could not lock the wheels before the modification, the previous formulas won't work because it was assumed that operating pressure is maintained with the new master cylinder. Instead, go through the brake-system calculations in Chapter 9. Most cars are not at the force limit, so the previous formulas should apply. But don't forget the usual

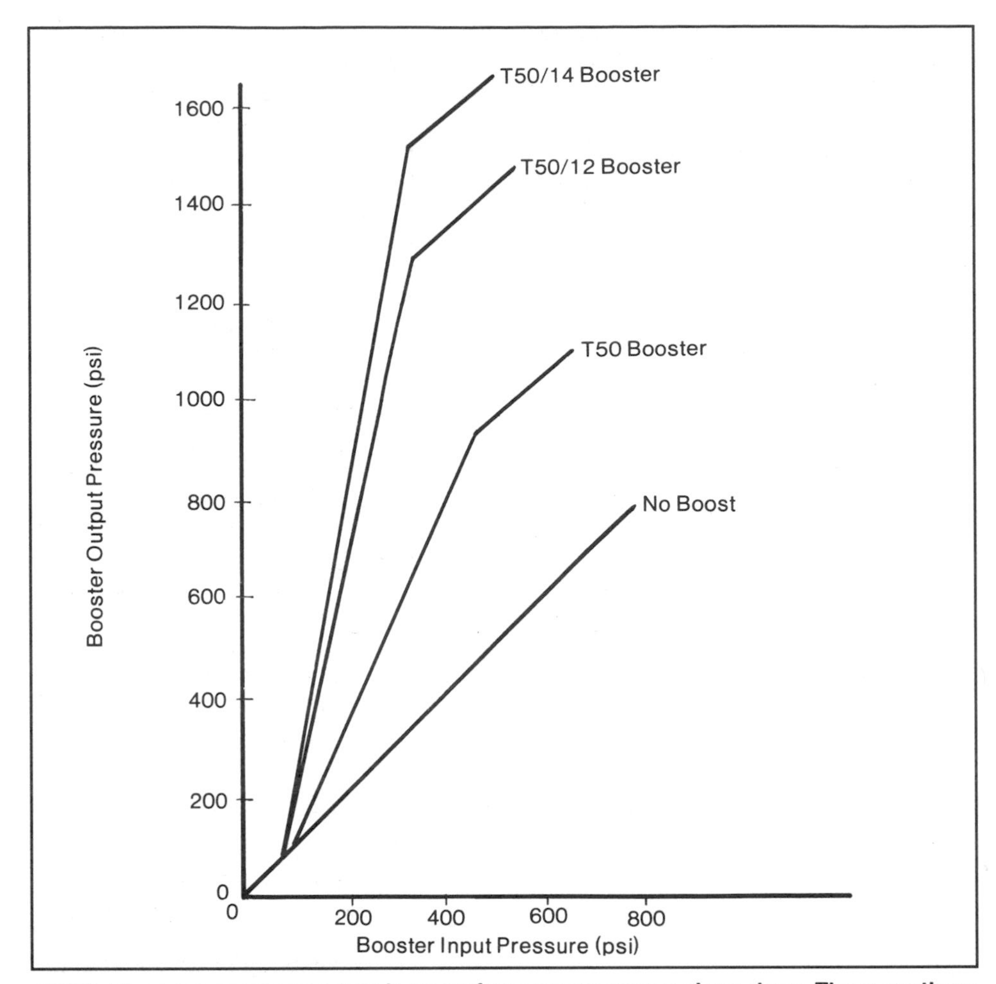

ATE T-50 series remote boosters have performance curves as shown here. There are three different models offered, each with a different boost. For most racing applications, only the T-50 unit would be used. More-powerful boosters would be useful on heavy road cars where more boost is required.

1000—1500-psi maximum safe operating pressure. If operating pressure gets too high, you cannot reduce master-cylinder diameter.

Adding a Power Booster—Converting to power brakes is a good way to reduce pedal effort. This method is used by most car manufacturers for road cars. Even light cars with disc brakes now come equipped with a power booster. With power brakes, you don't have to fight the problems of excessive pedal effort or travel. Instead, you can have the best of both worlds with power assist: low pedal travel and low pedal effort.

The easiest car to convert to power brakes is a road car that could be purchased with power brakes as an option. You may be able to purchase all the necessary parts from your dealer, or a salvage yard. Ask your dealership parts man what's necessary to convert to power brakes. This may require extra vacuum lines and a different pedal, master cylinder and pedal-support bracket. Get the entire picture before buying parts.

If power brakes weren't offered, you may have trouble fitting a booster. If the master cylinder is mounted under the floor or close to the engine, it may be impossible to fit a conventional vacuum brake booster. You might be able to install a smaller hydraulic brake booster, but a power-steering pump must be plumbed into or installed. This requires a lot of work.

Some brake boosters are designed as an accessory for existing brake systems. One such booster is the T-50, made by ATE in West Germany. This unit is designed to mount remotely from the master cylinder, rather than in tandem. It uses input pressure from the existing master cylinder and raises this pressure with a vacuum-operated piston to boost, or supply extra force. ATE provides data on this unit. Input pressure versus output pressure is shown by the nearby curve. This booster may be desirable if you have difficulty mounting a booster in the conventional location. Bendix makes a similar unit called the *Hydro-Vac.*

You may be able to measure pedal effort directly with a small fish scale. If there is a way to measure fluid pressure at same time, you can determine effect of power booster. Measure pedal effort and hydraulic pressure with booster working. Then, try it again with vacuum line to the booster disconnected. The difference between two pressures at a given pedal effort is effect of booster.

Most race cars don't use power brakes because:

- Power assist is not required on light race cars. If pedal effort is too high on a car weighing under 2500 lb, there's probably a brake-design problem.
- Power-brake equipment adds weight and complexity.
- Driver feel is not as precise as with unboosted brakes.

A heavy race car with disc brakes may require power brakes. For example, on a stock car, there may be no other solution if hard linings are used. Fortunately, most stock cars have enough room under the hood to mount a power-brake booster. This can be a simple change if stock pedals and master cylinder are used.

When adding a power booster, try to use the master cylinder that goes with the booster. If you don't, you'll have to check the stroke of the booster and master cylinder to be sure they are compatible. If master-cylinder stroke is more than the booster, the booster will not bottom the cylinder. However, if master-cylinder stroke is less, there's no problem.

You must investigate the inner workings of both the master cylinder and booster to determine their strokes. See the factory shop manuals for this.

Like any other master cylinder, one from a power-brake system must have sufficient displacement. So, if you're using the brakes from one car and a master cylinder from another, check

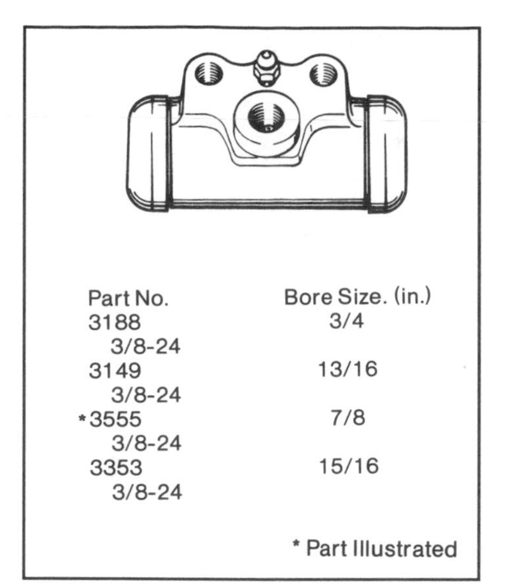

Part No.	Bore Size. (in.)
3188 3/8-24	3/4
3149 3/8-24	13/16
*3555 3/8-24	7/8
3353 3/8-24	15/16

* Part Illustrated

Drawing was taken from Bendix Brake Reference and Identification Catalog. It shows a wheel cylinder which comes in four different diameters. Many wheel cylinders are illustrated. This will show you what other size wheel cylinders may be available for your car. Go to an auto-parts store that sells Bendix parts. Drawing courtesy Bendix Corp.

master-cylinder displacement. If it bottoms, the brakes won't work. Displacement of the new master cylinder should be equal to or greater than the one designed for the brake system.

The big problem in adapting a brake booster to a special car is getting the correct pedal effort. If the booster supplies excessive boost, the brakes will be too sensitive. Too much boost can be dangerous. It's too easy to lock the wheels if pedal effort is low.

Pedal effort with power brakes is determined by the design of the brakes and booster, pedal ratio, and vacuum or power-steering-fluid pressure supplied to the booster. Calculating pedal force is complex. And, it is difficult to get all the required information.

To get this information, start by testing to determine brake-fluid pressure. Find a car with the power booster you plan to use. Install a fluid-pressure gage in the line to the front brakes. With the engine running, apply the brakes and record pedal effort and fluid pressure. You can use a small scale to measure pedal force. Try various forces and plot a graph of pedal effort versus hydraulic pressure.

If you know the hydraulic pressure for a given pedal effort, you can use the formulas in Chapter 9 to design your brake system. Correct for any difference in pedal ratio between your car and the one the power booster came from. If your car's pedal ratio is higher, fluid pressure on your car will be higher and vice versa.

Let's run through a sample calculation to correct for pedal ratio. Assume the pedal ratio is 8.0 on your car and 5.0 on the car the brake booster came from. Let's also assume that you tested the car with the brake booster and got 500-psi fluid pressure for 20-lb pedal effort. With the pedal ratio of 5.0, the force on the booster-input rod is:

$$\text{Force on booster} = R_P P_E$$

R_P = Pedal ratio
P_E = Pedal effort in pounds

$$\text{Force on booster} = (5.0)(20\text{ lb}) = 100\text{ lb}$$

Fluid pressure from the boosted cylinder is 500 psi with a 100-lb force on the booster. If you put this booster and cylinder on *your* car with its pedal ratio of 8.0, pedal effort will be lower for the same hydraulic pressure, providing vacuum or hydraulic pressure to the booster is the same.

Your car's pedal effort at 500-psi hydraulic pressure will be:

$$\text{Pedal effort} = \frac{F_B}{R_P}$$

F_B = Force on booster in pounds
R_P = Pedal ratio

$$\text{Pedal effort} = \frac{100\text{ lb}}{8.0} = 12.5\text{ lb.}$$

This compares to the 20-lb effort on the car with a pedal ratio of 5.0.

Larger Wheel Cylinders—If you have drum brakes on all four wheels, pedal effort can be reduced by replacing the wheel cylinders with larger ones. The effect of increasing wheel-cylinder diameter is the same as reducing master-cylinder size. Wheel-cylinder changes typically are more difficult than changing master-cylinder diameter because wheel cylinders are more trouble to adapt. And, you're working with more of them. However, the advantage of changing wheel cylinders is that fluid pressure is not affected.

Input pressure from the master cylinder is unchanged for a given pedal force. If the hydraulic system operates at high pressure, changing wheel cylinders is better than going to a smaller master cylinder. Regardless of pressure, larger wheel cylinders have the same effect as a smaller master cylinder. Both situations require more fluid displacement, so check for adequate master-cylinder displacement as already discussed.

The trouble with changing wheel cylinders is finding ones that will fit. Often, the mounting is different, the bleeder is in the wrong place, or other physical differences make the job impractical. If you decide to try changing wheel cylinders, first check wheel cylinders from a car by the same manufacturer as yours. For example, if you have a Chevrolet, first consider Chevrolet cylinders. If that fails, look at cylinders from other GM divisions. If you can't find a suitable cylinder, try other American cars using a similar drum-brake design.

Calculate the effect on pedal effort of changing wheel-cylinder diameters as follows. This assumes the car has drum brakes all around, all use the same-size wheel cylinders, and all cylinders are changed.

$$\text{New pedal effort} = P_{EO}\frac{D^2_{WO}}{D^2_{WN}}$$

P_{EO} = Old pedal effort in pounds
D_{WO} = Old wheel-cylinder diameter in inches
D_{WN} = New wheel-cylinder diameter in inches

Notice that the cylinder diameter is squared, so a small change in diameter has a major effect on pedal effort.

The previous formula only works if all the brakes use the same size wheel cylinders *and* all brakes are modified the same way. On most cars, this will not be true. Instead, a wheel-cylinder modification would usually fall into one of two categories:

- Disc brakes are used on the front and drums on the rear. Only the wheel cylinders on the rear brakes are modified.
- Four-wheel drum brakes are used, but only one pair of brakes has the wheel cylinders modified.

In either case, the wheel-cylinder modification affects both pedal effort and brake balance. Make sure you can deal with two changes at once before modifying the wheel cylinders on only one end of the car.

To see what a wheel-cylinder change does, it's easiest to determine the effect changing wheel-cylinder size has on brake torque at a given hydraulic pressure. Then, by using the formulas in Chapter 9, calculate the effect on pedal effort. This formula as-

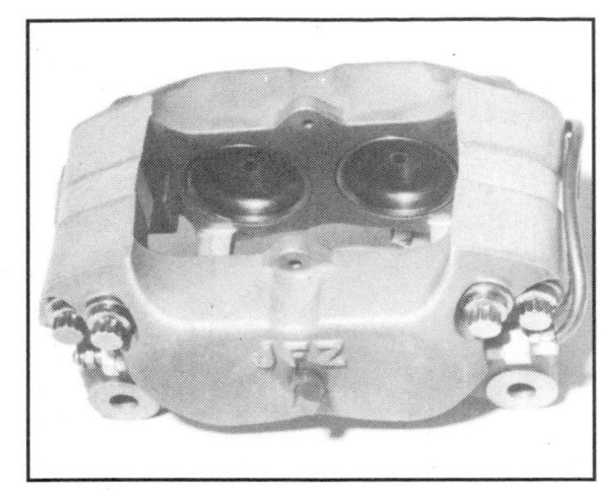

This double-piston caliper is manufactured by JFZ Engineered Products. If you want to double brake torque or cut pedal effort in half, install double-piston calipers in place of single-piston calipers. Of course, this also doubles pedal travel. Photo courtesy JFZ.

Installing two calipers per rotor and a slightly larger-diameter master cylinder can cut pedal effort and significantly improve lining life. This Formula 1 Grand Prix car uses two AP Racing calipers on each front rotor. On a track that requires little braking, one caliper can be removed to save weight. Notice temperature-indicating paint on rotor under nearest caliper. Photo courtesy AP Racing.

sumes brake balance is not changed after the modification. Readjusting brake balance usually reduces the effect of a wheel-cylinder change.

$$\text{New brake torque} = T_O \frac{D^2_{WN}}{D^2_{WO}}$$

in inch-pounds

T_O = Old brake torque in inch-pounds
D_{WN} = New wheel-cylinder diameter in inches
D_{WO} = Old wheel-cylinder diameter in inches

Changing only the rear-wheel cylinders only has less effect on pedal effort than if only the front ones are changed. This is because the front brakes do most of the work.

With disc brakes only on the front, changing the rear-wheel cylinders to reduce pedal effort is *not* worth the trouble. If the brakes were balanced before the modification, the new rear-wheel cylinders would cause the rear wheels to lock up first. There also would be a small reduction in pedal effort. If you use an adjustable proportioning valve to balance the brakes, rear-brake torque is *reduced.* Consequently, reducing rear-brake torque tends to cancel out the reduction in pedal effort gained by changing wheel cylinders. All you've done is spent time and money without significantly reducing pedal effort.

You should only change to larger rear-brake wheel cylinders if something is done to increase front-brake torque at the same time.

Larger Piston-Area Calipers—If you have disc brakes, pedal effort can be reduced by installing calipers with bigger pistons or with more pistons in each caliper. As long as total piston area is increased, this has the same effect as going to bigger drum-brake wheel cylinders.

Increasing caliper-piston area is a good modification on a system with high operating pressure. Pressure is reduced afterwards as is the deflection of brake hoses and other parts. However, increased piston area also increases brake-pedal travel. Therefore, you may also have to go to a longer-stroke master cylinder.

The problem with installing calipers that have larger-diameter pistons is that calipers usually have the largest pistons possible. If you find a caliper with larger pistons, it probably has larger pads. The problem here is that larger pads require a larger-diameter rotor. Therefore, finding a caliper with larger pistons not requiring a larger rotor generally is not usually possible.

Multiple-Piston Calipers—The usual modification for disc brakes is installing calipers with multiple sets of pistons. If your calipers have a single piston on each side of the rotors, you may be able to replace them with two- or three-piston sets. Racing calipers are sometimes made with more than one piston set. Depending on the number of pistons used, calipers are called *single-piston, dual-piston* or *triple-piston calipers.*

You may have trouble mounting new multiple-piston calipers even if you find one that fits your rotors. Check with one or more of the brake-caliper manufacturers listed in the Suppliers Index for calipers and mounting kits to fit particular cars. You may also have to purchase caliper-mounting hardware.

If you change to multiple-piston calipers, there's an added benefit besides reduced pedal effort. The calipers must be larger and heavier to contain multiple pistons. As a result, the calipers will run cooler than the smaller calipers they replaced. Fade will be less likely with the larger calipers. In addition, larger pads will last longer. Even though bigger calipers are expensive, you get value for the additional cost.

The new pedal effort for calipers with larger-diameter pistons can be calculated using the following formula. Note: This assumes that all four brakes have the same-size calipers and are changed the same amount:

Using low-profile tires will lower pedal effort slightly. If you put wider wheels and tires on a car, use lowest profile tire possible to end up with minimum pedal effort. This Fiat owner was able to lower the CG and put a lot of rubber on the road with low-profile tires.

$$\text{New pedal effort} = P_{EO}\frac{D^2_{EO}}{D^2_{CN}} \text{ in pounds}$$

P_{EO} = Old pedal effort in pounds
D_{EO} = Old caliper-piston diameter in inches
D_{CN} = New caliper-piston diameter in inches

When changing to multiple-piston calipers with the same diameter as the old pistons, use the following formula to find the new pedal effort. Again, it is assumed that all four brakes are being changed identically.

$$\text{New pedal effort} = P_{EO}\frac{N_{PO}}{N_{PN}} \text{ in pounds}$$

P_{EO} = Old pedal effort in pounds
N_{PO} = Number of pistons in old caliper
N_{PN} = Number of pistons in new caliper

For floating calipers, multiply the number of pistons by two. Each piston in a floating caliper does the job of two pistons in a fixed caliper.

When changing both piston diameter and the number of pistons, use both formulas. First, correct pedal effort for the change in piston diameter. Then, put that pedal effort into the formula for number of pistons and calculate final pedal effort.

Twin Calipers—If you have a race car with one caliper on each rotor it may be possible to mount two calipers on each rotor. This will reduce pedal effort by half if you change all four brakes to *twin calipers.* This modification has the same effect as changing the brakes from single to double-piston calipers. It may actually cost less. In addition, you can always remove one set of calipers if you run on a track that is easy on brakes. Remember that doubling caliper-piston area will double fluid displacement, so master-cylinder stroke or area may have to be increased to compensate.

If you increase master-cylinder area by a lesser percentage than caliper-piston area, you'll get a pedal-effort reduction. Also, the doubled lining area will significantly increase lining life. Therefore, you may run soft linings with twin calipers and hard linings with singles. Softer linings will further reduce pedal effort.

If you are thinking this is too good to be true, you're right. Adding a second caliper on a rotor restricts cooling. Because more of the rotor is covered, airflow through the rotor is restricted. Consequently, the rotor doesn't cool as well.

The previous formulas in this section apply only to a car that has the same-size brakes on all four wheels and to the same modification to all four brakes at the same time. This makes the formulas simple, but they may not apply in your situation. Most cars are modified differently front and rear, or don't have the same-size brakes front and rear.

Unfortunately, there are no simple formulas to predict exactly what happens to pedal effort when you modify only one pair of brakes. However, you can make some rough estimates. If you modify only the front brakes, you'll realize about two-thirds of the change in pedal effort that the formulas for changing all four brakes predict. For example, let's use one of the simple formulas to predict the new pedal effort for a modification to only the front brakes.

Let's say the formula predicts a change from a 90-lb pedal effort to a new pedal effort of 30 lb. The change in pedal effort is 60 lb. Because the rear brakes were not modified, the actual change is about two-thirds of this change. Thus the estimated change in pedal effort is (2/3) X 60 lb = 40 lb. The estimated new pedal effort after the modification is 90 lb − 40 lb = 50 lb.

If both front and rear brakes are the same size, but are not modified identically, the simple formulas can again be used to estimate new pedal effort. Use the formula twice—once for the front-brake modification and again for the rear-brake modification. Add the two answers and average them to get a rough estimate of the new pedal effort.

To get an accurate answer for a modification, you must use the formulas in Chapter 9. This means performing a number of calculations for both the old and new brake systems.

Reducing Tire Diameter—If you install tires with a smaller outside diameter or *roll-out,* pedal effort will be reduced. This is an easy change to make, but only a small reduction in pedal effort results. You will not want to spend money for new tires *only* to change brake-pedal effort. The amount of change won't be worth the cost. However, changing tires can be desirable for other reasons, such as lowering the car for improved handling.

Usually, putting smaller-diameter wheels on a car won't work. The brakes are usually designed to be as big as possible and still fit inside the wheels. Therefore, if you change tire diameter, you should do it with lower-profile tires. Tire designations refer to the ratio of height to width. For example, a 50-series tire has a section height that is 50% of its width. A 50-

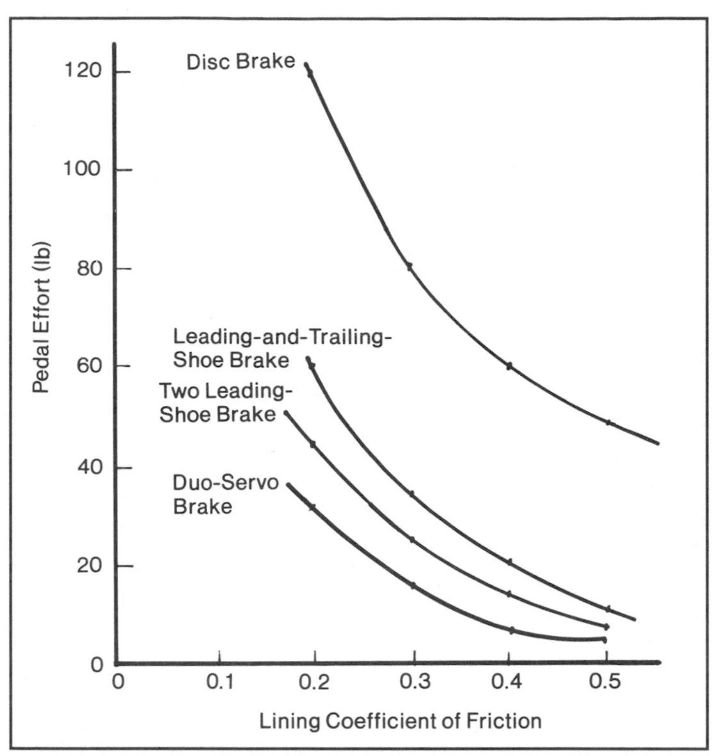

Graph shows how pedal effort would change if you change servo action of your brakes. Converting to drum brakes from disc brakes has the biggest effect, but I don't know why anyone would do such a thing. This graph assumes that all brakes are the same effective radius and have typical cylinder diameters. Purpose of this graph is to compare various types of drum brakes, because disc-brake pedal effort can be varied greatly with caliper design. This curve should be used only as a rough guide, because actual numbers will vary greatly from one car to another.

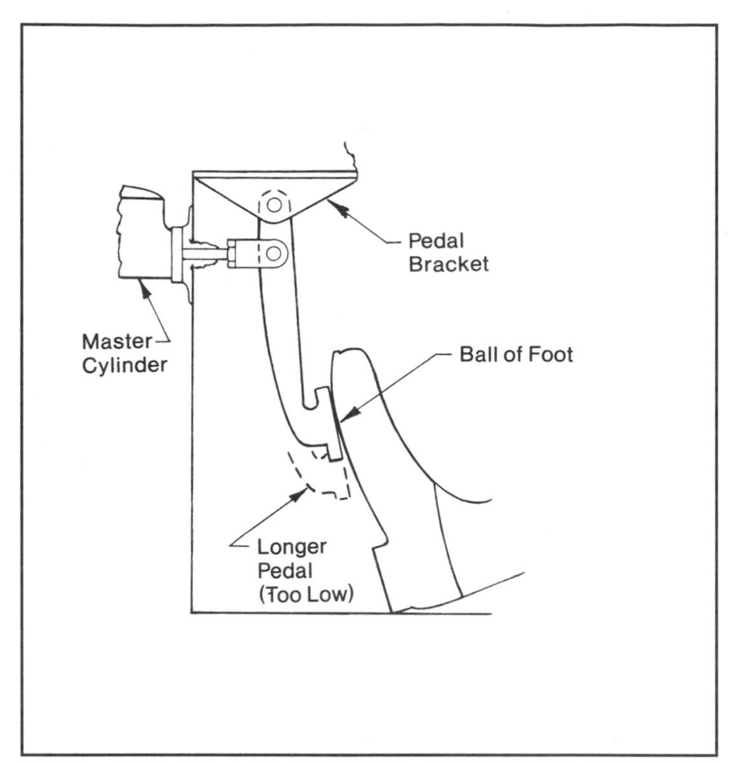

Most race-car brake pedals are designed to fit the driver's foot. If you changed pedal ratio by making the pedal arm longer, the pad would be in the wrong place. Changing the pedal bracket or master-cylinder location is nearly impossible on most cars, because mounting points are an integral part of the frame structure. Don't try changing pedal ratio unless your car is well suited to this modification.

This slalom racer uses VW drum brakes in a 15-in. wheel. On this car, it is possible to increase effective radius of brake. Usually, the brake on a race car fits close to the wheel rim. If larger-diameter Porsche drum brakes were installed, the pedal effort would be reduced by a noticeable amount.

series tire is lower than a 60-series tire of the same section width. Because tire sizes are confusing and tread widths vary, the best way to choose new tires is to measure them or compare their published specifications.

Tire diameter is not used in brake-system calculation. Instead, *rolling radius* of the tire is used. As discussed on page 94, it's the distance between the center of the wheel and the ground.

When changing tires, measure both tires. Measuring rolling radius is covered on page 94.

Once you have the rolling radius of both tires, calculate the new pedal effort:

$$\text{New pedal effort} = P_{EO}\frac{r_N}{r_O} \text{ in pounds}$$

P_{EO} = Old pedal effort in pounds
r_N = New-tire rolling radius in inches
r_O = Old-tire rolling radius in inches

As you can see, smaller tire rolling radius reduces pedal effort. By installing low-profile tires on your car, you can improve both braking and handling.

Larger-Diameter Drums or Rotors— If there is room in the wheels, changing to brakes with larger-diameter drums or rotors will reduce pedal effort. This modification will result in other improvements, too: The brakes will run cooler and pads or shoes may wear at a lower rate. Bigger brakes are a good modification if possible.

The critical dimension of a brake is its *effective radius,* as described in Chapter 9. This is the inside radius of a drum, or the rotor radius measured to the center of the pad. Increasing effective radius will reduce pedal effort.

The new pedal effort is calculated as follows:

$$\text{New pedal effort} = P_{EO}\frac{r_{EO}}{r_{EN}} \text{ in pounds}$$

P_{EO} = Old pedal effort in pounds
r_{EO} = Old brake effective radius in inches
r_{EN} = New brake effective radius in inches

Because effective radius is limited by existing wheel diameter, it can't be increased much. Therefore, it will give only small benefits and is expensive. However, the benefits will be great if you can make a large change in effective radius.

To find larger-diameter brakes for a road car, check the brakes used on cars similar to yours. Locate brakes that will mount easily on your car. Often, larger-diameter brakes from one model will fit another. Check brakes from a station wagon or truck.

They may fit a sedan suspension and be larger. Some manufacturers change brake sizes during a redesign, but retain the original mounting points. See your car dealer for information.

Increasing Servo Action—Drum brakes are designed with varying amounts of servo action. As you may recall from Chapter 2, the amount depends on the geometry of the internal parts of the brake. Leading-shoe brakes have high servo action; trailing-shoe drum brakes do not. If you put a dual-leading-shoe brake in place of a single-leading-shoe brake, this usually increases servo action and reduces pedal effort. A duo-servo brake has the most servo action. Replacing single-leading-shoe brakes with duo-servo brakes would cause the greatest reduction in pedal effort.

Changing servo action is difficult and expensive. Also, it is difficult to predict the resulting pedal effort. You must know the coefficient of friction and the exact geometry of the brake to accurately calculate servo action. The graph at left gives you a rough idea of what will happen. The curves in this graph apply to brakes of the same diameter, but with different amounts of servo action. These are rough estimates—do *not* use them for exact calculations.

If you want a big change in pedal effort and are willing to change the complete brake assembly, try to find brakes with a different servo action. It may be difficult to find a brake that will bolt onto your car. As with many other modifications discussed, start the search by looking at brakes used on cars by the same manufacturer.

The problem with increasing servo action is that the tendency to fade increases. However, if your car has no fade problems, increasing servo action may work. But, if you already experience fade, stay away from brakes with high servo action. Another disadvantage of increased servo action on front drum brakes is that it can cause the brake to pull.

To summarize, modifying servo action is expensive, somewhat unpredictable, and may cause fade and pull problems. It is not a good choice for lowering pedal effort.

Increasing Pedal Ratio—An obvious way to reduce pedal effort is to increase the pedal ratio. This means a different brake pedal with more leverage on the master cylinder. The pedal ratio is determined from the methods discussed in Chapter 6.

This Mercury requires adjustable brake balance for two reasons. Switching from street driving to racing increases weight transfer and requires more braking effort on the front wheels. Racing in the rain reduces weight transfer and requires less braking on the front wheels compared to racing in the dry. Photo by Steven Schnabel.

Pedal ratio is hard to change. Normally, simply extending the pedal won't work because the pedals are designed to fit the driver's foot while the driver is in his normal position. Therefore, a longer pedal arm may make the pedal pad too high or low for comfort. And, there's a danger that the extended pedal will bottom against the floor—if hanging pedals are used—and not match the position of the clutch and accelerator pedals. The only solution is to remove the pedal mechanism and redesign the whole thing, including relocating the master cylinder. Because the master cylinder generally is mounted on a reinforced part of the chassis or body, it is difficult to move.

The best time to modify pedal ratio is when the car is being designed. As for an existing car with pedals in the correct position, don't try to modify pedal ratio. It's too difficult. Instead, try a smaller-diameter master cylinder. The effect on the brakes is the same as a higher pedal ratio—more pedal movement and lower effort.

If you do manage to change the pedal ratio, calculate the effect on pedal effort:

$$\text{New pedal effort} = P_{EO}\frac{R_{PO}}{R_{PN}} \text{ in pounds}$$

P_{EO} = Old pedal effort in pounds
R_{PO} = Old pedal ratio
R_{PN} = New pedal ratio

You can see that a larger pedal ratio results in a lower pedal effort.

REDUCING LINING WEAR

Reducing lining wear is important for both street driving and racing. Unfortunately, making changes to reduce lining wear often has undesirable effects on brake performance. Before I discuss these effects, let's look at common ways of increasing brake-lining life:

- Increase cooling airflow.
- Use harder lining material.
- Increase lining area.

Following is a brief review of these modifications that are covered in this chapter.

Cooling modifications are important to all phases of brake performance, so they are top priority. On a road car, cooling improvements can cause increased lining wear if not done correctly. For example, removing splash shields from disc brakes can cause increased lining wear from increased dirt contamination and corrosion.

Racing a road car on racing tires requires adjustable brake balance. Street tires on this car are being changed to racing slicks for a slalom. Car will decelerate faster with slicks, and brake balance will have to be adjusted more towards the front. An adjustable proportioning valve could make balancing brakes fast and easy.

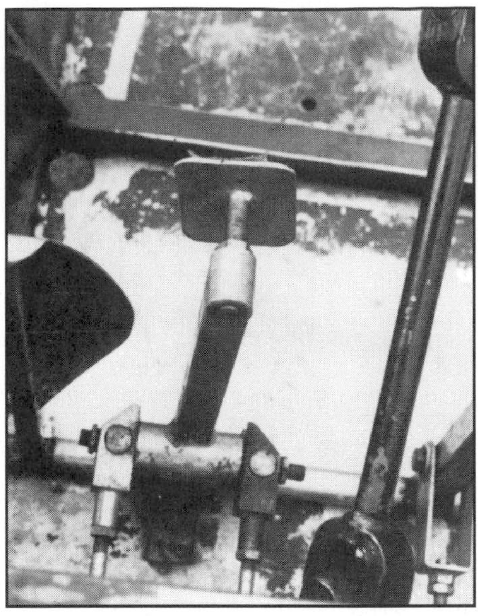

Typical balance bar used in a race car. If your car does not have a brake-balancing feature, balance bar could be adapted. It means changing pedals, linkage, and converting to dual master cylinders. Fitting an adjustable proportioning valve is much easier and cheaper.

If you change CG height of a car, you need to change brake balance. A tall car has a greater weight transfer than one with a low CG. I hope the owner of this truck never has to stop fast from high speed—or has to do anything else at high speed.

Tilton adjustable proportioning valve is used on this stock car instead of a balance bar. A tandem master cylinder is used for safety. The proportioning valve is mounted on drive-shaft tunnel where driver can easily reach lever. Valve is installed in brake line to rear brakes.

Harder lining material is a good way to reduce wear. Unfortunately, pedal effort increases with harder linings. Be sure you can operate the brakes at maximum deceleration after changing to harder linings. Hard linings combined with other modifications to reduce pedal effort may be the best overall compromise for reducing lining wear.

Increase lining area by installing wider drums or multiple-piston calipers to increase lining life. Temperature reduction also results. The combination of larger lining area and reduced temperature will have a major effect on wear. On disc brakes, multiple-piston calipers also reduce pedal effort.

BRAKE-BALANCE MODIFICATIONS

On most road cars, there's no way to adjust brake balance. Balance adjustment is usually a reasonable compromise, as designed. However, there may be reasons why modifying a car to allow balance adjustment is desirable.

If you drive your car on the street and occasionally race it, a balance adjustment may be necessary. Racing tires have much higher grip than street tires. Consequently, the additional weight transfer with race tires requires more braking at the front and less at the rear. The opposite is true on the street. If a proportioning valve is used, it partially compensates for tire-grip changes. But a more precise balance adjustment may be needed for racing.

If you drive on ice and snow, you may wish to change brake balance for this condition. The same goes for driving off-road or on unpaved roads. With reduced grip, the rear brakes do a greater share of the work. If you change brake balance, your car may have better braking on snow, ice or dirt.

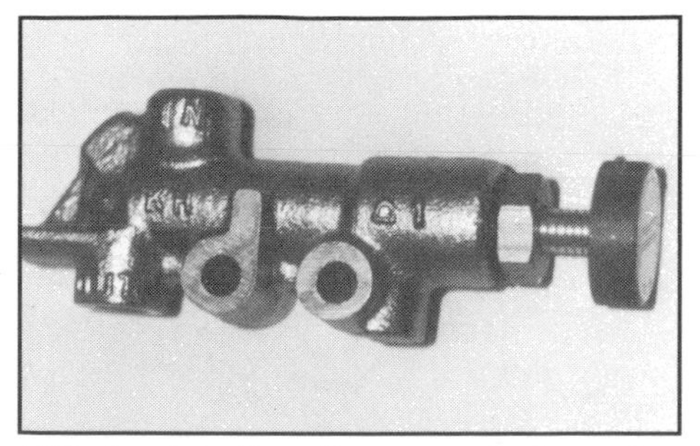

This handy adjustable proportioning valve is manufactured by Kelsey-Hayes. Adjustment is made by turning knob. Valve is available from aftermarket-parts suppliers such as Performance World in San Diego, CA and from many U.S.-car dealers through their high-performance-parts catalogs.

There are several ways to modify a car to allow brake-balance adjustments. Here are the easiest ways:

- Install an adjustable proportioning valve.
- Install a dual master-cylinder and balance-bar system.

Proportioning Valve—On a road car, the adjustable proportioning valve is best. It's much easier and cheaper than changing the pedals, master cylinder and linkage. A proportioning valve is virtually a bolt-on component. For a race car, use a balance bar—it gives a firmer pedal.

There are two types of adjustable proportioning valves—one can be adjusted from the cockpit and the other can't. A proportioning valve that can be adjusted in the cockpit is, by far, the most convenient. It has the advantage of allowing the driver to know exactly where the adjustment is. On the type that adjusts with a screw, you can't judge easily where the valve is set without counting the turns or taking a precise measurement.

The in-cockpit adjustable proportioning valve that I like is the Tilton Engineering CP 2611-3. This lever-operated valve has five settings and is designed for racing. The driver can make changes from the cockpit while at speed. For street use, you may wish to mount the valve where the lever can't be moved accidentally. This valve and the Alston Performance Brakes, Direct Connection and Kelsey-Hayes screw-type adjustable valve are discussed in more detail in Chapter 5.

Screw-type adjustable valves are available from several car makers. Porsche, Ford and Chevrolet installed adjustable proportioning valves on some models. They are also available

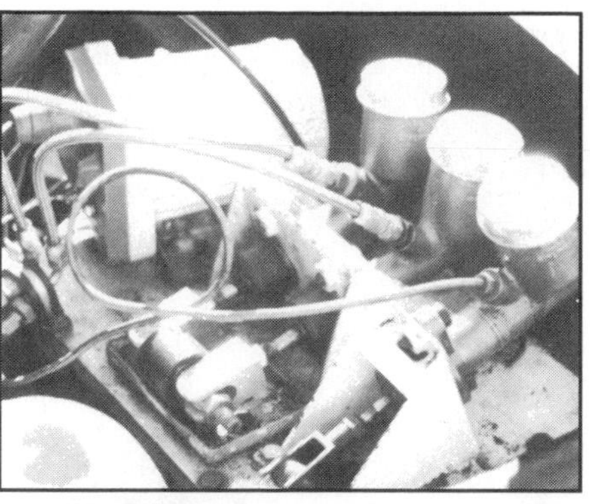

Triumph Spitfire was modified to add balance bar and dual master cylinders. Pedal was modified to install balance bar, and master-cylinder bracket was changed. This is an easy road car to modify, because master cylinders are easily accessible in engine compartment. Notice that flexible hoses are used on each master cylinder, resulting in extra pedal motion. Avoid using flexible hoses where movement isn't required.

from Alston Performance Brakes and Direct Connection. See your dealer or contact one of the brake-equipment suppliers listed in the Suppliers Index for this type of valve.

To briefly review what was discussed in Chapter 5, adjustable proportioning valves change the pressure-rise-reduction rate to the rear brakes *after* fluid pressure reaches a certain value. Changing the point at which the valve starts to operate is the adjustment. Above the actuating pressure, there is a *smaller increase* in pressure to the rear brakes than there would be without the valve. This effectively distributes more braking to the front brakes and less to the rear. However, its operation does not affect pressure to the front brakes.

A proportioning valve is installed in the line between the master cylinder and rear brakes. Its use is independent of the type of master cylinder. One can be used in brake systems using either single, tandem or dual master cylinders.

If you wish to race your car, a proportioning valve will work well. However, if you want more rear-wheel braking, you'll first have to modify the brake system, then reduce it with a proportioning valve. This should not be done by installing harder linings on the front brakes only, which results in balance changes with temperature. Instead, install larger wheel cylinders or calipers with more piston area at the rear, or change to a smaller-bore master cylinder for the rear brakes only.

Adjusting balance bar in hanging pedal. This mechanic highly recommends remote adjusters!

If your car has a proportioning valve and you have excessive rear braking under all conditions, install an adjustable valve in line between the stock valve and the rear brakes. Afterward, if there's too much front braking under some conditions and too much rear braking under others, remove the stock valve or gut it.

DIAGONALLY SPLIT BRAKE SYSTEMS

Brake-balance-adjustment devices can be added only if the front brakes are actuated by one hydraulic line and the rear brakes by a separate line. Diagonally split systems won't work with a balance bar or a single proportioning valve. Such a system is designed to have a diagonal pair of brakes operate separately from the opposite diagonal pair. This means the left-front brake is tied to the right-rear, and vice versa.

To modify a diagonally split brake system for balance adjustment, either install two adjustable proportioning valves or change the hydraulic system plumbing in a major way. The major change may require a different master-cylinder size and other changes to the system. Such a modification is too complex for most road-car owners. If two adjustable proportioning valves are used, they must be synchronized so they give the same output pressure at each input pressure.

Balance Bar—For an all-out race car, a balance bar is the best way to adjust brake balance. As a modification, adding a balance bar is difficult. You may have to design and fabricate the pedal and linkage. Or, you may be able to adapt a complete new pedal, bracket and balance-bar assembly from Neal Products or Tilton

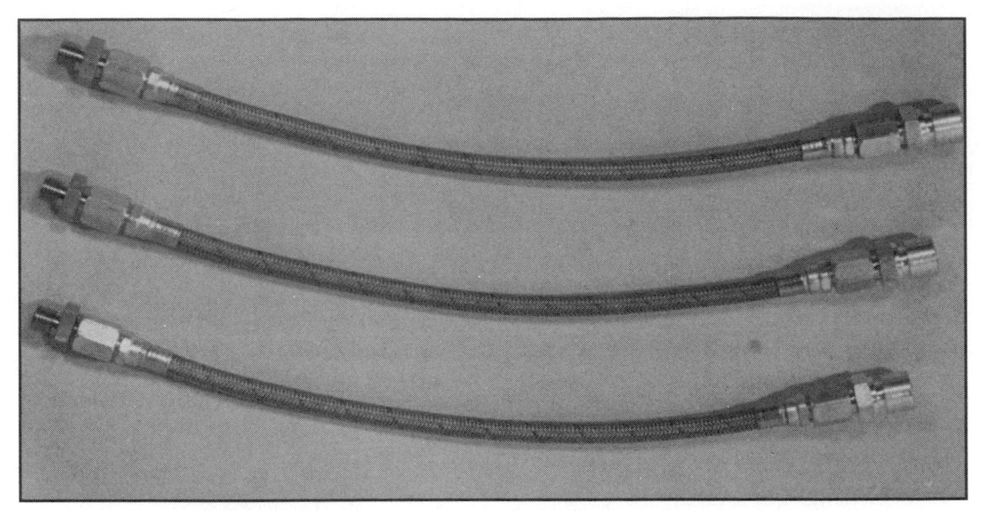

Steel-braided, high-pressure hoses are used to reduce deflections in brake system. They are stiffer than reinforced original-equipment rubber hoses, and are less likely to be damaged. Usually, adapter fittings are available from companies that sell hoses. Photo courtesy C & D Engineering Supply Ltd.

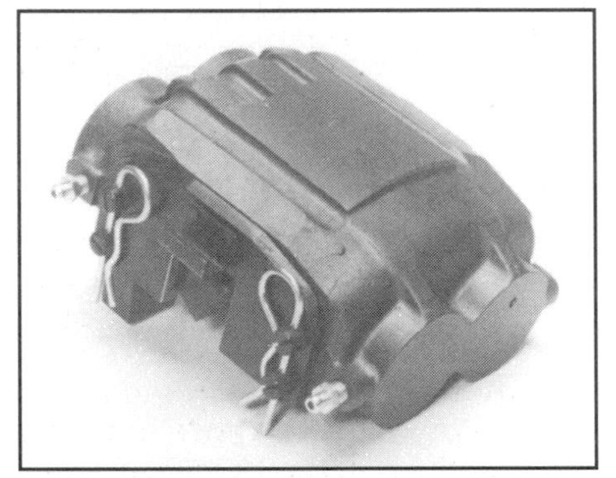

AP Racing caliper offers a stiff bridge to minimize deflection. This caliper is made for a 12-in.-diameter rotor, from 1.10- to 1.375-in. thick. If you can fit that size rotor to your racer, you probably won't find a stiffer caliper. Photo courtesy AP Racing.

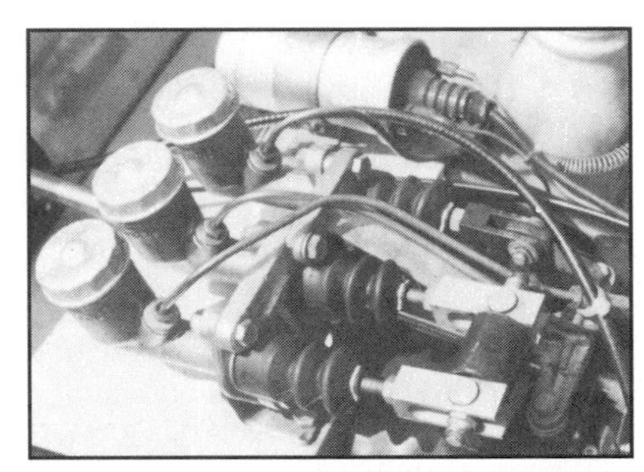

Reverse-mounted, hanging-pedal setup is easy to service.

Engineering. For a road car, an adjustable proportioning valve gives good results. On a road car modified for racing, converting to a balance bar may be impractical due to the cost and effort involved. An adjustable proportioning valve will also work for racing, but pedal travel is increased.

When changing from a tandem or single master cylinder to a dual arrangement with a balance bar, remember to recalculate pedal effort. Because the balance bar divides the pedal force into approximately two equal halves, the pressure in each brake system will be cut roughly in half. This will require double the pedal effort for the same deceleration if master-cylinder diameter is the same. For this reason, dual master cylinders are about three-fourths the diameter of a tandem or a single master cylinder.

REDUCE DEFECTIONS

Many brake systems have excessive deflections. This causes increased pedal travel and a spongy feel to the pedal. If you install a small-diameter master cylinder(s), consider reducing brake-system deflections at the same time. The small master cylinder increases required pedal movement. If deflections can be reduced, you can regain some of this extra movement.

Deflections in a brake system are cumulative. Each one is small. You couldn't see most deflections if you were looking at them as they occur. But, because they are cumulative and magnified by the pedal ratio, they can be significant. For instance, if the pedal ratio is 8.0, a deflection requiring 1/8 in. of movement at the master cylinder will give an additional 1 in. at the pedal pad.

Some of the causes of deflection that can be corrected are:

- Brake-hose expansion.
- Caliper flexing.
- Bending of master-cylinder mounting structure.
- Bending of pedal or linkage.
- Compression of gas in the brake fluid, or the brake fluid itself.

Modifications can be made to reduce each of these deflections. They can never be reduced to zero—only made smaller. Every object deflects under load. It is only a matter of how much it deflects.

Brake-Hose Expansion—The rubber hoses used on most road cars where flexing is necessary expand, or swell, under pressure. Additional fluid is required to fill an expanded hose. This swelling may add up to noticeable pedal movement.

The answer to hose deflection is to install stiffer hoses. Use steel-braided Teflon-lined ones. They are stiffer than fabric-reinforced rubber hoses, so they deflect less under pressure. Also, steel-braided hoses are usually smaller than rubber hoses. The smaller diameter further reduces hose swelling.

Steel-braided hoses are available in kits from brake-product outlets such as C & D Engineering, Earl's Performance Products, or WREP Industries.

Longer brake hoses deflect more than shorter ones; one that's twice as long deflects twice as much, and so on. On some cars, flex hoses are used where flexing is not required—say from a T-fitting to each brake on the rear axle. In some cases, a complete brake system may have been plumbed with flex hoses. Replace unnecessary flex hoses with steel tubing to reduce deflection in a brake system. Don't cheat by excessively reducing the length of the hoses to the wheels. If a hose doesn't have sufficient slack, it will likely fail with extreme wheel movement.

Caliper Flexing—A disc-brake-caliper body will deflect away from the surfaces of a rotor when the fluid is pressurized. This deflection is from bending of the bridge between the inboard and outboard sections of the caliper.

Caliper deflection can be measured with a dial indicator. If it deflects, additional piston movement and fluid flow is necessary to maintain brake pressure. The driver senses this as increased pedal stroke and a spongy feel.

The best way to cure caliper flex is to install more-rigid calipers. They can be bigger, wider or of another material. Cast-iron calipers are about twice the stiffness of aluminum calipers of the same size and design. Iron calipers aren't used on race cars due to higher weight, but they certainly are stiffer. On a road car, iron calipers are better.

Racing-caliper suppliers such as Hurst, JFZ or Tilton can help you select a stiffer caliper for a race car. Although expensive, better brakes will result. Remember that a caliper change may also require a mounting-bracket change.

Master-Cylinder-Mount Deflections—The master cylinder may "see" a force exceeding 1000 lb during hard braking. If the mount is flexible, additional pedal movement will be noticeable.

Check for master-cylinder deflection by remotely mounting a dial indicator and setting the indicator plunger against the master cylinder near its end. The indicator base shouldn't be mounted on or near the master-cylinder mount. Have someone step on the pedal while you observe the indicator dial. Check for deflections in different directions. Often an entire fire wall will flex.

If master cylinders are mounted to a sheet-metal fire wall, they may flex too much when driver pushes hard on pedal. To prevent front end of master cylinders from flexing upward, designer of this stock car added a tab to each frame tube. With a bolt passing through both master cylinders, flexing is greatly reduced. The driver feels this as a firmer pedal.

The answer is to stiffen the structure. The best method is a strut or several struts attached to or near the master-cylinder mount. Airheart and Alston master cylinders have additional mounting holes at both ends of the fluid reservoir for mounting struts. Thicker sheet metal doesn't do much. Beware of thin sheet metal loaded in bending. The brake-pedal pivot and the master cylinder should be attached to a common bracket if possible.

Pedal or Linkage Bending—Brake pedals are usually quite stiff because they are designed with a large safety factor. However, a pedal arm that is curved or bent to the side may twist and cause excessive deflection at the pad. If you have such a pedal arm, add metal to the pedal arm to make it into

Master-cylinder bracket on this car needs some modifications to stiffen the pedal. An angle bracket will deflect in bending—this can be eliminated with two 45° gussets. Bracket is mounted in center of thin sheet-metal panel. Tubular cross member is needed to minimize bending of floor panel. Cross member should be firmly attached to each frame side member.

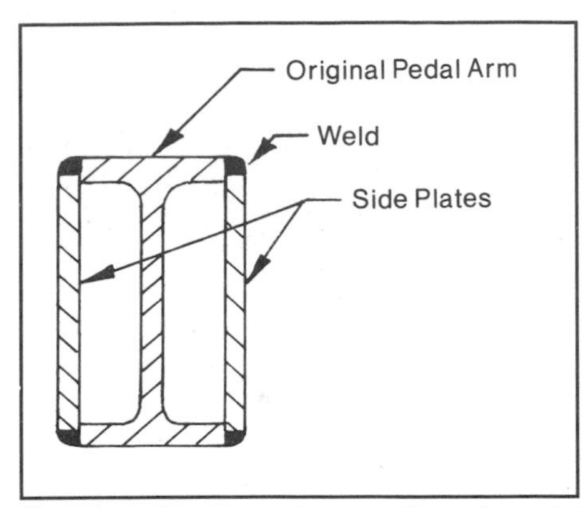

Drawing shows section cut through pedal arm. Original pedal arm is shaped like an I-beam cross section. To keep it from twisting, a side plate is welded on each side of the I-beam to make it a box-section. Resistence to twisting is much greater with this modification.

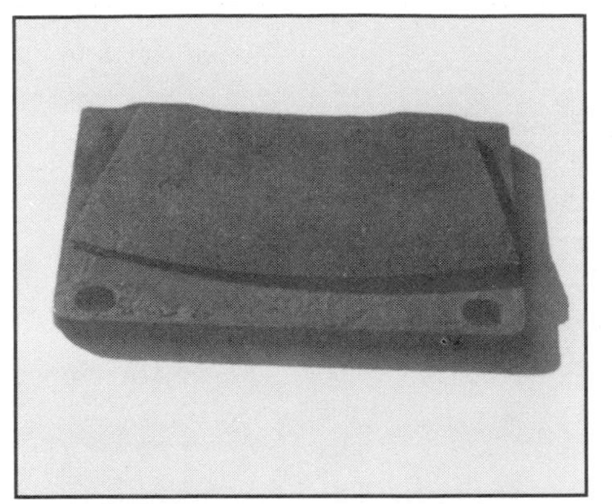

Pads normally wear on a taper because trailing edge of pad runs hotter than leading edge. Because wear increases with temperature, trailing edge wears faster. Tapered pads make pedal feel spongy because caliper or bracket has to flex to bring pads in contact with rotor. Some pads wear leading edges faster due to caliper design.

Bleeder on this race-car rear brake is below brake hose! Backing plates were mounted upside down on axle. By unbolting and repositioning backing plates, spongy pedal was cured forever. Bleeders must be at highest spot in cylinder to be effective in removing all air.

Checking brake-fluid level periodically is an important maintenance job on every car. If fluid reservoirs are hard to reach, consider a modification. These race-car reservoir caps are fitted with plastic breather tubes, which could be used as feed lines from a remote-mounted reservoir. Be sure reservoir-breather tubes breath from a clean, dry place such as inside a roll cage.

a tube. A closed—*boxed*—tubular structure, round or square, has much greater torsional strength than a flat plate, channel or I-beam. The accompanying drawing shows one method of how to do this.

Linkages usually don't deflect noticeably. However, if the system has a bell crank or anything subjected to bending or twisting, it may be a problem. A linkage must be as stiff as possible. Curved or bent pushrods should be strengthened or replaced with straight ones. This applies even if it means a major design change.

Compression of Air in Brake Fluid—If a brake system always acts as if it needs to be bled, maybe it does. Air tends to get trapped in high spots and corners of a hydraulic system. Check that the bleeders are positioned so all the air can get out. A bleeder should be at high spots in a hydraulic system. Some special cars may have the calipers mounted at an angle that trap a small amount of air. If you have such a car, remount calipers or remove them from their mounts and position the bleeder high for bleeding. Then remount calipers.

If a master cylinder is not mounted level, it can trap air. With this condition, try tilting the car with a jack when bleeding the brakes so no air can be trapped at the closed end of the master cylinder. Check the position of the cylinder with a level while the car sits on a level surface.

If the brake lines have a fitting at any high spot, there may be bubbles trapped inside the fitting. Examine the details of the fitting to be sure. Perhaps you can reroute the line to eliminate the high spot or, maybe, you can use another type of fitting that will not trap air. But, first, avoid high spots in the hydraulic system.

A spongy pedal can be caused by compression of the brake fluid. Brake fluid normally has little compression, but some fluids are better than others in this respect. For instance, at high temperatures, silicone fluids have a lower compression stiffness than glycol-based fluids. But, at low temperatures, silicone-fluid compressibility is usually not a problem.

If you have a race car with a spongy pedal, perhaps the fluid is the problem. If you are using silicone fluid, try a glycol-based fluid such as AP Lockheed racing fluid.

Deflections Due to Tapered Pads—A worn disc-brake pad usually is tapered. The lining thickness of a used pad is usually thicker at the leading edge than at the trailing edge. This results from the trailing edge of the

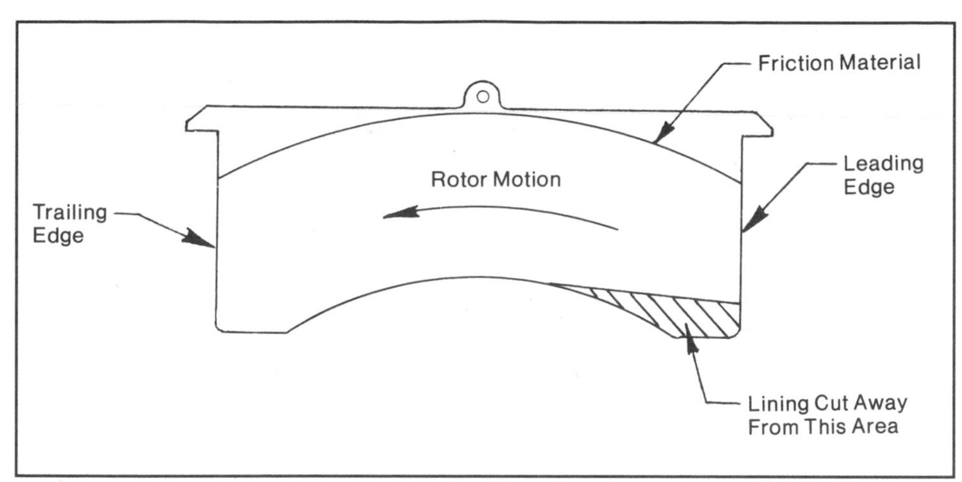

By removing some friction material on slow-wearing, leading edge of pad, taper wear can be reduced. If pad wears evenly, pedal travel will not increase as much as pad wear. Make sure both sides of the car are modified the same so braking will be even.

This caliper bracket is cut from a flat plate. It adapts a disc brake to a spindle from a car with drum brakes. This flat plate bracket should be 1/2-in. thick, if possible, to resist twisting.

pad running hotter than the leading edge, thus wearing more. Some caliper designs also cause the leading edge to wear more.

Once pads are worn tapered, the caliper assembly must deflect for the pad to contact the rotor fully. This increases pedal travel and effort. It can also damage the caliper by cocking the pistons in the bores. Although this can be minimized by changing the pads before they are worn much, another modification can help, too.

Remove material from the sides of the pad at its leading end to reduce pad area and increase wear rate locally. See the accompanying drawing. This helps balance the unequal wear rates. The amount of material to remove is found by trial and error. Start by removing about 25% of the width at the slow-wearing edge and see what happens. Make sure all pads and both sides of the car are modified equally to prevent pulling to the side during braking.

Tapered pads can also be caused by flimsy caliper-mounting brackets. If the bracket twists, this cocks the caliper and makes the pads taper. This taper wear due to twisting will be unequal on the inboard and outboard pads. One will show more wear on the trailing edge than the other one on the leading edge. If one pad tapers a lot compared to the other, caliper twist and higher trailing-edge temperature are involved.

Reduce caliper twist by using a stiffer caliper-mounting bracket. I suggest using at least a flat 1/2-in.-thick steel plate for a caliper bracket. If that thickness is impossible, use the thickest possible steel plate.

With sliding calipers, unequal inner-to-outer lining wear can be caused by high slide friction. Clean the slides and lubricate them with high-temperature grease between the inner and outer caliper sections.

IMPROVING STABILITY DURING BRAKING

Stability refers to how straight the car travels during braking. When the brakes are applied hard, an unstable car may swerve, pull to the side, or even spin out. There are many things that affect stability, but the following items are the most likely.

Stable braking depends on balanced braking at both sides of the car. If braking is different from right to left, the car will be unstable. Start by checking brake diameters, lining condition, and mechanical condition of the brakes to be sure they are equal on both sides. Also check tire rolling radius to be sure this is equal on each side. Measure with a tape wrapped around the circumference of each tire—it is more accurate than measuring radius.

The car's suspension must work well for stability. It's essential that the wheels don't steer or change toe-in when the brakes are applied. For tips on setting up a chassis, read my HP Book, *How to Make Your Car Handle*.

Excessive front *scrub radius* can cause instability during braking. Scrub radius is the distance from the point the kingpin axis intersects the ground to the center of the tire patch.

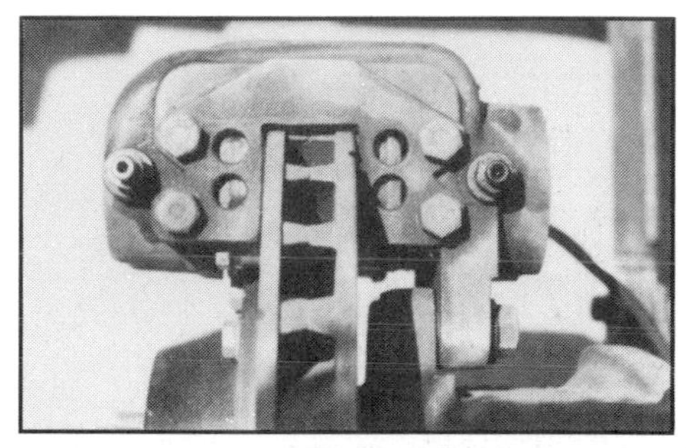

Bolted pad retaining plate increases caliper-bridge strength. Studs with quick-release clips are used in place of bolts when quick pad changes are required.

Widening the track by installing wheels with increased offset increases scrub radius by the same amount. Cars should not have a scrub radius greater than about 2 in., although many road and race cars have been designed with more.

If stability problems are caused by rear-wheel lockup, that can be eliminated by balancing the brakes. See the earlier part of this chapter dealing with modifications for balancing brakes. Adding an adjustable proportioning valve can be the answer to rear-wheel lockup.

Drum brakes with high servo action, such as duo-servo brakes, are notorious for causing pulls and instability. If this is a problem, be sure the drums, linings and wheel cylinders are in good condition. If this doesn't help, try harder linings or disc brakes. Instability with duo-servo drum brakes is a major reason why modern American cars use front disc brakes exclusively.

Trouble-Shooting Guide

Trouble-shooting brakes is too big of a subject for complete coverage in this book. However, the subject of maintenance and testing would not be complete without a trouble-shooting guide. Briefly listed below are some possible causes of brake trouble. Cures are covered in this book, the factory manual, and in other publications on brake repair and maintenance.

This reference guide to trouble-shooting may help you with a quick solution at the test track:

Problems	**Possible Cause**
Low pedal—will pump up	Rotor Runout Loose wheel bearings Air in hydraulic system Drum brakes out of adjustment
Low pedal—won't pump up	Bad seals in caliper Badly worn pads Rotor runout Leak in hydraulic system Loose wheel bearings Excessive free-play in brake linkage Balance bar too far off-center
Changing brake-pedal travel	Slop in wheel bearing or suspension Balance-bar failure Balance bar too far off center
Spongy pedal	Air in hydraulic system Deflection of caliper or mount Deflection of hoses Brake fluid too hot Badly worn linings or pads Concave or convex lining wear Deflection of master-cylinder mount Master cylinder too small Pedal ratio too high Drum-brake-shoe arc incorrect Distorted brake shoes or backing plate Old brake fluid Cracked brake drum
Brake pedal not returning	Master-cylinder reservoir not vented No clearance in brake pushrod Binding in pedal pivots or pushrod
Front or rear brakes locking	Too much front- or rear-brake balance Failure in opposite brake system
One brake locking	Caliper piston seizing in cylinder Wheel cylinder seizing Bad brake-shoe return spring Loose or distorted backing plate Oil or fluid leak into brake Loose caliper mount Excessive weight on other wheel Twist in car chassis or suspension Mismatched rotors, linings or drums
Pedal effort too low	Master cylinder too small Pedal ratio too high Too much servo action on drum brakes Linings too soft Too much power assist Defective booster
Pedal effort too high	Master cylinder too large Pedal ratio too low Linings too hard Racing linings too cold Power-assist failure

	Brakes wet Glazed linings Oil or fluid on linings Drum-brake-shoe arc incorrect Brakes too hot
Vibrating pedal	Excessive rotor runout or thickness variation Damaged wheel bearings Cracked drum or rotor Bent axle Drum warped or eccentric Brakes not releasing Vented-rotor fins rusted out so side plates deflect under pressure
Swerving under hard braking	Front suspension bottoming Toe-in adjustment wrong Bump-steer problems Shock-absorber failure Shock-absorber adjustment dissimilar Brake balance way off Caster or camber uneven Chassis or suspension twisted Worn steering or suspension pivots Tire sizes dissimilar Excess front-wheel scrub radius
Pedal goes to floor	Air in hydraulic system Leaking seal in master cylinder Leak in hose or tube Leak in caliper or wheel cylinder Tapered pad wear Drum brake not adjusting Electric current passing through fluid
Brakes grab or lock	Drum-brake-shoe arc incorrect Loose or distorted backing plate Contaminated linings Brakes wet Racing linings too cold Failed brake-shoe return spring Drum-brake linings burned up
Brakes not releasing	Blocked master-cylinder port Binding pedal pivots No free play in pushrod Seized caliper or wheel cylinder Aged or overheated caliper seals Swollen seals—incorrect fluid Caliper slides corroded and stuck Heel drag on cup-type seals Parking brake on or maladjusted Power booster faulty Distorted shoes or backing plate
Brakes squeal	Wear in brake shoes or attachments Worn pads Cold linings Need new brake-pad anti-squeal shims Need anti-squeal compound behind pads Need to chamfer ends of linings Brakes wet
Rapid lining wear	Wrong surface finish on rotor or drum Lining too soft Cracks in drum or rotor Adjustment too tight Brakes not releasing Inadequate brake cooling

Trouble-Shooting

This road-car hydraulic-brake-system diagnostic procedure may help you solve some perplexing brake-system problems:

PRELIMINARY ACTION

1. Clean master-cylinder cap.
2. Fill master-cylinder reservoir.
3. Turn ignition switch to start to prove out instrument-panel brake light.

PEDAL-FEEL CONDITIONS & APPROPRIATE ACTIONS

1. Pedal to the floor quickly—light on:
   Perform action A.
      If pedal holds, perform Action B.
      If pedal does not hold, perform Action C.
2. Pedal to the floor slowly (creeps)—light on:
   Perform Action C.
3. Pedal firm—long stroke—no light:
   Perform Action A.
      If pedal pumps up—perform Action B.
      If pedal does not pump up—perform Action D.
4. Pedal spongy—no light:
   Perform Action A.
      If pedal pumps up—perform Action B.
      If pedal does not pump up—perform Action G.
5. Pedal Normal—Goes to floor at times:
   If disc brakes—perform Action E.
   Perform Action F.

ACTIONS

A. Pump pedal as fast as possible 30 times and hold pedal down on last stroke.
B. Have an assistant remove master-cylinder cap and observe reservoir-fluid levels as the pedal is released as quickly as possible. A spout in either chamber indicates the possibility of air in the system. Bleed the brakes. If problems remain, adjust the brakes.
C. Have an assistant observe master-cylinder-fluid levels as the brake pedal is slowly depressed. A level that rises 1/16 in. on pedal application and drops to original level on release—not including the spout on apply—indicates an internal bypass leak in a tandem-type master cylinder. A reservoir that overflows after long use while the other reservoir drops also indicates an internal bypass. Rebuild or replace master cylinder. A reservoir that drops progressively with each pedal stroke indicates an external pressure leak. Check all lines and fittings of the affected system—front or rear—and repair as required.
D. Adjust brakes. Check for loose wheel bearings and adjust to specifications if required.
E. Check for loose wheel bearings and adjust to specification if required.
F. If condition occurs only after long periods of idleness, replace the hydraulic stoplight switch and flush front-system brake fluid. A shorted switch can pass current through the brake fluid. Fluid subjected to electric current for a long period of time will bubble.
G. Check caliper slides, pins and pistons for sticking. Check disc-brake linings for tapered wear. Check drum brakes for bent shoes or backing plates.

Suppliers List

Suppliers are listed two different ways. First, they are listed by company name in alphabetical order—then the products are listed alphabetically. If you are looking for a particular product, look first in the product list. The number behind each item refers to the number of the company(ies) shown in the alphabetical supplier list.

Every year new companies are formed and some go out of business. Thus, you should not consider this list to be the last word. If you want a complete and current list of companies supplying a particular brake-related product, refer to ads in the latest trade and racing publications. The following supplier list is merely a guide to get you started on your shopping task.

ALPHABETICAL LIST OF SUPPLIERS

1. **Aeroquip Corp.**
300 S. East Ave.
Jackson, MI 49203

2. **Alfred Teves Technologies, Inc.**
(ATE Brakes)
2718 Industrial Row
Troy, MI 48084

2A. **Alston Industries**
6291 Warehouse Way
Sacramento, CA 95826

3. **AMMCO Tools, Inc.**
Wacker Park
North Chicago, IL 60064

4. **AP Racing**
Leamington Spa
Warwickshire, England CV31 3ER

5. **Automotive Products USA, Inc.**
1864 Northwood Dr.
Northwood Industrial Park
Troy, MI 48084

6. **Bendix Automotive Aftermarket**
1904 Bendix Dr.
Jackson, TN 38301

7. **C & D Engineering Supply Ltd.**
Rt. 2, Box 47-1
Fort Valley, CA 31030

7A. **Chrysler Corporation**
Direct Connection
P.O. Box 1718
Detroit, MI 48288

8. **Columbia Motor Corp.**
140 West 21st St.
New York, NY 10011

9. **Contemporary Chassis Design**
15506 Vermont St.
Paramount, CA 90723

10. **CRC Chemicals**
885 Louis Dr.
Warminster, PA 18974

11. **Deist Safety, Ltd.**
641 Sonora Ave.
Glendale, CA 91201

12. **Delco Moraine Div.**
General Motors Corp.
1420 Wisconsin Blvd.
Dayton, OH 45401

13. **Dick Guldstrand Enterprises**
11924 W. Jefferson Blvd.
Culver City, CA 90230

13A. **Dillon Enterprises**
66820 SR 23
North Liberty, IN 46554

14. **Dow Corning Corp.**
Box 0994
Midland, MI 48640-0994

15. **Dwyer Instruments, Inc.**
P. O. Box 373
Michigan City, IN 46360

15A. **EIS Automotive Corp.**
129 Worthington Ridge
P.O. Box 1315
Berlin, CT 06037

16. **Earl's Performance Products**
825 E. Sepulveda
Carson, CA 90745

17. **Edco Specialty Products**
3411 W. MacArthur Blvd.
Santa Ana, CA 92703

17A. **Ford Motorsport**
Ford Motor Co.-SVO
17000 Southfield Rd.
Allen Park, MI 48101

18. **Friction Products Co.**
920 Lake Road
Medina, OH 44256

19. **ELSCO Inc.**
1843 E. Adams St.
Jacksonville, FL

20. **Frankland Racing Equipment**
P.O. Box 278
Ruskin, FL 33570

21. **Girling Ltd.**
200 Manchester Ave.
Detroit, MI

22. **Girlock**
c/o Tilton Engineering
McMurray Rd. & Easy St.
Buellton, CA 93427

23. **Girlock Ltd.**
36-40 Harp. St.
Belmore, N.S.W. 2192
Sydney, Australia

24. **Grey Rock Div.**
Raybestos Manhattan Corp.
P.O. Box 9140
Bridgeport, CT 06603

25. **Harwood Performance Sales, Inc.**
11501 Hillguard Rd.
Dallas, TX 75243

26. **Henry's Engineering Co.**
P.O. Box 629
Prince Frederick, MD 20678

27. **Huffaker Engineering, Inc.**
22 Mark Dr.
San Rafael, CA 94903

28. **Hurst Performance, Inc.**
Airheart Brakes
50 West Street Road
Warminster, PA 18974

29. **JFZ Engineered Products**
9761 Variel Ave.
Chatsworth, CA 91311

29A. **Ja-Mar Off-Road Products**
29300 3rd St.
San Rafael, CA 94903

30. **Kano Laboratories**
1000 S. Thompson Ln.
Nashville, TN 37211

31. **Kelsey-Hayes Co., Inc.**
Kelsey Axle & Brake Division
5800 W. Donges Bay Road
Mequon, WI 53092

32. **Kelsey-Hayes Co., Inc.**
Kelsey Products Division
38481 Huron River Dr.
Romulus, MI 48174

33. **Lakewood Enterprises**
4566 Spring Rd.
Cleveland, OH 44131

34. **Lamb Components**
150 W. College Way
La Verne, CA 91750

35. **Lee Manufacturing Co.**
Division of Rolero
P.O. Box 7187
Cleveland, OH 44128

37. **Lucas Industries, Inc.**
Girlock
Girling Brake & Hydraulic Systems Group
5500 New King St.
Troy, MI 48098

38. **Marsh Instrument Co.**
3501 W. Howard St.
Skokie, IL 60076

39. **McLeod Industries**
1125 N. Armando
Anaheim, CA 92806

40. **Midwest Race Engineering**
503 S. 7th St.
Boonville, IN 47601

41. **MICO, Inc.**
1911 Lee Blvd.
North Mankato, MI 56001

42. **Morak Brakes, Inc.**
9902 Ave. D
Brooklyn, NY 11236

43. **Mueller Fabricators**
10872 Stanford Ave.
Lynwood, CA 90262

Suppliers List

44. **Nationwide Electronic Systems, Inc.**
1536 Brandy Parkway
Streamwood, IL 60103
45. **Neal Products, Inc.**
7170 Ronson Rd.
San Diego, CA 92111
46. **Omega Engineering, Inc.**
Box 4047, Springdale Station
Stamford, CT 06907
47. **Performance World**
3550 University Ave.
San Diego, CA
48. **Plasma Technology Inc.**
1754 Crenshaw Blvd.
Torrance, CA 90501
49. **Professional Racers Emporium**
1463 E. 223rd St.
Carson, CA 90745
50. **R.A.C.E. Pads**
8 S. 582 John St.
Big Rock, IL 60511
51. **Ron Minor's Racing Specialties**
6511 N. 27th Ave.
Phoenix, AZ 85017
52. **Servex**
1520 10th Ave.
Seattle, Washington 98122
53. **Simpson Safety Equipment, Inc.**
22630 S. Normandie Ave.
Torrance, CA 90502
54. **Speedway Engineering**
13040 Bradley Ave.
Sylmar, CA 91342

54A. **Speedway Motors**
P.O. Box 81906
Lincoln, NE 68501

55. **Stainless Steel Brakes Corp.**
11470 Main Rd.
Clarence, NY 14031

55A. **Stop & Go Products**
1835 Whittier, Suite A5
Costa Mesa, CA 62627

55B. **Tempil Division**
Big Three Industries, Inc.
2901 Hamilton Boulevard
South Plainfield, NJ 07080

56. **Tilton Engineering, Inc.**
McMurray Rd. & Easy St.
P.O. Box 1787
Buellton, CA 93427
57. **Troutman, Ltd.**
3198 'L' Airport Loop Dr.
Costa Mesa, CA 92626
58. **Velvetouch Div.**
S.K. Wellman Corp.
200 Egbert Rd.
Bedford, OH 44146
59. **Weevil Ltd.**
206 Queens Court
Ramsey, Isle of Man
British Isles
60. **White Post Restorations**
White Post, VA 22663
61. **Wilwood Engineering**
4580 Calle Alto
Camarillo, CA 91360

61A. **Winters Performance Products**
2819 Carlisle Rd.
York, PA 17404

62. **WREP Industries, Ltd.**
140-C Shepard
Wheeling, Ill. 60090
63. **Yankee Silicones, Inc.**
P.O. Box 1089
Schenectady, NY 12301

PRODUCT INDEX

Numbers refer to those shown in alphabetical suppliers list.

Adjustable Proportioning Valve 2A, 7A, 17A, 32, 47, 56, 61
Adjusters 5, 6
Anti-Noise Compound 10

Balance Bars 29A, 45, 56, 61A
Bleeder Screws 5, 6
Brake Fluid 4, 5, 6, 7, 15A, 45, 47, 51, 54A, 55A, 56, 61, 62
Brake-Locking Valves 29A, 37
Brake Tubing 6, 45

Calipers 2, 2A, 4, 5, 6, 12, 13A, 28, 29, 29A, 32, 51, 54A, 55, 56, 60, 61, 61A
Caliper Brackets 28, 56
Cleaning Fluids 10, 51
Cutting Brakes 29A, 45
Cylinder Restoring 55, 60

Decelerometers 3
Dragster Brakes 2A, 28, 29, 31, 61
Drums 5, 15A, 29

Fast-Fill Master Cylinders 41
Finned Drums 5
Fittings 1, 4, 5, 6, 7, 16, 45, 51, 52, 56, 62
Flaring Tools 6

Gage Protectors 41

Hand-Operated Brake 29A, 45
Hats 2A, 4, 28, 29, 56, 61A
Hoses 1, 4, 5, 6, 7, 16, 62

Lining Material 4, 5, 6, 8, 13A, 24, 50, 51, 54A, 55A

Master Cylinders 2A, 4, 5, 6, 12, 13A, 28, 29A, 37, 45, 51, 54A, 55A, 56, 61, 61A, 62
Metric Adapter Fittings 6, 7, 45, 62
Moisture Barriers 45, 56

Pads 2A, 4, 5, 6, 7, 12, 13A, 15A, 18, 28, 29, 29A, 32, 50, 51, 54A, 55A, 61A
Parachutes 11, 53
Pedals 2A, 13A, 29A, 45, 51, 51, 54A, 56, 61, 61A
Pedal Brackets 2A, 13A, 29A, 45, 54A, 56, 61A
Penetrating Oil 30
Power Boosters 2, 5, 6
Pressure Bleeders 6, 15A
Pressure Gages, Air 15
Pressure Gages, Hydraulic 28, 38, 41
Pressure Switches 41
Proportioning Valves 2A, 7A, 17A, 31, 47, 56
Pushrods 56
Pyrometers/Thermometers 4, 44, 46, 51, 56

Racing Brakes 2A, 4, 13A, 28, 29, 51, 54A, 55A, 56, 61A
Rebuilding Kits 4, 5, 6, 15A, 28, 29, 45, 55, 56, 62
Remote Balance-Bar Adjuster 13A, 45, 54A, 56, 61A
Reservoirs 5, 45, 51, 55A, 56
Reservoir Extensions 45, 51, 55A, 56
Residual-Pressure Valves 28, 45
Return Springs 5, 6
Rotors 2A, 4, 5, 6, 12, 13A, 15A, 28, 29, 29A, 32, 51, 52, 54A, 55A, 61A

Shoes 5, 6, 15A, 18, 32, 52, 55
Silicone Fluid 15A, 47, 51, 55, 62, 63
Sprint-Car Brakes 28, 29, 54A, 55A, 56, 61
Steel-Braided Hoses 1, 4, 7, 16, 52, 62
Stock-Car Brakes 13A, 28, 29, 54A, 55A, 56, 61, 61A
Switches 6

Temperature Indicators 4, 46, 51, 55B, 56, 59
Tools 3, 6

Water-Cooling System 56
Wheel Cylinders 5, 6

METRIC CUSTOMARY-UNIT EQUIVALENTS

Multiply:		by:		to get:	Multiply: by:		to get:
LINEAR							
inches	X	25.4	=	millimeters(mm)	X	0.03937	= inches
feet	X	0.3048	=	meters (m)	X	3.281	= feet
yards	X	0.9144	=	meters (m)	X	1.0936	= yards
miles	X	1.6093	=	kilometers (km)	X	0.6214	= miles
inches	X	2.54	=	centimeters (cm)	X	0.3937	= inches
microinches	X	0.0254	=	micrometers(Mm)	X	39.37	= microinches
AREA							
inches^2	X	645.16	=	millimeters^2(mm^2)	X	0.00155	= inches^2
inches^2	X	6.452	=	centimeters^2(cm^2)	X	0.155	= inches^2
feet^2	X	0.0929	=	meters^2(m^2)	X	10.764	= feet^2
yards^2	X	0.8361	=	meters^2(m^2)	X	1.196	= yards^2
acres	X	0.4047	=	hectacres(10^{4}m^2) (ha)	X	2.471	= acres
miles^2	X	2.590	=	kilometers^2(km^2)	X	0.3861	= miles^2
VOLUME							
inches^3	X	16387	=	millimeters^3(mm^3)	X	0.000061	= inches^3
inches^3	X	16.387	=	centimeters^3(cm^3)	X	0.06102	= inches^3
inches^3	X	0.01639	=	liters (l)	X	61.024	= inches^3
quarts	X	0.94635	=	liters (l)	X	1.0567	= quarts
gallons	X	3.7854	=	liters (l)	X	0.2642	= gallons
feet^3	X	28.317	=	liters (l)	X	0.03531	= feet^3
feet^3	X	0.02832	=	meters^3(m^3)	X	35.315	= feet^3
fluid oz	X	29.57	=	milliliters (ml)	X	0.03381	= fluid oz
yards^3	X	0.7646	=	meters^3(m^3)	X	1.3080	= yards^3
teaspoons	X	4.929	=	milliliters (ml)	X	0.2029	= teaspoons
cups	X	0.2366	=	liters (l)	X	4.227	= cups
MASS							
ounces (av)	X	28.35	=	grams (g)	X	0.03527	= ounces (av)
pounds (av)	X	0.4536	=	kilograms (kg)	X	2.2046	= pounds (av)
tons (2000 lb)	X	907.18	=	kilograms (kg)	X	0.001102	= tons (2000 lb)
tons (2000 lb)	X	0.90718	=	metric tons (t)	X	1.1023	= tons (2000 lb)
FORCE							
ounces—f(av)	X	0.278	=	newtons (N)	X	3.597	= ounces—f(av)
pounds—f(av)	X	4.448	=	newtons (N)	X	0.2248	= pounds—f(av)
kilograms—f	X	9.807	=	newtons (N)	X	0.10197	= kilograms—f

TEMPERATURE

°F -40 0 32 40 80 98.6 120 160 212 00 240 280 320 °F

°C -40 -20 0 20 40 60 80 100 120 140 160 °C

Degrees Celsius (C) = 0.556 (F - 32)

Degree Farenheit (F) = (1.8C) + 32

ACCELERATION

feet/sec^2	X	0.3048	=	meters/sec^2(m/s^2)	X	3.281	= feet/sec^2
inches/sec^2	X	0.0254	=	meters/sec^2(m/s^2)	X	39.37	= inches/sec^2

Multiply:		by:		to get:		Multiply: by:		to get:

ENERGY OR WORK (Watt-second=joule=newton-meter)

Multiply:		by:		to get:		by:		to get:
foot-pounds	X	1.3558	=	joules (J)	X	0.7376	=	foot-pounds
calories	X	4.187	=	joules (J)	X	0.2388	=	calories
Btu	X	1055	=	joules (J)	X	0.000948	=	Btu
watt-hours	X	3600	=	joules (J)	X	0.0002778	=	watt-hours
kilowatt-hrs	X	3.600	=	megajoules (MJ)	X	0.2778	=	kilowatt-hrs

FUEL ECONOMY & FUEL CONSUMPTION

Multiply:		by:		to get:		by:		to get:
miles/gal	X	0.42514	=	kilometers/liter(km/l)	X	2.3522	=	miles/gal

Note:
235.2/(mi/gal)=liters/100km
235.2/(liters/100km)=mi/gal

PRESSURE OR STRESS

Multiply:		by:		to get:		by:		to get:
inches Hg (60F)	X	3.377	=	kilopascals (kPa)	X	0.2961	=	inches Hg
pounds/sq in.	X	6.895	=	kilopascals (kPa)	X	0.145	=	pounds/sq in
inchesH_2O(60F)	X	0.2488	=	kilopascals (kPa)	X	4.0193	=	inches H2O
bars	X	100	=	kilopascals (kPa)	X	0.01	=	bars
pounds/sq ft	X	47.88	=	pascals (Pa)	X	0.02088	=	pounds/sq ft

POWER

Multiply:		by:		to get:		by:		to get:
horsepower	X	0.746	=	kilowatts (kW)	X	1.34	=	horsepower
ft-lbf/min	X	0.0226	=	watts (W)	X	44.25	=	ft-lbf/min

TORQUE

Multiply:		by:		to get:		by:		to get:
pound-inches	X	0.11298	=	newton-meters (N-m)	X	8.851	=	pound-inches
pound-feet	X	1.3558	=	newton-meters (N-m)	X	0.7376	=	pound-feet
pound-inches	X	0.0115	=	kilogram-meters (Kg-M)	X	87	=	pound-feet
pound-feet	X	0.138	=	kilogram-meters (Kg-M)	X	7.25	=	pound-feet

VELOCITY

Multiply:		by:		to get:		by:		to get:
miles/hour	X	1.6093	=	kilometers/hour(km/h)	X	0.6214	=	miles/hour
feet/sec	X	0.3048	=	meters/sec (m/s)	X	3.281	=	feet/sec
kilometers/hr	X	0.27778	=	meters/sec (m/s)	X	3.600	=	kilometers/hr
miles/hour	X	0.4470	=	meters/sec (m/s)	X	2.237	=	miles/hour

COMMON METRIC PREFIXES

mega	(M)	=	1,000,000	or	10^6	centi	(c)	=	0.01	or	10^{-2}
kilo	(k)	=	1,000	or	10^3	milli	(m)	=	0.001	or	10^{-3}
hecto	(h)	=	100	or	10^2	micro	(μ)	=	0.000,001	or	10^{-6}

Conversion chart courtesy Ford Motor Company.

BOLT TORQUE BY GRADE

SAE Grade Number	1 or 2	5	8	Special Alloy*
Manufacturer's marks may vary. Bolt grades up through grade 8 are shown at right.				
Capscrew Diameter (inch) and Minimum Tensile Strength (psi)	To 1/2—69,000 To 3/4—64,000 To 1—60,000	To 3/4—120,000 3/4 to 1—115,000	To 1—150,000	To 1—185,000
Bolt size & thread pitch	**Torque (foot-pounds) for plated bolts with clean, dry threads.**			
1/4-20	6	10	12	14
-28	7	12	15	17
5/16-18	13	20	24	29
-24	14	22	27	35
3/8-16	23	36	44	58
-24	26	40	48	69
7/16-14	37	52	63	98
-20	41	57	70	110
1/2-13	57	80	98	145
-20	64	90	110	160
9/16-12	82	120	145	200
-18	91	135	165	220
5/8-11	111	165	210	280
-18	128	200	245	310
3/4-10	200	285	335	490
-16	223	315	370	530
7/8-9	315	430	500	760
-14	340	470	550	800
1-8	400	650	760	1130
-14	460	710	835	1210

For lubricated threads, reduce torque values by the following amount: 45% when using anti-seize compound; 40% with grease or heavy oil; 30% with graphite; or 25% with white lead. *Torque Values shown are for high nickle-chrome alloy Bowmalloy cap screws. Other high-alloy cap screws may not have the same values. Always use manufacturer's specifications if they do not agree with values listed in this chart.

DECIMAL EQUIVALENTS

0.0156—1/64	0.2656—17/64	0.5156—33/64	0.7656—49/64
0.0312—1/32	0.2812—9/32	0.5312—17/32	0.7812—25/32
0.0469—3/64	0.2969—19/64	0.5469—35/64	0.7969—51/64
1/16—0.0625	5/16—0.3125	9/16—0.5625	13/16—0.8125
0.0781—5/64	0.3281—21/64	0.5781—37/64	0.8281—53/64
0.0937—3/32	0.3437—11/32	0.5937—19/32	0.8437—27/32
0.1094—7/54	0.3594—23/64	0.6094—39/64	0.8594—55/64
2/16—0.1250	6/16—0.3750	10/16—0.6250	14/16—0.8750
0.1406—9/64	0.3906—25/64	0.6406—41/64	0.8906—57/64
0.1562—5/32	0.4062—13/32	0.6562—21/32	0.9062—29/32
0.1719—11/64	0.4219—27/64	0.6719—43/64	0.9219—59/64
3/16—0.1875	7/16—0.4375	11/16—0.6875	15/16—0.9375
0.2031—13/64	0.4531—29/64	0.7031—45/64	0.9531—61/64
0.2187—7/32	0.4687—15/32	0.7187—23/32	0.9687—31/32
0.2344—15/64	0.4844—31/64	0.7344—47/64	0.9844—63/64
4/16—0.2500	8/16—0.5000	12/16—0.7500	16/16—1.000

Chart courtesy Acme Rivet & Machine Corp.

Index